CU00708812

Sponsors' Endorsement

In times of change, commitment to making our env[i] which to live contributes to society's stability and enhances the quality of life. Syfrets believes in these values and, together with two of the major charitable trusts it administers, has facilitated the publication of this volume by making a grant to the Southern African Nature Foundation. SANF is the local arm of WWF – World Wide Fund for Nature.

The contributors are:

Nedcor Community Development Fund
Syfrets Charitable Trust
Lorenzo & Stella Chiappini Charitable and Educational Trust
Rowland & Leta Hill Trust.

The trusts were set up by farsighted people who recognised we should live not by bread alone but in harmony with Nature if we are to fulfil our destinies. It is clear the responsibility for conservation funding will rest largely with the private sector. Regrettably there are too few of these Trusts to cater for the calls made on them. There is a real need for enlightened people to come forward with more bequests of this nature. In Rowland Hill's words:

> To some the assets of our joint estate may seem modest, but the said assets were acquired from a modest salary, and would have been much less had it not been for our faith in the future of our country . . . When others panicked in our many times of crisis, real or imagined, the Testator invested in sound equities at the lower prices prevailing at such times of uncertainty, he being himself convinced that South Africa would go forward, as it, indeed, has done. In handing back to the country and its people what they gave us during our lifetime, we trust only that the awards . . . will induce . . . pride in our countrymen and women in their civic progress . . .

The SANF heartily endorses this outstanding publication. The WWF family – the largest non-governmental conservation organisation in the world – has adopted the marine biome as one of the top three priority biomes for attention during the 1990s. This book will help to increase awareness of and appreciation for the amazing diversity of life in the marine environments of southern Africa.

Funding was also received from Sea Fisheries, Department of Environment Affairs.

The authors and publishers especially appreciate a donation from Mr & Mrs A. J. van Ryneveld; and a private donation 'in memory of those who loved these shores'.

WWF

**SUIDER-AFRIKAANSE NATUURSTIGTING
SOUTHERN AFRICAN NATURE FOUNDATION**

Two Oceans

A Guide to the Marine Life of Southern Africa

G. M. Branch, C. L. Griffiths,
M. L. Branch, L. E. Beckley

Photographic Credits

The bulk of the invertebrate photographs (Plates 1–98) were taken by Charlie Griffiths and George Branch, and most of the photographs of plants (Plates 146–165) by George Branch. The authors owe a special debt to Dennis King for contributing the majority of the fish photographs (Plates 99–135). Without his exceptional talents, the fish section of the book would never have been feasible. Many other people gave freely of their photographs, often of rare or difficult-to-photograph animals, or of such exquisite quality that we could never hope to match them:

Mike Anderson-Reed: 99.1
The *Argus*: 145.2
Amos Barkai & Ken Findlay: 76.5
Peter Best: 143.1, 144.1
Pat Berjak & Norman Pammenter: 164.1d, 164.2d
Colin Buxton: 113.1
Cape Bird Club: 139.3
Cape Nature Conservation (Zelda Wahl): 142.3
Simon Chater: 104.1, 104.4, 108.4, 113.6, 114.6, 115.7, 118.6, 123.3, 124.3
Geremy Cliff: 99.5, 101.3, 109.5, 118.5, 122.3, 122.4, 124.5, 124.6, 125.4, 144.4
Phillip Coetzee: 6.3, 8.1, 8.5, 8.8, 97.4, 97.5, 97.7, 112.3, 130.3, 130.9
Tertius Coetzee: 141.6
Martina Compagno-Roeleveld: 86.5
Allan Connell: 25.4, 110.2
Ben Durham: 145.4
Bruce Dyer: 144.2
Dane Gerneke: 86.4
Phil Hockey: 138.3, 138.6, 139.2, 141.3
Deon Horstman: 35.1

Donald Kinross: 139.4
Mandy Lombard: 136.1
Pierre Malan: 145.5
Mike Meyer: 143.2a, 143.2b
MIKOMTEK: Map on p. viii
Mike Mittelmeyer: 145.1
Natal Sharks Board: 99.4
Natal Parks Board: 136.2, 136.3j
Rene Navarro: 141.2, 141.4
Oceanographic Research Institute: 99.3, 99.6, 110.7, 118.2, 127.1, 132.3
Sue Painting: 80.7, 95.2
Sue Painting & Wally Shave: 13.5
Kim Prochazka: 131.1, 132.5
Jack Randall: 100.1, 101.4, 103.3, 122.2, 125.2, 125.7, 133.1
Koos Reddering: 123.4
Bernhard Riegl: 2.5, 8.7, 9.4, 10.5, 11.1d, 11.7d, 12.5d, 12.6, 12.7, 35.2
Mike Schleyer: 1.2, 6.4, 7.1, 7.2, 7.4, 7.5, 7.6, 7.9, 7.10, 7.11, 9.2, 10.8, 11.1, 12.1, 12.2, 88.4, 89.5, 98.1, 98.8
Sea Fisheries Research Institute: 112.4, 113.2, 115.2, 115.5, 115.8, 125.6
Jurgen Seier: Front cover
Malcolm Smale: 87.3, 114.5, 124.4, 144.3
Rob Tarr: 71.3, 72.8, 72.10
Rudi van der Elst: 103.6, 113.7, 118.4
Claudio Velasquez: 43.7, 137.2, 137.3a, 137.4, 138.5, 138.7, 139.1, 139.5, 141.1a, 141.1b, 141.5
Gary Williams (South African Museum slide collection): 16.4, 41.5, 54.8
Alan Whitfield: 103.1, 104.2, 104.6, 129.2

Struik Publishers
(a division of New Holland Publishing
(South Africa) (Pty) Ltd)
Cornelis Struik House
80 McKenzie Street
Cape Town 8001

New Holland Publishing is a member of Johnnic Communications Ltd.

Visit us at **www.struik.co.za**
Log on to our photographic website
www.imagesofafrica.co.za for an African experience

ISBN 978 1 77007 633 4

First published in 1994 by David Philip Publishers
Published in 2007 by Struik Publishers

Cover design: Fresh Identity

Typesetting and Reproduction by Hirt & Carter Cape (Pty) Ltd
Printed and bound by Kyodo Printing Co (S'Pore) Pte Ltd

Front cover: Underwater scene at 'Coral Gardens' on the west coast of the Cape Peninsula. (Photo by Jurgen Seier)

Acknowledgements

A Core Programme Grant to G.M. Branch, C.L. Griffiths and J.G. Field from the Foundation for Research Development funded much of the background research for this book, particularly the costs of the literally thousands of photographs from which the final selection was culled. We also gratefully acknowledge the support of our institutions, the Zoology Department of the University of Cape Town and the Oceanographic Research Institute. Special thanks to their technical staff, who were called on (often at short notice) to produce bizarre items of collecting and photographic equipment. Natal Parks Board arranged accommodation at Sodwana, and Willem and Sharon Prinsloo, kindred free spirits, provided friendship, potjiekos and accommodation for impoverished biologists. The shell plates could not have been produced with such exquisite detail without the loan of a 4x5 camera by Tim Timlin, Peninsula Technikon, and the use of the collections and photographic studio of the South African Museum. For the latter, our thanks are due to Michelle van der Merwe and Liz Hoenson.

Many people helped to collect animals and plants, including Frans Kriel (Sea Fisheries Research Institute), Wendy Robertson (ORI), Anne Drummond (Zoology, Pietermaritzburg), Prof. Anton McLachlan, An de Ruyck and Gary Dobkins (University of Port Elizabeth), Graham Brill (Irvin and Johnson), Janda Maybank of Struisbaai, and Mike Mittelmeyer (De Beers Marine). A crate of beer has been set aside to acknowledge the enthusiasm of a special group of people: the students in the Coastal Ecology Unit at the University of Cape Town – Bernhard Riegl, Rodrigo Bustamante, Jan Korrubel, Kim Prochazka, Lisa Kruger, Yves Lechanteur, Bruce Tomalin, Yvonne Dempster, Bruce Emanuel, and Claudio Velasquez.

Several authorities gave of their time and expertise to help us check identifications or sections of the text: Martina Compagno-Roeleveld (cephalopods), Prof. Jan Heeg (zoanthids), Bernhard Riegl (corals), Dr Phil Hockey (birds), Dr Peter Best and Eva Plaganyi (cetaceans), Drs Yvonne Chamberlain and Derek Keats (coralline algae), Prof. John Bolton and Dr Rob Anderson (seaweeds), Drs Dick Kilburn and Dai Herbert (shelled molluscs), Simon Chater and Dr Phil Heemstra (fish), Dr Dave Pollock (rock lobsters), Drs Gary Williams and Mike Schleyer (octocorals), Dr Mark Gibbons (jellyfish), Wendy Robertson and Dr Pat Backwell (some of the crabs), Dr Naomi Millard (hydrozoans), and Dr Patti Wickens (seals). Particular thanks to Bruce Emanuel for the important task of checking the distribution ranges of all the invertebrates. Special thanks to our children.

About the Authors

PROFESSOR GEORGE BRANCH is world-renowned for his research on rocky shores. Together with his wife, Margo, he wrote *The Living Shores of Southern Africa,* the first book on the ecology of marine life in the region. It became a best-seller and won the UCT book award. Professor Branch has been awarded Fellowships from the University of Cape Town and the Royal Society of Southern Africa, the Gilchrist Medal for marine science, the Gold Medal of the Zoological Society, and a Distinguished Teachers Award in recognition of his contribution to marine research and education. He contributed to fisheries policy, development of subsistence fisheries, the rights of recreational fishers and the formation of Marine Protected Areas.

PROFESSOR CHARLES GRIFFITHS is director of the Marine Biology Research Institute at the University of Cape Town and is internationally known for his research on crustaceans, mussels, linefish and predator-prey interactions on rocky shores. He is a co-author, with his wife and D. Thorpe, of the pocket guide *Seashore Life,* and with M. Picker and A. Weaving of *Field Guide to the Insects of South Africa.* His natural history photographs have been widely published in books and magazines around the world.

DR LYNNATH BECKLEY is a dynamic fish biologist who is currently a Senior Lecturer in Marine Science at Murdoch University, Western Australia. She has published widely on many aspects of the biology of fishes, but is equally well known among yachting enthusiasts as an author and commentator on ocean yachting, and is a superb navigator.

MARGO BRANCH is a biologist with wide interests and has been involved in seven natural history books including *Trees of Southern Africa, Seaweeds of South Africa,* and monographs on various groups of lilies. She co-authored *A Field Guide to Mushrooms of South Africa,* and wrote and illustrated *Explore the Seashore* and *Explore the Cape Flora and Fauna,* which are widely used by groups concerned with environmental education among children. She is presently developing coastal environmental education posters.

Contents

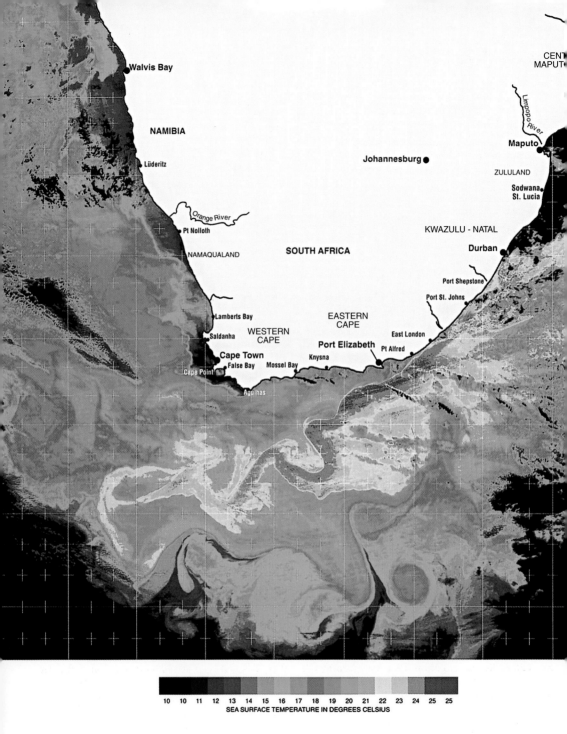

| 10 | 10 | 11 | 12 | 13 | 14 | 15 | 16 | 17 | 18 | 19 | 20 | 21 | 22 | 23 | 24 | 25 | 25 |
SEA SURFACE TEMPERATURE IN DEGREES CELSIUS

Satellite photograph of the coast of Southern Africa

The mighty Agulhas Current drives down the East Coast of southern Africa, bringing warm water (red) from the tropics. On the West Coast, cold, nutrient-rich upwelled water (blue-black) drifts northwards. Five biogeographic provinces can be recognised. The cold temperate Namib Province runs from northern Namibia to Lüderitz; the cold temperate Namaqua Province from there to Cape Point; the warm temperate Agulhas Province from Cape Point to northern Transkei; the subtropical Natal Province from there to southern Moçambique; central Moçambique is tropical, but the precise boundaries of this province have not been defined.

Introduction

There are about 900 species of birds in southern Africa, and no less than 15 books available to identify them. There are over 12 000 species of marine plants and animals, but not a solitary field guide is currently available for all the diverse groups in the rich fauna and flora of this region. That is why this book was written. J.H. Day's *A Guide to Marine Life of South African Shores,* published 35 years ago, is dated in its information and has been out of print for many years. Thirty years ago, *The Living Shores of Southern Africa* was published to communicate the exciting research that had been done on sea life, particularly on coastal ecosystems. The present field guide has a different, and complementary, purpose: to allow the ready identification of the most common forms of marine life that inhabit our coasts – including the invertebrates, fish, reptiles, mammals and plants.

Southern Africa (stretching from northern Namibia to southern Moçambique) has a particularly rich marine fauna and flora – over 12 000 species, or almost 6% of all the coastal marine species known world-wide. About 31% of these are endemic, occurring nowhere else. Given this rich assemblage, it is impossible to cover every species in one book. Instead, we have chosen to focus on the most frequently encountered species that live in the intertidal zone and in shallow subtidal waters that can readily be explored by scuba-divers. Not all groups of animals or plants are covered in the same depth. The emphasis falls on groups that are diverse, most often encountered, and poorly covered in other field guides. In particular, the focus is on the life of open-coast beaches and rocky shores, although aquatic estuarine animals and plants are also covered. The coverage of birds is limited to the coastal shoreline. Selecting which fish should be included was another difficult task. The focus falls on species commonly seen in tide pools, by divers, or frequently caught by anglers. Special attention has been given to the smaller rock-pool fish that have largely been ignored in other popular field guides.

Finding Your Way Around the Guide

The outline of contents on pp. vi-vii includes a pictorial guide to the major groups of animals and plants in the text. In the sections dealing with each group of organisms there are brief introductory paragraphs that outline the characteristics and key features used to identify the species. We have kept technical terms to a minimum, but where they are necessary they are defined in the introductory paragraphs and also in the Glossary (p. 350).

Almost all the species are illustrated with a colour photograph. In a few cases, line drawings or detailed photographs show details necessary for identification. These are indicated by 'd' after the figure number. Where juveniles are shown, the figure number is followed by a 'j'. On the page facing the photograph, each species is described under the following headings:

IDENTIFICATION: key features used to recognise the species.

SIZE: gives an indication of the size a species normally achieves; in the case of fish, maximum sizes and weights are given. Strictly speaking, size should always be given in millimetres, but this convention becomes clumsy when dealing with larger individuals. We have opted to express sizes for species smaller than 100 mm in millimetres, those of medium-sized ones (10–100 cm) in centimetres, and larger animals and plants in metres.

BIOLOGY: notes on where the species can be found, its diet, any particular relationship it has with other species, and other interesting aspects of its biology.

MAPS: accompany the text for each species and show its geographic range.

RELATED SPECIES: some species are dealt with less fully, only being given in sufficient detail to allow them to be distinguished from relatives or similar species which are described more fully.

Identification of an animal or plant should first involve comparison with the photographs: species with similar features have been grouped together to simplify comparison. Having matched the specimen with a photograph, one should read the accompanying text to check if it fits the description. Resist the temptation to 'force' a name on an animal or plant if it does not conform to the description: sometimes you will discover species that are not in the guide and may even be undescribed new species. For those enthusiasts wishing to pursue these rarities further, detailed scientific monographs on particular groups are listed in the References (p. 352).

Currents and Tides

Part of the reason for southern Africa's rich and varied fauna and flora is the extreme contrast between the water masses on the East and West coasts. The Agulhas Current, one of the most powerful currents in the world, sweeps warm water from the subtropics down the East Coast. In the region of East London, the continental shelf widens so that the Agulhas Current is further offshore and the coastal waters become slightly cooler. Eventually the Agulhas Current swings back on its tracks (retroflects) to flow eastwards. Conditions on the West

Coast are quite different. It is chilled by northward-drifting cold water. Wind blows the surface waters offshore, and deep water upwells near the coast to replace it. Because this water comes from depths where it is too dark for plant life to grow, it is not only cold but also rich in nutrients. Reaching the sunlit shallows, these nutrient-rich waters fertilise microscopic floating plant life known as phytoplankton. Both phytoplankton and seaweeds are far more productive on the West Coast than on the South and East coasts, and fuel more productive foodchains, culminating in the lucrative fisheries that are concentrated in this region. Although productivity is high on the West Coast, it supports many fewer species than the East Coast, which is particularly diverse because a large suite of tropical Indo-Pacific species contributes to its fauna and flora.

With the changes in temperature around the coast come accompanying changes in the marine life. For example, the East Coast boasts a large array of crabs and many species of corals that are absent from cooler waters. The West Coast, on the other hand, has prolific kelp forests that are absent from the warmer, less productive South and East coasts. Because of these faunal and floral changes, we can recognise five distinct biogeographic provinces, as illustrated on p. viii.

One of the most frequent questions asked is where the influence of the warm Indian Ocean ends and that of the cold Atlantic begins. Even scientists will disagree on the issue, because the retroflection of the Agulhas Current moves back and forth, setting a boundary that lies between Cape Point and Cape Agulhas but is difficult to define precisely. However, if we believe what the animals are telling us, Cape Point is the dividing line, for there is a decisive difference between the fauna north-west and south-east of Cape Point.

Another factor affecting the distribution of seashore plants and animals on a smaller scale is the tides. Twice a day the tide rises and falls, reaching its greatest range during the spring-tides that occur during full and new moons. At this time the gravitational forces of the sun and moon act together to drive the tide higher and lower than normal. Neap tides, when the tidal range is at its smallest, occur during the first and third quarters of the moon, when the gravitational pull of the two bodies act in opposition. The best time to visit the shore is at about 10 a.m. on days immediately following full or new moons: predictably (and obligingly), low spring-tides always fall at these times in southern Africa.

At the high-tide level, the shore is seldom submerged; at the low-tide level it is almost continually bathed. Between these extremes lies the intertidal zone; and below it the subtidal zone. In the intertidal zone there is a gradient of physical stress. Solar heating and water loss impose their greatest stresses on the high-shore. As a result, few species survive there. Tiny snails (*Afrolittorina* species) are often the only abundant occupants, and give their name to this region of the shore, which is termed the Littorina Zone. Below that is a zone dominated by barnacles, the Upper Balanoid Zone, named after the barnacle *Balanus*. Lower still, seaweeds or mussels or both become important, in the Lower Balanoid Zone. These zones (or their equivalents) are found on shores throughout most of the world. In southern Africa there are, additionally, three unique zones. The East Coast has a high-shore band of oysters; the South and North-west coasts have low-shore belts dominated, respectively, by the limpet *Scutellastra cochlear* (the Cochlear Zone) and the limpet *Scutellastra argenvillei* (the Argenvillei Zone). These unique zones are of special interest because the organisms that create them have a powerful influence on other life. The oysters provide shelter to many species and a food-source for others. The limpets are so concentrated that they virtually prevent any seaweeds from becoming established in 'their' zones.

Exploitation and Conservation

Southern Africa has a long, diverse and, by global standards, still relatively pristine coastline. Its shores offer a wide range of renewable natural resources that can, and should, be shared and enjoyed by fishermen, collectors and nature-lovers for generations to come. Each of these resources is, however, limited in that it can provide only a certain yield. Repeated removal of catches in excess of this maximum sustainable yield will inevitably lead to the collapse of the resource – as history has all too frequently demonstrated. Virtually all exploitable marine species in southern Africa, as elsewhere, are under increasing exploitation pressure. This results not only from growth of the human population (southern Africa already has less than 10 cm of coastline per person) but from the growing mobility of the population and popularity of such sports as angling and diving. There is also an increasing reliance of rural people on the sea for subsistence. As the number of individuals exploiting any resource increases, it is imperative that each person's catch is restricted, such that the total take remains below the sustainable limit. A variety of control measures has been instituted to achieve this. These regulations are compiled on the basis of available information on both the biology of the species concerned and current catch rates. Regulations are thus subject to periodic change, both as better biological data become available and as the number of resource-users changes. The tables in Appendices 1 & 2 (pp. 348–349) summarise the most important regulations as they stood at the time of going to press. Readers should also familiarise themselves with local by-laws and the locations of any marine reserves in the areas where they wish to collect. Note that seashore regulations in KwaZulu-Natal differ markedly from those in the Western, Eastern and Northern Cape provinces. The following free pamphlets are available and should be consulted for more detailed information about current regulations:

A Guide to the Coastal Fishing Regulations of Natal – Ezemvelo KwaZulu-Natal Wildlife, P.O. Box 662, Pietermaritzburg 3200.

Marine Conservation: Do's and Don'ts – Chief Directorate, Sea Fisheries, Private Bag X2, Roggebaai, 8012.

It is not the intention of regulations to spoil the enjoyment of resource users, but rather to conserve the shore so that it can continue to be enjoyed, and exploited, into the future.

The Classification and Naming of Species

Each species in this volume is accorded both its formal scientific or 'Latin' name and a common name. Both systems have their assets. Well-established common names exist for most familiar forms, and many users find these more descriptive and easier to pronounce and remember than the scientific names. Paradoxically they also tend to be more stable over time. On the other hand common names lack any formal status, and different names are applied to the same animal not only in each of many languages, but often also from region to region, making them an inaccurate means of communication. Worse still, the same common name has sometimes been applied to a number of (sometimes unrelated) species, while no common names exist for thousands of smaller or less abundant species. Unique scientific names, on the other hand, are established for every known species and these are universally recognised – even in Arabic, Russian or Japanese texts! Although we have attempted to allocate standard common names to each species, we encourage users to adopt the more powerful scientific terminology. A brief explanation of how this operates and is correctly used is given below.

Scientific Names

The system of scientific names, or scientific nomenclature, is best thought of as a hierarchical 'address' system in which each species is positioned according to its relationship with others. At the broadest level the animal kingdom is divided into phyla, or groups which share a similar overall body plan. Each phylum is then subdivided into more closely related classes, which in turn contain orders, families, and finally genera and species. Every species is allocated a pair of names (rather like the names of a person at an address). The first word of this bino-

mial, which is always written with an initial capital letter, identifies the genus, while the second, which appears in lower case, identifies the individual species. Closely related species will thus share the same generic name. The names of the genus and species are always printed in italics or, if handwritten, are underlined. For example, the classification of the well-known pear or ear limpet is as follows:

Phylum: Mollusca
Class: Gastropoda
Order: Patellogastropoda
Family: Patellidae
Genus: *Scutellastra*
Species: *cochlear*

Note that this scientific name, like many others, is descriptive of the species, the term *Scutellastra* referring to the star-shaped shell of some limpets, and *cochlear* to the 'ear' shape of this particular species. A further name and date are sometimes given after the binomial and identify the person who first named the species and when. If the name and date appear in brackets, this indicates that the original generic name has changed. In the above example, the full description is thus *Scutellastra cochlear* Born, 1778.

Occasionally variants within a species are recognised, and these may be given a third or subspecies name. Names may also be abbreviated, most commonly by reducing the generic name to an initial when it is used repeatedly (e.g. *Scutellastra cochlear* subsequently becomes *S. cochlear*). Members of a genus may also be referred to inclusively by the notation 'spp.' (e.g. *Scutellastra* spp.) or a single unnamed species as 'sp.' (*Scutellastra* sp.).

Why Names Change

One of the most frustrating and misunderstood aspects of scientific nomenclature is that scientific names may change over time. The reasons for this relate both to the hierarchical nature of the 'address' system and to strict rules of precedence, which dictate that each species may have only one valid name, which must be the earliest one given to the species. As additional species are discovered and more is learnt about their relationships to one another, scientists may decide to amalgamate existing genera (resulting in replacement and loss of the younger generic names), subdivide existing ones (resulting in erection of new generic names for those species split off from the original group), or simply to transfer a species from one genus to another. Research may equally reveal that two or more existing species are in reality simply variants of a single species (resulting in the invalidation of the more recent name and application of the older one to both forms). Alternatively, a single name may have been applied to what are in fact separate species, necessitating the creation of a new name for one of these.

Classification System

*Groups not covered in this book

Kingdom Animalia

PHYLUM PORIFERA   Sponges

PHYLUM CNIDARIA
  Class Hydrozoa
        Order Hydroida   Sea firs or hydroids
        Order Milleporina   Allopora
        Order Siphonophora   Bluebottles
  Class Scyphozoa   Bell-shaped jellyfish
  Class Cubozoa   Box-shaped jellyfish
  Class Anthozoa
     Subclass Octocorallia
        Order Alcyonacea   Soft corals
        Order Gorgonacea   Sea fans
        Order Pennatulacea   Sea pens
     Subclass Zoantharia
        Order Actiniaria   Sea anemones
        Order Coralliomorpha   Corynactis
        Order Zoanthidea   Zoanthids
        Order Scleractinia   Corals

PHYLUM CTENOPHORA   Comb jellies

PHYLUM PLATYHELMINTHES
  Class Turbellaria   Free-living flat worms

PHYLUM NEMERTEA   Ribbon worms

PHYLUM NEMATODA*   Round worms

PHYLUM SIPUNCULIDA   Peanut worms

PHYLUM ANNELIDA   Segmented worms
  Class Oligochaeta*   Earthworms
  Class Hirudinea*   Leeches
  Class Polychaeta   Bristle-worms

PHYLUM ARTHROPODA
Subphylum Hexapoda
  Class Collembola   Springtails
  Class Insecta   Insects
Subphylum Chelicerata
  Class Pycnogonida   Sea spiders
  Class Arachnida
        Order Araneae   Spiders
Subphylum Crustacea
  Class Ostracoda   Seed shrimps
  Class Copepoda   Copepods
  Class Cirripedia   Barnacles
  Class Malacostraca
        Order Phyllocarida   Nebalia
        Order Stomatopoda   Mantis shrimps
        Order Tanaidacea   Tanaids
        Order Cumacea   Cumaceans
        Order Isopoda   Isopods
        Order Amphipoda   Amphipods

        Order Mysidacea   Opossum shrimps
        Order Euphausiacea   Krill
        Order Decapoda   Prawns, lobsters, crabs

PHYLUM BRYOZOA   Moss or lace animals

PHYLUM BRACHIOPODA   Lamp shells

PHYLUM MOLLUSCA
  Class Polyplacophora   Chitons
  Class Bivalvia   Bivalve molluscs
  Class Scaphopoda   Tusk shells
  Class Gastropoda
     Subclass Prosobranchia   Snails, limpets
     Subclass Opisthobranchia   Sea slugs
  Class Cephalopoda   Octopus, squid

PHYLUM ECHINODERMATA
  Class Asteroidea   Starfish
  Class Crinoidea   Feather stars
  Class Ophiuroidea   Brittlestars
  Class Echinoidea   Sea urchins
  Class Holothuroidea   Sea cucumbers

PHYLUM CHORDATA
Subphylum Tunicata
  Class Ascidiacea   Sea squirts
Subphylum Vertebrata
Superclass Agnatha   Hagfish
Superclass Pisces
  Class Chondrichthyes
        Subclass Elasmobranchii   Sharks, rays
        Subclass Holocephali   St. Joseph shark
  Class Teleostomi   Bony fish
Superclass Tetrapoda
  Class Amphibia*   Frogs
  Class Reptilia
        Order Cheloni   Turtles
        Order Squamata   Snakes & Lizards
  Class Aves   Birds
  Class Mammalia
        Order Cetacea   Whales, dolphins
        Order Pinnipedia   Seals
        Order Carnivora   Otters

Kingdom Plantae

DIVISION CHLOROPHYTA   Green algae

DIVISION PHAEOPHYTA   Brown algae

DIVISION RHODOPHYTA   Red algae

DIVISION SPERMATOPHYTA   Seed plants
  Subdivision Angiospermae   Flowering plants

THE SEA

Harbinger of eternity,
Time's time-keeper.
She is, in all her vast diversity,
the reflection of man's soul.

Porifera : Sponges

Sponges are primitive, sedentary animals which lack a mouth, digestive tract or any other conventional organ. They consist of a few types of cells, forming tissues that are supported by a skeleton made of a fibrous material called collagen or spongin (as in the familiar bath sponge), or of lime or silica spicules, or of a combination of these. The type of skeleton and form of the spicules are important features in the classification of sponges. Water and small food particles enter the sponge through numerous tiny pores which dot the body surface and exit via one or a few larger openings, or oscula, which are often raised, like turrets. Collar cells line the body cavity and beat their hair-like flagellae to generate the water flow and filter out food particles through their net-like collars. Sponges reproduce either by budding or by producing planktonic larvae. Although sponges are conspicuous on rocky reefs, particularly in temperate areas, the southern African species are extremely poorly known, and many remain unnamed. As a result, the following account includes only those few forms to which names can be attached.

1.1 Encrusting turret sponge   *Haliclona oculata*

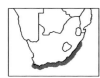

IDENTIFICATION: Oscula raised on turrets, which can be very tall and tubular in more sheltered locations. Colour purple. SIZE: 50 cm or more across, turrets up to 10 cm long. BIOLOGY: Encrusting sheets growing on the walls of intertidal pools or on subtidal reefs. RELATED SPECIES:

1.2 *Haliclona tulearensis* (Sodwana northwards) is bright yellow with conical turrets and a pocked surface.

1.3 *Haliclona* sp. is a conspicuous blue to mauve form common in rock pools on the Natal coast, but appears to be unnamed.

1.4 Crumb-of-bread sponge   *Hymeniacedon perlevis*

IDENTIFICATION: Thick, encrusting, ochre-yellow sheet with a rough, lumpy surface texture. Oscula on short, rather flattened turrets. SIZE: Typically 10 cm across. BIOLOGY: The most abundant intertidal sponge, found in crevices and on shaded rock surfaces up to mid-tide level. RELATED SPECIES:

1.5 *Hymeniacedon* sp. is a similar but bright red species, and is abundant in Natal and Moçambique. Its identification is still uncertain.

1.6 Golf ball sponges   *Tethya* spp.

IDENTIFICATION: Easy to identify to genus by virtue of the spherical body form. Surface usually with warty knobs. Several species probably represented. SIZE: Typically 5–10 cm diameter. BIOLOGY: Often occur in large colonies on shallow subtidal reefs.

1.7 Hairy tube-sponges   *Sycon* spp.

IDENTIFICATION: Small, erect, tube- or vase-shaped sponges. Surface white and covered in hair-like spicules. A crown of longer, stiffer spicules surrounds the osculum at the apex. SIZE: Typically 10 mm. BIOLOGY: Generally growing attached to hydroids, bryozoans or other marine growths in gullies and caves or on vertical rock-faces.

1.8 Branching ball-sponge   *Leucosolenia* sp.

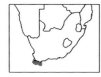

IDENTIFICATION: Small, white hemispherical to spherical sponges which often send out 'roots' or extensions from which other balls bud. SIZE: Balls 10–20 mm diameter. BIOLOGY: Usually under loose rocks near low tide. The solid appearance is a deception as this sponge is formed of a network of tubes (as in 1.9) which have become fused such that only a few openings remain.

1.9 Tube-sponges   *Leucosolenia* spp.

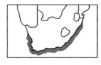

IDENTIFICATION: Small, elongated whitish tubes with a large osculum at the end of each. Bases united to form a colony. SIZE: Tubes 20 mm long. BIOLOGY: Found in groups growing amongst algae or encrusting animals on the walls of pools or under overhangs.

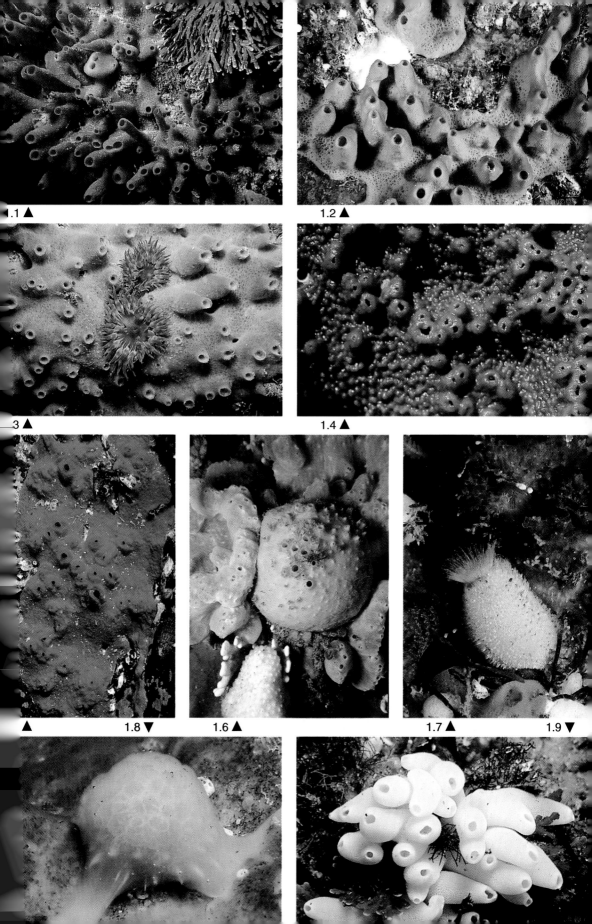

1.1 ▲   1.2 ▲

3 ▲   1.4 ▲

▲   1.8 ▼   1.6 ▲   1.7 ▲   1.9 ▼

2.1 **Teat sponge** *Polymastia mammillaris*

IDENTIFICATION: A tough brown encrustation covered in regularly-spaced teat-like projections which carry both the inhalant and exhalant apertures: their tips are dotted with tiny holes. The flesh is firm and leathery to the touch. SIZE: Up to 30 cm across, and about 25 mm thick. BIOLOGY: Spreads over flat rock surfaces in gullies and sandy pools, and often grows beneath rocks. In areas where the colonies are scoured by the flailing of kelp plants, the 'teats' are short and flat and separated by 'valleys', giving the sponge a convoluted surface a little like the texture of a brain.

2.2 **Vented sponge** *Spirastrella sp.*

IDENTIFICATION: A flat, encrusting orange sponge with numerous, large exhalant openings, which project to form short thin-walled columns. SIZE: 20 cm across. BIOLOGY: Grows in pools and on shallow reefs in more tropical seas. The specimen photographed here carries two nudibranchs, *Jorunna zania* (85.1), which may feed on it.

2.3 **Orange wall-sponge** *Spirastrella spinispirulifer*

IDENTIFICATION: A massive sponge with an orange-to-red exterior 'skin' and yellow 'flesh'. Usually grows as a smooth, fairly regular wall without obvious oscula. SIZE: Typically 20 cm thick, 20–30 cm wide, and as much as 2 m long. BIOLOGY: The largest and most conspicuous sponge on reefs 10–30 m deep in the Western Cape. A chemical extracted from this sponge shows promise as an anti-cancer drug and is currently undergoing clinical tests.

2.4 **Tree sponge** *Clathria sp.*

IDENTIFICATION: A red-brown tubular sponge with an erect tree-like growth form. Oscula not obvious. SIZE: About 10 cm in height. BIOLOGY: Found on shallow reefs and occasionally in deep pools or under overhangs in the low intertidal. It is quite often overgrown by a zoanthid (*Parazoanthus* sp.), as shown in Pl. 5.2.

2.5 **Cup sponge** *Ircinia sp.*

IDENTIFICATION: A very large grey-brown sponge with a characteristic cup or goblet shape and rough ridged external sculpturing. SIZE: Up to 1 m across. BIOLOGY: A conspicuous resident of deeper tropical reefs. The cup-like shape may be a means of concentrating food particles, which are likely to settle in the 'cup'. Like the coral *Turbinaria* (12.1), the 'cup' may also serve to concentrate silt at one point, so that the whole colony is not smothered.

2.6 **Crumpled sponge** *Axinella weltneri*

IDENTIFICATION: A very hard, almost woody red sponge with a strongly ridged surface, a rough, spiny surface texture and no oscula. The growth form is upright, resembling crumpled corrugated cardboard. SIZE: 5–10 cm across. BIOLOGY: Occurs on shallow tropical reefs, in gullies, and occasionally in rock pools. Tends to grow on vertical surfaces, possibly to avoid the effects of siltation.

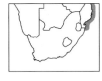

2.7 **Black stink-sponge** *Ircinia sp.*

IDENTIFICATION: Forms a thick, flat black mat with rounded edges. Surface texture firm but slippery to the touch. When rubbed it leaves a strong pungent smell on the fingers. Oscula hardly raised, but surface textured by a network of coarse spongin fibres. SIZE: Average 1–2 cm thick and 10–20 cm across. BIOLOGY: An encrusting species found in deep rock pools or caves low in the intertidal or on shallow reefs. Sand-grains and other debris are often embedded with the tissues. The smell is a reliable means of identification.

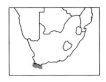

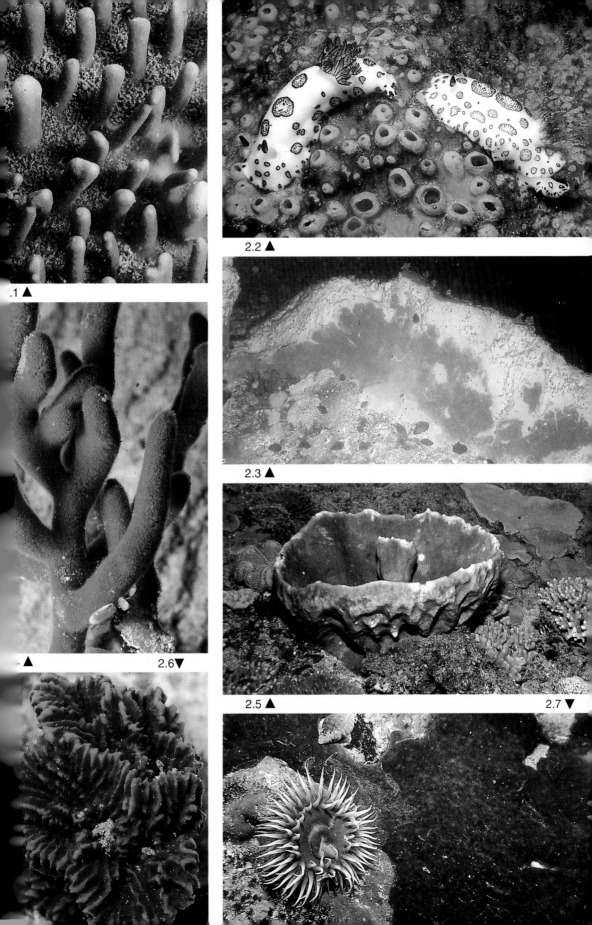

.1 ▲

2.2 ▲

2.3 ▲

2.6▼

▲

2.5 ▲

2.7 ▼

Cnidarians

The phylum Cnidaria includes the jellyfish, anemones, corals and hydroids. All have a simple structure. Their bodies can be visualised as being sac-like, with only two cell layers, an outer 'skin', or ectoderm, and an inner lining to the gut, called the endoderm. The centre of the body is taken up by a cavity that serves as the gut, but the mouth is the only entrance into this gastric cavity, there being no anus. There are no specialised organs for respiration or excretion.

Despite their simplicity, cnidarians have their own unique features, and are very successful when viewed in terms of their diversity and the fact that many are familiar because of their abundance. They were also amongst the earliest forms of multicellular life to evolve on earth, having arisen at least 650 million years ago. Their most obvious unique feature is the possession of highly specialised stinging cells (nematocysts). These vary in shape and function, but usually house a coiled thread-like sting that can be rapidly discharged in attack or defence to penetrate the skin of the target and release toxins into its body. The painful effects of a sting from a bluebottle are well known to bathers.

The phylum is divided into four classes. The Hydrozoa are the least familiar because most species are small. They have a complex life cycle involving alternation between two types of body plan. The asexual phase consists of a tubular body that is attached to the substratum at the lower end and bears a ring of tentacles around the mouth at the opposite end. Such individuals are called polyps. They asexually bud off the second stage, which resembles a miniature jellyfish, termed a medusa. Medusae are shaped like umbrellas, with a pulsing bell that propels them through the water, and a central trailing manubrium that carries the mouth below the bell. Medusae are the sexual phase in the life cycle, and produce sperm and eggs that fuse to yield planktonic larvae. Ultimately these settle and give rise to polyps. A few species have solitary polyps, but in most cases the polyps divide repeatedly to form colonies that are often feather-like in appearance. An extreme example of division of labour within a colony occurs in the bluebottle: one individual forms the float, others serve only as defensive stinging tentacles, yet others are for digestion, and some serve solely as organs of reproduction.

The anemones are a more familiar example of the second class, the Anthozoa. They have only a polyp stage in the life cycle, and have much larger and more complex polyps than the Hydrozoa. Their gastric cavities are divided vertically by sheets that extend inwards from the body wall, and increase the surface area for digestion. While anemones are solitary, many anthozoans also form colonies of polyps, as are found in the sea fans. Corals are also colonial anthozoans, and secrete a hard, calcium carbonate skeleton that supports a relatively thin skin of living tissue.

The remaining classes, Scyphozoa and Cubozoa, have developed the medusa stage at the expense of the polyp stage, which is reduced to a short-lived larval stage. Both classes consist of jellyfish, some of which attain remarkable sizes, exceeding a metre in diameter. The Scyphozoa comprise the jellyfish familiar to most people, most of which have numerous long, trailing tentacles around the margin of the bell. The Cubozoa are box-shaped and have only four, very long tentacles. The group includes one species which has probably the most toxic venom of any animal in the world – the sea wasp, which does not occur in southern African waters.

Actiniaria : Sea Anemones

Simple solitary animals lacking a hard skeleton, but supported by internal water pressure. Body hollow, cylindrical, attached at the base by a flat adhesive disc. The mouth is ringed by tentacles. These are armed with stinging cells but are harmless to humans. Prey captured with the tentacles is stuffed through the mouth into the digestive cavity. Reproduction can be sexual or by simple division of the body.

3.1 Strawberry anemone   *Corynactis annulata*

IDENTIFICATION: The white knobs on the tips of the tentacles and exquisite semi-transparent pink colour are distinctive. SIZE: 10 mm diameter. BIOLOGY: More closely related to the corals than the true anemones, this small colonial species occurs in clusters under ledges in low-shore pools and in extensive sheets on shallow reefs. Probably feeds on small planktonic organisms.

3.2 Striped anemone   *Anthothoe chilensis*

IDENTIFICATION: Delicate green, pink or yellowish-brown, sometimes very pale. Column smooth, and vertically striped. Shoots sticky white defensive threads through the body wall when disturbed. SIZE: 20 mm diameter. BIOLOGY: Abundant in rock pools and under boulders from mid-tide down. Harbours symbiotic algae in its tissues. Fed upon by the sea slug *Aeolidiella indica* (83.6) Previously *A. stimpsoni*.

3.3 Plum anemone   *Actinia equina*

IDENTIFICATION: Plum red with a glossy, smooth column and a row of bright-blue bead-like spherules hidden under the outer row of tentacles. SIZE: Typically 20 mm diameter. BIOLOGY: Survives out of water high on the shore by closing up tightly when exposed at low tide, trapping water within the body cavity. Often found hanging like ripe fruit in shady gullies and overhangs. The young are brooded within the body cavity and are born as fully-developed young anemones.

3.4 False plum anemone   *Pseudactinia flagellifera*

IDENTIFICATION: Column smooth, orange or red. Cannot readily retract tentacles or close up; 3–5 rows of bubble-like orange vesicles and one row of spherules just outside tentacles. Tip of tentacles often mauve. SIZE: 50–100 mm. BIOLOGY: Large and conspicuous, found singly or in small family groups in pools or on shallow reefs. Can move slowly about, inflating vesicles to attack and sting unrelated individuals, thus maintaining its territory. Preys mostly on molluscs and crustaceans. Has the most potent venom known for an anemone, even affecting humans if tentacles contact cuts in the skin. RELATED SPECIES: *Pseudactinia varia* (Cape Town to East London) is smaller (20 mm), has only 1–2 rows of vesicles, and is usually confined to intertidal pools.

3.5 Knobbly anemone   *Bunodosoma capensis*

IDENTIFICATION: Column with small non-adhesive knobs (papillae). These are sometimes brilliantly pigmented to contrast with the background colour, which can be white, pink, orange, red, blue or purple. Spherules present. SIZE: Generally 20–40 mm diameter. BIOLOGY: A strikingly variable species abundant in sand-free areas; often found attached to mussels or red-bait. Sand never adheres to the column (compare with 3.6).

3.6 Sandy anemone   *Aulactinia reynaudi*

IDENTIFICATION: Column covered in flattened sticky knobs, to which sand and gravel particles adhere. Over 300 short tentacles, no spherules. Colours include brown, green, pink and blue. SIZE: Up to 80 mm diameter. BIOLOGY: Often crowded into sandy gullies, around the bases of boulders or the margins of pools. Feeds on dislodged mussels, sea urchins, whelks and other animals tumbled by the waves. Juveniles common in mussel beds. Previously known as *Bunodactis reynaudi*.

3.3 ▲

3.4 ▼

▲   3.2 ▼

▼

3.6 ▼

4.1 **Long-tentacled anemone** *Anthopleura michaelseni*

IDENTIFICATION: Column with large sticky pads, to which shells and gravel adhere. Dark stripes usually radiate from mouth to the 96 long pink-to-brownish tentacles. Spherules present. SIZE: Reaches about 70 mm diameter. BIOLOGY: Lives partially buried in sand in rock crevices and sandy pools from mid-shore downwards.

4.2 **Giant anemone** *Heteractis magnifica*

IDENTIFICATION: Forms flat sheets covered with hundreds of tentacles which waft in currents. All the tentacles are identical, uniform in diameter and longer than 30 mm. Column brightly coloured, often red, purple or green, and with longitudinal rows of papillae that are the same colour as the column or slightly lighter. SIZE: Often reaches a width of 50 cm. BIOLOGY: Attaches itself to rocks or dead coral heads, and characteristically displays itself very prominently. Carnivorous, but plays host to anemone or clown fish (Pl. 126). May cause irritation to human skin and raise welts. Previously called *Radianthus ritteri*. RELATED SPECIES: *Stichodactyla mertensii* (Sodwana northwards) is gigantic, reaching 1 m in diameter, and is flat and sprawling. Its upper surface is densely covered with tentacles that are uniform in diameter and nearly all short (10–20 mm), although patches of tentacles may be longer. Lives on coral reefs, buried in sand among coral debris or in seagrass beds on sheltered tropical sandbanks. Hosts anemone-fish and commensal shrimps. *Entacmea quadricolor* (Sodwana northwards) is not as common and usually hides in holes and crevices with only its tentacles protruding. The tentacles are distinctive, being very long (100 mm) and having small onion-like bulges near their tips.

4.3 **Brooding anemone** *Halianthella annularis*

IDENTIFICATION: A pale, slender anemone distinguished by having only 24 elongated, transparent tentacles. SIZE: Reaches only about 10 mm diameter. BIOLOGY: Found in sheltered areas, particularly under boulders. A fold in the skin part-way up the column is used to brood the young, which emerge as fully-formed young anemones.

4.4 **Violet-spotted anemone** *Anthostella stephensoni*

IDENTIFICATION: Column yellow with bright violet spots. Scarlet lines radiate from the mouth to the 48 short, blunt tentacles. Column dotted with mauve spots. SIZE: About 30 mm diameter. BIOLOGY: A strikingly-coloured but little-known species usually found singly or in small groups in sandy sites near or below low tide. Previously placed in the genus *Anthostella*.

4.5 **Natal anemone** *Anemonia natalensis*

IDENTIFICATION: Column smooth and red-brown, with one row of spherules beneath the 70–100 cream-to-brown tentacles. Upper surface plum-coloured with pale lines radiating onto tentacles. SIZE: About 20 mm diameter. BIOLOGY: The most common rock-pool anemone on the KwaZulu-Natal coast. Like many anemones, it has the ability to undergo asexual division of the body to create clones of identical individuals: often forms small groups as a result.

4.6 **Colonial anemone** *Gyractis excavata*

IDENTIFICATION: Column pale and dotted with purple papillae, to which coarse gravel adheres. Outer row of tentacles white with brown tips. Oral disc green, often brilliantly so around the mouth, from which brown lines radiate. Short brown tentacles sprout irregularly from the disc surface. SIZE: About 20 mm diameter. BIOLOGY: Its colonies form sheet-like expanses in intertidal pools along the KwaZulu-Natal coast. Easily mistaken for zoanthids, but differs in that the individuals are not joined at their bases. Previously called *Actinoides sultana*.

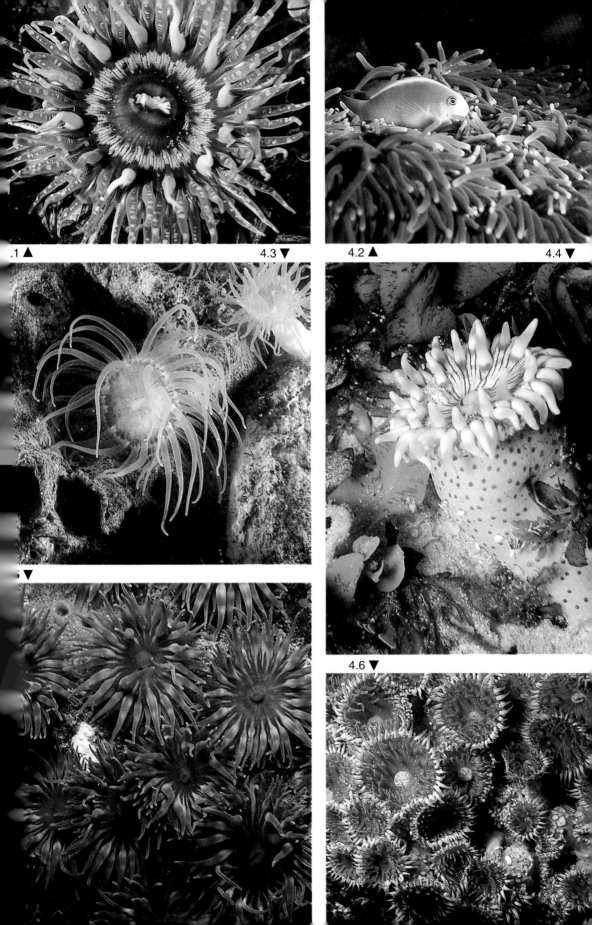

.1 ▲

4.3 ▼

4.2 ▲

4.4 ▼

4.6 ▼

Zoanthidea : Zoanthids

Zoanthids are anemone-like, their polyps forming upright, hollow columns crowned by tentacles around the mouth. Unlike anemones, they are colonial, their polyps being joined basally by a sheet-like coenenchyme, forming continuous carpets on subtropical and tropical shores. Zoanthids capture tiny prey but, like corals, most depend on microscopic symbiotic algae (zooxanthellae) in their tissues for much of their nutrition.

5.1 **Cape zoanthid** *Isozoanthus capensis*

IDENTIFICATION: Orange to pink, column coated with coarse sand; about 30 tentacles. SIZE: Polyps 12 mm tall, 5 mm diameter. BIOLOGY: Seldom abundant and forming small colonies of 20–50 individuals, the Cape zoanthid occurs under overhangs in shallow water and is the only zoanthid found on the South Coast. Lacking zooxanthellae, it feeds on plankton. RELATED SPECIES:

5.2 *Parazoanthus* **sp.** (Cape Peninsula) is never sand-coated and usually grows on sponges. Its mouth is mounted on a central finger-like projection of the oral disc.

5.3 **Knobbly zoanthid** *Isaurus tuberculatus*

IDENTIFICATION: Column tall, curving to one side and usually distinctly knobbly; grey to brown. Tentacles seldom visible. SIZE: Polyps 30 mm tall, 7 mm diameter. BIOLOGY: Occurs in deep pools and in the shallow subtidal zone. Forms extensive colonies, but is seldom common. Previously known as *I. spongiosus*.

5.4 **Columnar sandy zoanthid** *Protopalythoe nelliae*

IDENTIFICATION: Polyps tall and only thinly connected at their bases. Column brown, embedded with fine sand grains, giving a sandpapery texture; trumpet-shaped when expanded. The disc at the top ranges from bright green to brown. SIZE: 30 mm tall, 10-15 mm wide. BIOLOGY: Abundant in low-shore pools, especially in areas periodically covered by sand.

5.5 **Squat sandy zoanthid** *Palythoe natalensis*

IDENTIFICATION: Polyps short and squat, joined together by a thin basal sheet that forms grooves between adjacent colonies. Tissues embedded with fine sand, giving the column a slightly rough feel. Over 40 tentacles. SIZE: Polyps 15 mm tall, 10 mm diameter. BIOLOGY: Never grows on rocks exposed at low tide, but occurs in pools even quite high on the shore.

5.6 **Violet zoanthid** *Zoanthus sansibaricus*

IDENTIFICATION: Polyps tall (height 2–3 times width) and connected by a thin coenenchyme. Like all *Zoanthus* species, the column is smooth, never embedded with sand. Column usually violet to grey, disc green (often vivid), tentacles violet to dull green. SIZE: Polyps 10 mm tall, 4 mm diameter. BIOLOGY: Common low on the shore, particularly around pools. Intolerant of sand. RELATED SPECIES: *Zoanthus parvus* is also tall but distinctly smaller: 8 mm tall and 3 mm wide. Disc bright green, column pink.

5.7 **Durban zoanthid** *Zoanthus durbanensis*

IDENTIFICATION: Very similar to *Zoanthus natalensis*, being short and squat (height roughly equal to width); column powdery-grey; tentacles any grade between green and chocolate brown; oral disc green to grey. SIZE: Polyps 10 mm tall, 8 mm diameter. BIOLOGY: Same as *Zoanthus sansibaricus*.

5.8 **Green zoanthid** *Zoanthus natalensis*

IDENTIFICATION: Polyps short and squat, smooth-textured and joined together by a relatively thick basal coenenchyme. Grass-green, with darker green tentacles, colour darkening when the animal contracts. SIZE: Polyps 5 mm tall, 8 mm diameter. BIOLOGY: Grows on exposed rocks from low- to mid-tide levels, often higher on the shore than other zoanthids. RELATED SPECIES: 5.9 *Zoanthus eyrei* has a lime-green rim to its column and oral disc, and bright-green to khaki-green tentacles and a grey pharynx.

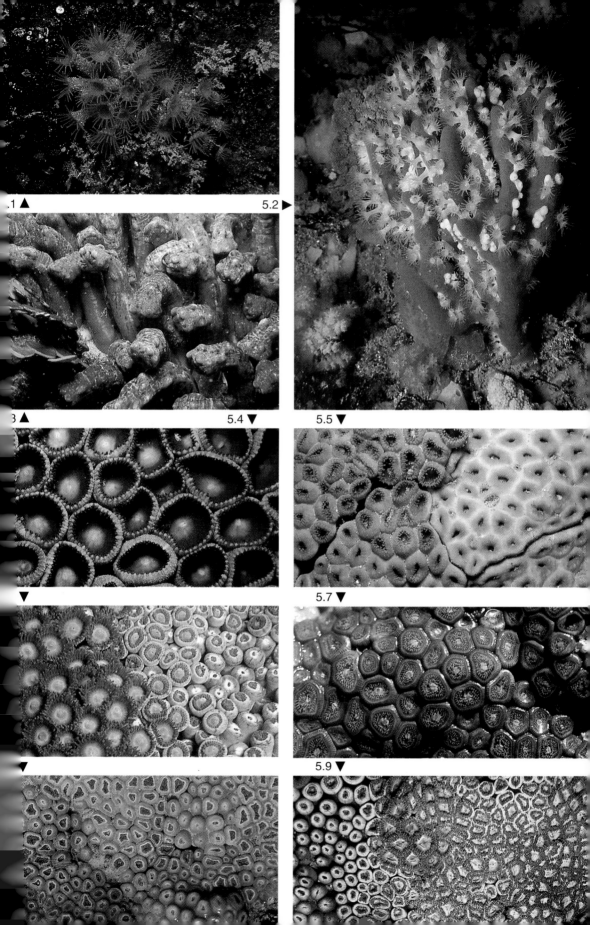

5.1 ▲

5.2 ▶

5.3 ▲

5.4 ▼

5.5 ▼

5.6 ▼

5.7 ▼

5.8 ▼

5.9 ▼

Alcyonacea : Soft Corals

Soft corals have no internal skeleton but form colonies of polyps, each carrying eight distinctive feathery (pinnate) tentacles. They are microcarnivores, using their tentacles (which contain stinging cells) to capture planktonic animals. To confirm the identification of different species, the spicules in their tissues need to be microscopically examined. However, most species can be recognised by external features.

6.1 Purple soft coral   *Alcyonium fauri*

IDENTIFICATION: Commonly vivid purple (see Pl. 24.3), less often yellow, pink or red and orange. Colonies encrust the substratum, but thrust out lobes covered with polyps. SIZE: Colony height 30 mm, polyp diameter 4 mm. BIOLOGY: Abundant in shallow subtidal zones. Its earlier (more descriptive) scientific name was *Parerythropodium purpureum* (meaning equal-sized red legs and purple colour).

6.2 Valdivian soft coral   *Alcyonium valdiviae*

IDENTIFICATION: A conspicuous stout trunk with stubby lobe-like branches that become almost globular when the colony is contracted. Colony colour extremely variable: white, pink, orange or brick-red, with white polyps. West Coast specimens usually bright orange. SIZE: 90 mm in height. BIOLOGY: A common inhabitant of rock walls and reefs between 14 and 18 m depths.

6.3 Variable soft coral   *Alcyonium variabile*

IDENTIFICATION: Shaped like a mushroom bud, with a short stalk and a round, unbranched head covered with polyps. Background colour extremely variable: red, purple, orange, yellow, pink or white, or combinations of these. SIZE: 50 mm length. BIOLOGY: One of the most common soft corals, regularly seen by divers. Wide depth-range: 13–468 m.

6.4 Stalked soft coral   *Xenia crassa*

IDENTIFICATION: Upright stalk that expands to swollen head resembling the shape of a mushroom bud. Head covered with polyps that are all identical and are incapable of withdrawing. Polyps borne on slender, banded stems and have eight obviously feathery tentacles. SIZE: Colony 30 mm, polyps 10 mm. BIOLOGY: This species is confined to the subtidal, but other species in the genus occur intertidally.

6.5 Cave-dwelling soft coral   *Carijoa* sp.

IDENTIFICATION: Pinky-orange. Attached by a narrow stalk that divides into short branches, each ending in a pink polyp with the eight pinnate tentacles. SIZE: Colony 30 mm, polyps 5 mm. BIOLOGY: Hangs down from the roofs of caves or rock overhangs; confined to the extreme low-shore. Resembles a bunch of orange grapes when the tentacles are withdrawn.

6.6 Blue soft coral   *Anthelia flava*

IDENTIFICATION: Forms small, flat carpets. Polyps tall and cannot withdraw, even when disturbed. Tentacles soft, floppy, royal blue to grey-blue. SIZE: Polyps 10 mm. BIOLOGY: Common in intertidal pools. RELATED SPECIES:

6.7 *Anthelia glauca* is twice the size, pale creamy-grey; recorded in 9–19 m off Sodwana.

6.8 *Clavularia* sp. has root-like stolons with upright branches ending in tiny (2 mm) bright-blue polyps which withdraw into their bases if touched.

6.9 Cauliflower soft coral   *Capnella thyrsoidea*

IDENTIFICATION: A large, soft, floppy colony; side-branches terminate in bunches of polyps. Colour variable, often translucent whitish-cream with transverse, opaque-white 'stretch-marks'. SIZE: Colonies up to 30 cm, polyps 2 mm. BIOLOGY: Hangs from vertical rock faces in depths of 10–240 m; often seen by divers. During the day, especially in clear water, it contracts tightly and resembles stalks of cauliflower.

1 ▲

6.2 ▲

▲   6.6 ▼

6.4 ▲   6.7 ▼

6.5 ▲   6.9 ▼

6.8 ▼

7.1 Thistle soft coral *Dendronephthya sp.*

IDENTIFICATION: Tree-like, with soft branches ending in white spikes and bright pink, yellow or red polyps, resembling exploding fireworks. SIZE: Colony 20 cm. BIOLOGY: Lives on tropical reefs at 12–50 m. The name is derived from *dendron*, tree, and *Nephthys*, an Egyptian goddess. RELATED SPECIES:

7.2 *Dendronephthya sp.*: similar but not as vivid, and its polyps lack spiky tips.

7.3 Sun-burst soft coral *Malacacanthus capensis*

IDENTIFICATION: A tough cylindrical, brown stalk crowned by a radiating ball of bright yellow polyps (autozooids), in-between which are tiny dot-like siphonozooids which serve to re-inflate the colony after it contracts. SIZE: Colony 15 cm tall when expanded. BIOLOGY: Occurs subtidally, in relatively deep water (10–40 m). When disturbed, the entire crown can fold inwards and be withdrawn into the top of the stalk.

7.4 Finger-lobed soft coral *Sinularia leptocladus*

IDENTIFICATION: Hard and rubbery. Forms flat growths with finger-like lobes. Covered by short polyps, which are all identical. Uniformly dull brown or grey. SIZE: Colonies 30 cm wide, polyps 1 mm. BIOLOGY: Occurs subtidally (10–30 m). RELATED SPECIES:

7.5 *Sinularia dura* has radiating twisted 'walls'.

7.6 *S. gyrosa* has upright radiating knobs, and *S. abrupta* radiating, wall-like ridges.

7.7 *S. heterospiculata* has distinctly spiny, knobbled branches.

7.8 Fleshy soft coral *Sarcophyton trocheliophorum*

IDENTIFICATION: Attached by a broad whitish stalk that mushrooms into a fleshy, funnel- or cup-shaped head which is soft and has two types of polyps: large, tentaculate polyps (autozooids), and tiny polyps (siphonozooids) that lack obvious tentacles and dot the surface. Small colonies are mushroom-shaped, larger ones are thrown into lobes. SIZE: Colonies 10–50 cm; large polyps 5 mm. BIOLOGY: Abundant at Sodwana, 2–30 m. RELATED SPECIES: *S. glaucum* is similar, but its polyps differ in being unable to retract fully and its stalk is yellow (bright pollen-yellow when cut open).

7.9 Blanching soft coral *Cladiella kashmani*

IDENTIFICATION: Colony leathery, attached by a broad creeping base, which gives rise to numerous short stalks and stumpy lobes. Dark chocolate brown or dark khaki-coloured when undisturbed. Dramatically turns almost white when touched, due to the withdrawal of the polyps. All the polyps are similar in size and structure. SIZE: Colony 30 cm in diameter, 10 cm tall. BIOLOGY: Common from Sodwana Bay northwards in moderately deep water (10–20 m).

7.10 Dimorphic soft coral *Lobophytum ¢rassum*

IDENTIFICATION: Low and creeping, with radiating ridges. Soft. Yellow-brown with pale polyps. *Lobophytum* differs from *Sinularia* in having two or more types of polyps: large autozooids (0.5 mm diameter) and tiny siphonozooids that dot the surface. It shares this feature with *Sarcophyton*, but has smaller polyps and is not stalked like *Sarcophyton*. SIZE: Colony 30 cm. BIOLOGY: Subtidal, 5–40 m. RELATED SPECIES:

7.11 *Lobophytum venustum* forms a flat crust with smooth folds. Green microalgal balls often dot its surface. *L. depressum* is smooth and flat, grey-to-green and soft. *L. patulum* may be smooth or has low humps, and is grey and stiffer to the touch.

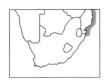

7.12 Organ-pipe coral *Tubipora musica*

IDENTIFICATION: Builds a bright-red skeleton in the form of parallel upright tubes, from which project daisy-like white polyps with eight flattened tentacles. SIZE: Colonies up to 25 cm tall, 'organ pipes' 1.5 mm in diameter, polyps 5 mm across the tentacles. BIOLOGY: Dead skeletons are a familiar sight washed up on tropical shores, but live animals are seldom seen, though strikingly beautiful. Subtidal, down to 50 m.

1 ▲　　　　　　　　　7.2 ▲　　　　　　　　　7.3 ▲

▲　　　　7.8 ▼　　　　7.5 ▲　　　　7.6 ▲　　　　7.7 ▲　　　　7.9 ▼

7.10 ▼　　　　　　　　　7.11 ▼　　　　　　　　　7.12 ▼

Gorgonacea & Pennatulacea : Sea Fans & Sea Pens

Gorgonians (sea fans) form tree-like colonies, often arranged like a fan. They are distinguished by a stiff central rod of gorgonin (a horn-like protein), covered with small polyps, each with eight feathery tentacles. Old skeletons, resembling miniature dead trees, often wash ashore. Pennatulaceans (sea pens) have a fleshy body covered with polyps and a soft, unbranched peduncle that anchors the colony in mud or sand.

8.1 Sinuous sea fan   *Eunicella tricoronata*

IDENTIFICATION: Branches characteristically flattened and sinuous, all tending to lie in one plane. Vivid orange-yellow. When the polyps are fully expanded they give the colony a fuzzy appearance because their tentacles obscure the outline of the branches. SIZE: Colony 30–40 cm tall, branches 7–10 mm wide. BIOLOGY: Occurs on subtidal rocky reefs in fairly deep water, 10–40 m, below wave surge.

8.2 Nippled sea fan   *Eunicella papillosa*

IDENTIFICATION: Arises as a single stock which forks at fairly regular intervals, usually forming a flat fan. Branches cylindrical, densely covered with nipple-like protuberances from which polyps extend. Uniform pale orange or creamy colour. SIZE: Colony 15 cm tall, branches 2 mm wide. BIOLOGY: Grows at depths of 2–20 m in caves or on sides of boulders sheltered from strong water movement.

8.3 Flagellar sea fan   *Eunicella albicans*

IDENTIFICATION: Colony divides near the base and then forms very long, thin, whip-like branches, conspicuously flattened near the base and often with longitudinal lines on the surface. SIZE: Colonies 30–40 cm tall, branches 3–5 mm wide. BIOLOGY: Grows profusely on deeper reefs (10–20 m) where wave action is minimal.

8.4 Palmate sea fan   *Leptogorgia palma*

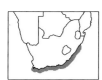

IDENTIFICATION: Flattened central stem gives rise to lateral branches. White polyps adorn the bright-red colony, but can withdraw into slits. SIZE: 2 m tall; branches 3–10 mm wide. BIOLOGY: Forms fantastic underwater forests at 10–100 m. Grows only 15 mm per year; large colonies are over 100 years old and easily exterminated by souvenir-hunting divers. It is eaten by a sponge crab (47.2) and a topshell (66.1). Previously called *Lophogorgia flammea*. RELATED SPECIES:

8.5 *Homophyton verrucosum* has sparse, round, finger-like branches that are covered with polyps. Colour extremely varied.

8.6 Multicoloured sea fan   *Acabaria rubra*

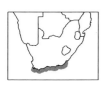

IDENTIFICATION: Stems cylindrical, forking irregularly, often fusing and tangled; surface dotted with knobs housing polyps. Colours diverse: white polyps on red stems, yellow on white, orange on yellow, or plain red. SIZE: 10 cm tall; branches 1 mm wide. BIOLOGY: Grows beneath overhangs, often tangled with bryozoans. Previously called *Melithaea africana* or *Wrightella coccinea*. RELATED SPECIES: *Astromuricacea fusca* (see 6.4) is similar; its main branches give off slightly flattened lateral branches in one plane.

8.7 Radial sea pen   *Eleutherobia aurea*

IDENTIFICATION: Colony plump and sausage-shaped; peduncle fleshy, thick. Polyps evenly distributed around the central axis. Colour extremely variable; often yellow but may be cream, red-brown or mauve. Colony covered with soft, prickly protuberances, but only visible when the polyps withdraw. SIZE: 12–150 mm long. BIOLOGY: Occurs at 12–300 m, often insinuating its peduncle into sandy crevices on rocky reefs.

8.8 Feathery sea pen   *Virgularia schultzei*

IDENTIFICATION: Colony very long and thin, with a brittle central axis. Polyps arranged in groups of 15–35 on flat leaf-like expansion on either side of the central 'stem'. Colour usually white or cream. SIZE: Colony 20 cm long, 10 mm wide. BIOLOGY: Grows in dense colonies, projecting vertically out of sand.

Scleractinia : Hard Corals

Corals consist of anemone-like individuals (polyps) which produce a limestone skeleton. Some are solitary, but most form colonies with massive skeletons (accumulating to form reefs over 1000 m thick). The surface of the skeleton forms small craters or projections (corallites), each housing one polyp. Usually the upper surface of each corallite is divided by vertical radiating plates (septa), which may be joined by tiny bridges (synapticula). Identification depends on these skeletal features, revealed by cleaning and bleaching. Colonial corals house symbiotic single-celled algae (zooxanthellae) in their tissues. The algae gain fertilising nitrogen; in return they supply food and help build the skeleton. This association limits reef corals to sun-lit, warm waters.

9.1 Noble coral *Allopora nobilis*

IDENTIFICATION: A single thick 'trunk' divides into cylindrical branches. Tiny polyps project from star-shaped pin-pricks in the skeleton. Skeleton bright pink, often with pale tips. SIZE: Colony 20 cm, polyps 1 mm. BIOLOGY: *Allopora* is not a true coral, belonging to the Hydrozoa (order Milleporina). Lacking zooxanthellae, it can occur in cool temperate waters and depths of 5–100 m, often in caves or beneath overhangs. Growth is very slow; large colonies may be over 100 years old. Protected in South Africa: permit required for collection.

9.2 Staghorn corals *Acropora* spp.

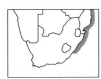

IDENTIFICATION: Corallites small and porous, projecting as distinct cups; an apical corallite always caps each branch (9.2d). Colony form varies with species, but is generally branching, bushy or resembles flat-topped trees; a few species form tables or plates. Colour variable; brown, whitish-pink, pale green or purple-tipped. SIZE: Colonies up to 40 cm, corallites <3 mm. BIOLOGY: 30 closely-related species occur in southern Africa. All occur in relatively calm water due to their brittle branches. Skeleton light; colonies thus fast-growing; early colonisers of disturbed areas.

9.3 Knob-horned corals *Pocillopora verrucosa*

IDENTIFICATION: Upright, hemispherical colonies; branches flattened, with knob-like projections. Corallites small and sunken into skeleton. Skeleton not porous. Projecting tentacles make live colonies appear fuzzy. Colour rich brown to light purple, often with a beautiful blue-brown sheen. SIZE: Colony 25 cm, corallites <2 mm. BIOLOGY: Common in shallow water, even in intertidal pools; extends down to 40 m. Ends of branches may form tiny caves, encapsulating commensal crabs. RELATED SPECIES:
9.4 *Stylophora pistillata* has flat branches but no knobs.
9.5 *Pocillopora damicornis* has much larger knobs, half the branch-width.

9.6 Mushroom coral *Fungia scutaria*

IDENTIFICATION: Solitary, consisting of a huge, single, flat polyp which is oval or elongate. Skeleton resembles an upturned mushroom, having a central groove and radiating vertical septa which are finely toothed. Live animals bright yellow, green or brown, with yellow stripes. SIZE: 20 cm. BIOLOGY: Juveniles are attached by a short stalk, but soon break free; adults lie loose on the bottom. Often collected by divers; illegal removal from Sodwana reserve is a recognised problem. Occurs from 4 to 30 m.

9.7 Cup coral *Balanophyllia bonaespei*

IDENTIFICATION: Solitary, but often lives in groups. Bright orange, tentacles almost transparent and beaded in appearance. Skeleton columnar, upper surface divided by radiating septa which have prickly, toothed inner edges (9.7d). SIZE: 10 mm diameter, 15 mm height. BIOLOGY: Occurs subtidally, often in caves or under dark overhangs, in depths of 5–150 m. RELATED SPECIES: *Balanophyllia annae* (known only from Still Bay) is almost identical but has smooth septa. **9.8 *Caryophyllia* spp.:** larger skeletons (15–20 mm diameter) with uniform longitudinal stripes on the column, and obvious pali (finger-like projections) in the centre of the disc (9.8d).

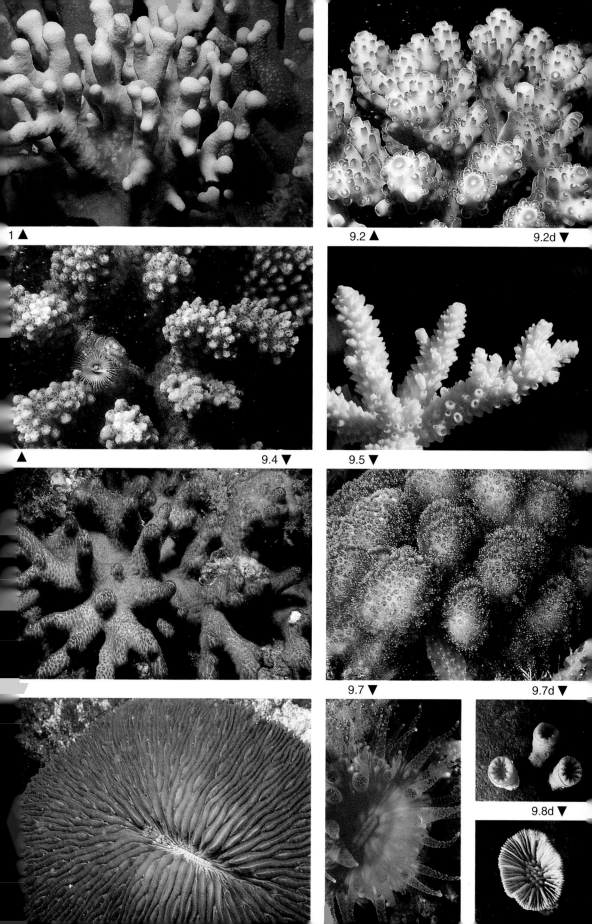

1 ▲   9.2 ▲   9.2d ▼

▲   9.4 ▼   9.5 ▼

9.7 ▼   9.7d ▼

9.8d ▼

10.1 Honeycomb corals *Favites* spp.

IDENTIFICATION: Hemispherical and boulder-like; completely covered with moderately large corallites, which are sunken into the surface. The polyps do not project and have short tentacles, which are seldom visible by day. Colour extremely varied, including bright green, brown, grey, russet. The walls of touching corallites are fused together, forming a honeycomb pattern (10.1d). Adjacent polyps thus share a common wall. The septa are well developed but lack synapticula (tiny bridges). SIZE: Colony width 30 cm, corallites 12 mm wide. BIOLOGY: Abundant, from intertidal pools down to 15 m. There are many species in the genus, all difficult to separate. RELATIVES: The genus *Goniastrea* is very similar, but its corallites are smaller and its septa always have thickened lobes near the centre, lacking in *Favites*. Rare in South Africa, but common in Moçambique.

10.2 False honeycomb corals *Favia* spp.

IDENTIFICATION: Very similar to *Favites*, with hemispherical colonies, but distinguished because the walls of adjacent corallites are distinctly separated from one another by a gap or, if they are touching, by at least a groove (10.2d). Colours again extremely variable. SIZE: Colony diameter 30 cm, height 15 cm, corallite diameter 12 mm. BIOLOGY: As for *Favites*.

10.3 Irregular honeycomb coral *Anomastrea irregularis*

IDENTIFICATION: Forms creeping hummocks. Corallites cover the whole surface and are similar to those of *Favites* but are half the size and their septa (10.3d) are joined by synapticula (tiny bridges). Polyps form a thin coating, so that the shape of the skeleton is clearly visible. Living colonies usually brown, sometimes dull green. SIZE: Colony 15 cm, corallites 6 mm. BIOLOGY: One of the commonest corals in intertidal pools.

10.4 Spiny honeycomb coral *Acanthastrea echinata*

IDENTIFICATION: Colonies irregular but generally domed and boulder-like. Corallites heavily spined, especially on the inner edges of the septa (10.4d). Typically the surface of live animals is thickly fleshy, concealing the skeleton, and textured with tiny bubbles. Colour variable, often khaki-green with paler centres. SIZE: Colony 25 cm, corallites 20 mm wide. BIOLOGY: Never abundant, but occurs from intertidal pools down to 25 m.

10.5 Labyrinthine brain coral *Platygyra daedalea*

IDENTIFICATION: Forms large flat slabs or hemispherical colonies; surface convoluted like a brain. The polyps are joined in rows, and the corallites housing them form long grooves separated by ridges (10.5d). Live animals have a striking network of green grooves and brown ridges, or green ridges on a khaki or white background. Tentacles not evident. SIZE: Colonies 5–100 cm, corallite grooves 5 mm wide. BIOLOGY: Occurs subtidally. Named after the mythical Daedalus, inventor of the labyrinth that housed the Minotaur. RELATED BRAIN CORALS:

10.6 *Oulophyllia crispa* is similar to *Platygyra daedalea*. Its tentacles are also not evident, but it is often more convoluted and its corallites much larger (10–12 mm groove width) with sharp ridges and finely-toothed septa (10.6d).

10.7 *Coscinaraea* spp. have a skeleton that is less obviously brain-like, the corallites being almost separate from one another, the grooves less distinct and the ridges gently rounded (10.7d). The septa are densely packed, giving the impression of a sieve. (*Koskinon* is Greek for 'sieve'.)

10.8 *Gyrosmilia interrupta* also has wide grooves (12 mm) but can be distinguished by its extended tentacles and the large, smooth skeletal septa, which are alternately large and small (10.8d).

10.9 *Symphyllia valenciennesi* is a giant brain coral, forming massive colonies: its corallites are 30 mm wide and coarsely toothed on their inner edges (10.9d). All brain corals are tropical, extending from Zululand northwards.

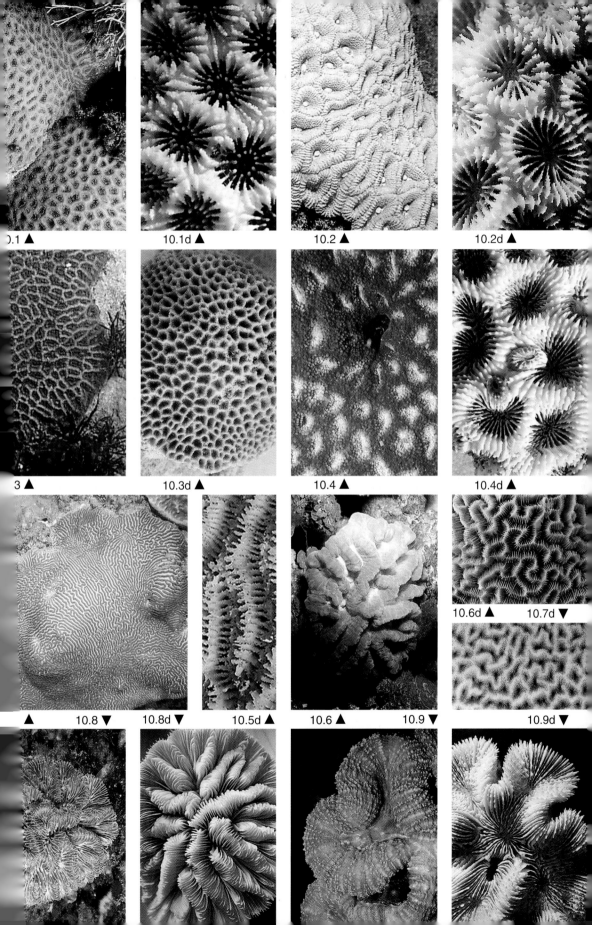

0.1 ▲ 10.1d ▲ 10.2 ▲ 10.2d ▲

3 ▲ 10.3d ▲ 10.4 ▲ 10.4d ▲

10.6d ▲ 10.7d ▼

▲ 10.8 ▼ 10.8d ▼ 10.5d ▲ 10.6 ▲ 10.9 ▼ 10.9d ▼

11.1 Diurnal coral *Goniopora djiboutensis*

IDENTIFICATION: Forming low boulder-shaped or encrusting colonies of moderate size. Polyps large and project boldly (by as much as 20 mm), resembling bunches of pale flowers. Skeleton completely covered by corallites which are closely positioned but still separated by a groove (11.1d), similar to the arrangement in *Favites*, but much smaller. SIZE: Colony 25 cm wide, corallites 4 mm wide. BIOLOGY: Unique in being the only coral which fully expands its polyps during the day. Aggressive: stings other corals to death.

11.2 Porous corals *Porites* spp.

IDENTIFICATION: Massive boulder-like colonies, covered densely with small sunken corallites (11.2d) which have well-developed septa (unlike *Montipora*, which can develop colonies superficially resembling those of *Porites*). Polyps with tiny brown tentacles, scarcely emerging from their corallites, but giving the colony a fuzzy appearance even though the skeletal structure is always evident. SIZE: Colony may be several metres across, corallites 1–2 mm. BIOLOGY: Common in shallow sheltered water, forming massive coral heads.

11.3 Peacock coral *Pavona decussata*

IDENTIFICATION: Forms flat thin blades from which twisted secondary plates jut at right-angles. This may produce a structure a little like a peacock's tail, hence the name (*pavo* is Latin for 'peacock'). The corallites (11.3d) are distinctive, lacking definite boundaries because the septa extend uninterrupted from one corallite to the next. SIZE: Colony 15 cm diameter, plates 2 mm thick. BIOLOGY: Abundant at Inhaca Island in shallow sheltered lagoons. Plates often break off, settle on the sandy bottom, and sprout fresh upright plates. RELATED SPECIES: Several other species occur in southern Africa. **11.4 Pavona maldivensis** has a massive boulder-like skeleton (70 cm in diameter), and the septa on its corallites radiate in a daisy-like manner.

11.5 Warty corals *Montipora* spp.

IDENTIFICATION: Large, flat, undulating colonies. If the skeleton is examined closely (11.5d), the corallites are poorly defined, very simple in structure, and their septa rudimentary. However, the surface is distinctive, being dotted with irregular warty protuberances which distinguish the genus. SIZE: Colony up to 30 cm diameter, 'warts' 2–3 mm. BIOLOGY: Restricted to the subtidal zone.

11.6 Plate coral *Leptoseris* sp.

IDENTIFICATION: Forming flat plate-like sheets, the septa clearly visible on the upper surface and radiating outwards. Like *Pavona*, the septa extend from one corallite to the next so that the boundaries between corallites are undefined. However, the corallites are much larger and more spaced out than those of *Pavona*. SIZE: Plates 40 cm diameter, corallites about 10 mm apart. BIOLOGY: Occurs only in deep waters (20–40 m). Common in Sodwana Bay.

11.7 Many-eyed coral *Astreopora myriopthalma*

IDENTIFICATION: Forms encrusting hillock-like sheets; densely covered with polyps which do not extend but are recessed in the skeleton. Usually pale purple. Superficially like *Porites*, but the skeleton is highly porous, the surface covered with fine spines, and the corallites consist of simple holes which either lack septa or have very reduced, simple septa (11.7d). SIZE: Colony very large: metres across; corallites 2 mm. BIOLOGY: Found only subtidally.

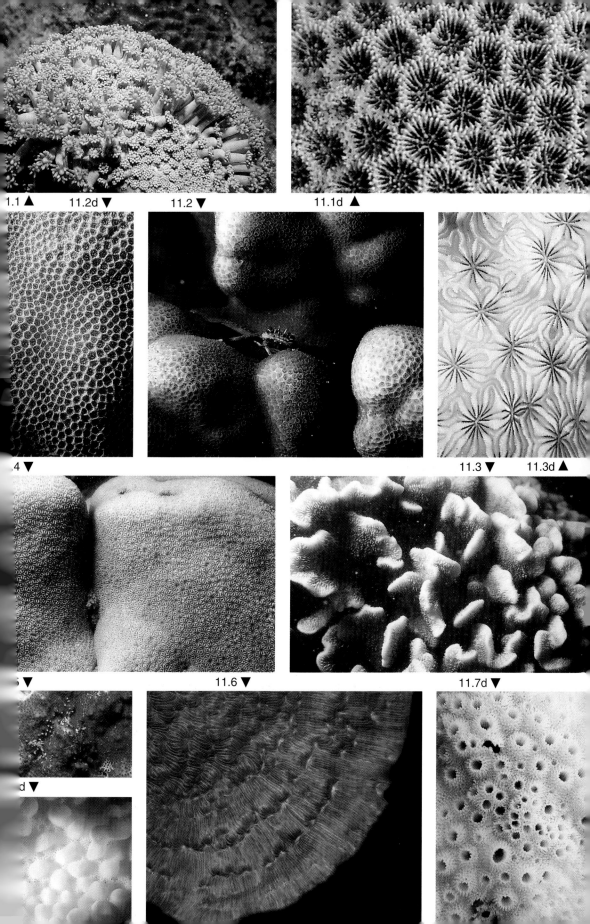

1.1 ▲   11.2d ▼   11.2 ▼   11.1d ▲

4 ▼   11.3 ▼   11.3d ▲

5 ▼   11.6 ▼   11.7d ▼

d ▼

12.1 **Turbinate coral** *Turbinaria mesenterina*

IDENTIFICATION: Colony attached by a distinct 'trunk' that expands into a flat table-like or bract-like plate, often spiralling or funnel-shaped. Corallites (12.1d) widely spaced and sunken into the skeleton. Septa simple, smooth and all of similar size. SIZE: Colony 30 cm, corallites 3–4 mm. BIOLOGY: Occurs subtidally. The funnel shape may be an adaptation to concentrate silt at the bottom of the funnel rather than letting it smother the whole colony; in turbulent waters the funnel voids silt. The name is derived from *turbo*, Latin for a whirlpool or spinning top.

12.2 **Spiky coral** *Galaxea fascicularis*

IDENTIFICATION: Colony rounded, surface distinctly spiky because the corallites are elevated on columns and their septa project sharply outwards from the apex (12.2d). The tentacles range in colour from brown to bright green, are flattened and tapering and usually have a distinct terminal white knob. They extend well beyond the septa. SIZE: Colonies 15–20 cm, corallite columns 4 mm diameter. BIOLOGY: Occurs sub-tidally in depths of 1–15 m.

12.3 **Small-coned coral** *Hydnophora microconos*

IDENTIFICATION: Colony flat and encrusting. The corallites are unusual in that the septa project from the surface and taper abruptly to a central ridge, forming tiny cone-like structures (12.3d). In the living animals the cones are surrounded by polyps that lie in the valleys between the cones, their mouths being flanked by short stubby tentacles. SIZE: Whole colony up to 40 cm, corallites 7 mm. BIOLOGY: Restricted to the subtidal zone, at depths of 1–25 m.

12.4 **Star-like coral** *Plesiastrea versipora*

IDENTIFICATION: Colonies encrusting and relatively small, densely covered with pale grey polyps with darker tentacles. The skeleton is distinctive, the corallites (12.4d) being almost circular and projecting slightly, crowded but definitely separated from each other by small gaps. SIZE: Colonies 15–20 cm, corallites 5–7 mm. BIOLOGY: Occurs in intertidal pools and down to 20 m subtidally. Previously referred to as *Solenaster*.

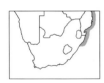

12.5 **Prickly-pored coral** *Echinopora hirsutissima*

IDENTIFICATION: Encrusting, the surface misshapen and rough because the corallites are irregularly scattered, differ in size and vary in the extent to which they project. The corallites of dead specimens resemble snowflakes, their outer surfaces covered with delicately-spined ridges. Septa toothed on their inner edges (12.5d). SIZE: Colony 10–20 cm, corallites 8 mm. BIOLOGY: Subtidal, but tolerant of a wide range of conditions. Grows well in sheltered areas where it is often covered by a thin blanket of silt, but survives equally well in very exposed situations.

12.6 **Turret coral** *Dendrophyllia aurea*

IDENTIFICATION: Colony small, consisting of a few cylindrical branches terminating in large corallites, each mounted by a single, very large, golden polyp. SIZE: Colony 50 mm tall, branches 10 mm wide. BIOLOGY: Polyps normally withdrawn by day. At night when they expand they are larger than those of most other corals (30 mm across the tentacles) and are big enough to be confused with anemones. RELATED SPECIES: *Dendrophyllia* species are very similar to *Tubastrea*, but pairs of their septa fuse together along their inner margins – a feature never seen in adult *Tubastrea* spp. Both genera feed voraciously on zooplankton, and lack zooxanthellae in their tissues.

12.7 ***Tubastrea diaphana*** is coated with a thin film of tissue, giving a green sheen over an orange-brown background colour. Its tentacles are not extended by day. *T. micranthus* is a tree-like, deep-water species, deep black-green, much admired by divers for its stunning beauty.

2.1 ▲

12.1d ▲

12.2 ▲

12.2d ▲

.3 ▲

12.3d ▲

12.4 ▲

12.4d ▲

5 ▲

12.6 ▼

12.5d ▲

12.7 ▼

Scyphozoa : Jellyfish

Jellyfish are bell-shaped gelatinous creatures with a simple body structure (described as a medusa). Viewed from above, they are round and their organs radiate out from a central stomach, so that they have a radial symmetry. Apart from a brief larval stage, jellyfish are planktonic, swimming by pulsing their bodies, and jetting water from beneath their bells. Most jellyfish are carnivores, stunning prey with stinging cells on the tentacles that fringe the bell, and passing it to the frilly mouth (manubrium) which hangs down from the centre of the bell. Jellyfish are an important source of food for some species of turtles.

13.1 Box jellyfish *Carybdea robsoni*

IDENTIFICATION: Bell very deep and almost cube-shaped, with a very long whip-like tentacle hanging from each of the four lower corners of the bell. SIZE: Bell diameter 40 mm, tentacles up to 70 cm long. BIOLOGY: Often occurs in swarms. Tentacles armed with potent stinging-cells which can inflict a painful sting, although not nearly as virulent as that of its relative, the lethal sea wasp *Chironex*, which does not occur in southern African waters.

13.2 Frilly-mouthed jellyfish *Chrysaora* sp.

IDENTIFICATION: Bell shallow, transparent blue with a more opaque white edge, 24 long marginal tentacles and 16–32 short stubby tentacles. Manubrium hangs down as an elaborate, frilly structure, which extends to form four trailing 'tails' when fully expanded. SIZE: 12 cm diameter. BIOLOGY: Less common than *Chrysaora hysoscella* (13.4); also feeds on relatively large planktonic animals.

13.3 Root-mouthed jellyfish *Eupilema inexpectata*

IDENTIFICATION: Large jellyfish with a smooth domed bell and no tentacles; usually translucent white or blue. The manubrium does not have a single mouth, but is dotted with numerous pore-like entrances into the gut. SIZE: The largest jellyfish known, averaging 30 cm diameter, but individuals up to 1.5 m have been recorded. BIOLOGY: Commonly washed ashore. Unusual in that they lack tentacles, rhizostomid jellyfish can feed only on microscopic prey. They have an elaborate manubrium with many tiny pores in it instead of the usual single mouth, and this acts as a sieve, allowing the jellyfish to filter water, trapping tiny prey in the process.

13.4 Red-banded jellyfish *Chrysaora hysoscella*

IDENTIFICATION: Bell relatively shallow, margin scalloped and armed with 24 long, hollow, marginal tentacles (damaged and scarcely visible in the photograph). The top of the bell is strikingly marked with dark purple-red radiating bands. SIZE: Up to 15 cm diameter. BIOLOGY: Feeds on larger planktonic animals including fish larvae. Often abundant in the cold waters of the West Coast, it is thought to consume significant numbers of fish larvae, including those of anchovy and pilchard. Sometimes it concentrates in sheltered bays in enormous numbers. A commensal hyperiid amphipod (a shrimp-like animal with very large eyes) often lives embedded in its tissues.

13.5 Night-light jellyfish *Pelagia noctiluca*

IDENTIFICATION: Bell hemispherical; surface characteristically warty. Margin carries eight tentacles, each flanked by two small lappets. Manubrium very long, and divided into four frilly lobes. SIZE: Bell diameter 90 mm. BIOLOGY: Abundant in the Mediterranean, forming enormous swarms over the past 15 years. The causes of these outbreaks are unknown, but there is concern about their potential effect on fish stocks. Not previously recorded from southern African waters. Likely to occur around the whole coast, but only known with certainty from False Bay.

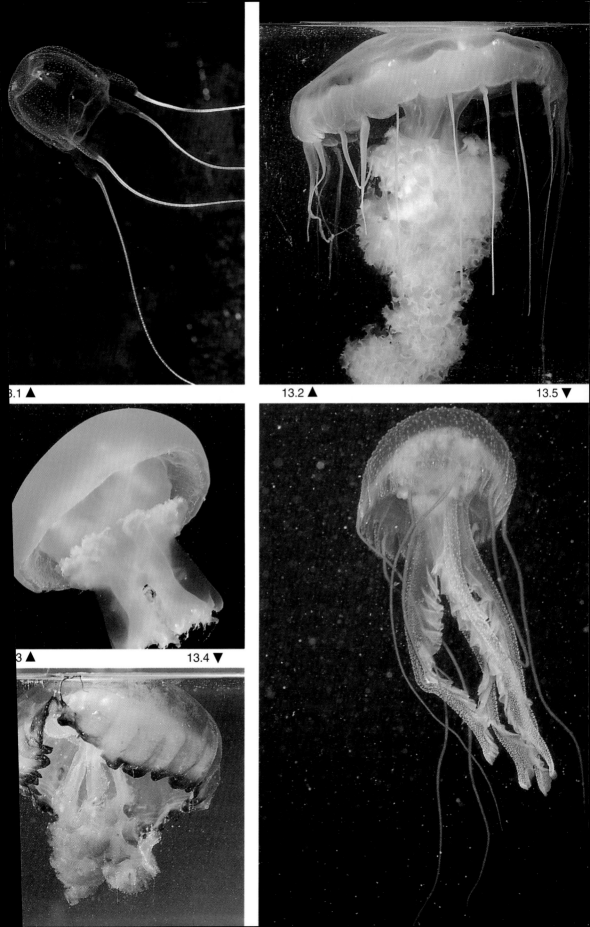

13.1 ▲ 13.2 ▲ 13.5 ▼

13.3 ▲ 13.4 ▼

Hydrozoa : 'Sea Firs' or Hydroids

Hydroids form colonies of numerous individuals (polyps), and are often tree-like or feather-like. The polyps have a ring of stinging tentacles around the mouth, used for defence and to capture microscopic animals. In some species the colony has an external skeletal sheath (perisarc) and each polyp is protected in a cup-like housing (the hydrotheca). The numbers of teeth on the margin and the shape of the hydrothecae distinguish species, and line drawings of these microscopic features are provided. Some polyps are reduced to a single, long, stinging tentacle and housed in tubular nematothecae. Sac-like reproductive structures on the colony (gonothecae) form medusae (miniature jellyfish) which reproduce sexually, yielding larvae that initiate the next generation of polyps. About 286 species of hydroids occur in southern Africa.

14.1 High-spined commensal hydroid   *Hydractinia altispina*

IDENTIFICATION: Forms an orange coat on shells of the whelk *Nucella squamosa*. Polyps (14.1d) naked (lacking a sheath and not housed in hydrothecae) and interspersed with chitinous spines and tiny ball-like reproductive individuals (gonozooids). SIZE: Polyps 2 mm. BIOLOGY: Occurs only on *N. squamosa* and protects its host, repelling at least some of its predators; found as deep as 24 m. RELATED SPECIES: *Hydractinia marsupialis* (Saldanha to Port Elizabeth) forms a white coat on the whelk *Nassarius speciosa*. *Hydractinia kaffraria* (Breede River–Durban Bay) occurs in estuaries on *N. kraussianus*.

14.2 Sea-fern hydroid   *Pennaria disticha*

IDENTIFICATION: Colony fern-like, with a central stem that gives off branches alternately on the left and right, each branch bearing about six polyps on the upper surface. Polyps naked (not housed on hydrothecae) but attached by a short stalk; a ring of tentacles circles the base and knobbed tentacles are scattered over the surface (14.2d). SIZE: Colony 60 mm tall. BIOLOGY: Attaches to rocks in pools and down to 20 m. Common on ships' hulls and wrecks.

14.3 Bushy hydroids   *Eudendrium* spp.

IDENTIFICATION: Colony branching haphazardly, forming a small bush. Stem enclosed in a sheath which is ringed (annulated) at the origins of branches. Polyps (14.3d) not housed in hydrothecae. Tentacles less than 40 in number, forming a single whorl around the mouth. SIZE: Colony 17 cm tall. BIOLOGY: The seven species in the genus are difficult to tell apart. They grow on rocks, ships' hulls, algae, or other hydroids.

14.4 Tubular hydroid   *Tubularia warreni*

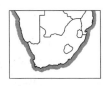

IDENTIFICATION: Stems long, unbranched and upright, encased in a skeletal sheath. Each carries a single apical polyp. Polyps spectacular, pink and white with short tentacles ringing the mouth and a second ring of much larger tentacles, below which hang yellow grape-like bunches of reproductive sporosacs. SIZE: Stems 100 mm long, polyps 10 mm across. BIOLOGY: Grows commonly on ships' hulls and dock piles, even in harbours where pollution is severe. Feeds on tiny planktonic crustaceans. A giant version, 25 cm tall and with polyps 30 mm across, occurs in False Bay. RELATED SPECIES:
14.5 *Zyzzyzus solitarius* (Saldanha to Natal) is only 15 mm tall, solitary and does not have a firm sheath around its stem; it grows embedded in sponges.

14.6 Thin-walled obelia   *Obelia dichotoma*

IDENTIFICATION: A creeping, root-like base gives rise to upright, unbranched stems. Colour pale pink to white. Polyps protected by hydrothecae, which have ringed stalks and arise directly from the stem; margins of hydrothecae smooth or gently undulating, never toothed. Reproductive bodies flask-shaped and three times the size of hydrothecae (14.6d). SIZE: Stems 10 mm long. BIOLOGY: World-wide and is common on ships, dock piles and seaweeds. RELATED SPECIES:
14.7d *Obelia geniculata* (Lüderitz to Cape Agulhas) has a markedly thickened stem below each hydrotheca. Common on kelp and on ships' hulls.

14.1 ▲

14.1d

14.2 ▲

14.2d

14.3 ▲

14.3d

14.4 ▼

14.5 ▲

14.6 ▼

14.6d

14.7d

15.1 Smoky feather-hydroid *Lytocarpus filamentosus*

IDENTIFICATION: Upright, feathery stems; mottled black, smoky-grey and white. Reproductive bodies protected in flat, circular structures. Hydrothecae roughly cup-shaped and have a projecting tooth above a central tubular nematotheca (15.1d). SIZE: Colonies 15 cm. BIOLOGY: Common in KZN; often grows on algae. RELATED SPECIES: **15.2 *Macrorhynchia philippina***, the fire hydroid, forms branching, pale-white, feathery colonies; its hydrothecae are similar, but lack the projecting tooth above the central tubular nematotheca (15.2d). Associated with coral reefs in KwaZulu-Natal and Moçambique. Capable of inflicting a fierce, fiery sting.

15.3 Rusty feather-hydroid *Thecocarpus formosus*

IDENTIFICATION: Creeping rootlets give rise to densely-packed undivided feathery stems that are rusty red-brown. Hydrothecae (15.3d) with several blunt teeth around the margin, and a long, obvious spur, a short nematotheca on either side, and a more obvious tubular one below the spur. Reproductive individuals gathered in a structure resembling a slender pine-cone (corbulae). SIZE: Stems 10–30 mm long. BIOLOGY: The most frequently-seen hydroid in KwaZulu-Natal; often grows on seaweeds, and is apparently able to penetrate their tissues with its rootlets.

15.4 Plumed hydroid *Plumularia setacea*

IDENTIFICATION: A low, creeping base, with pale, upright, unbranched plume-like stems. The cup-like hydrothecae have a smooth margin, and are flanked by two lateral nematothecae and a lower central nematotheca (15.4d). Gonothecae (reproductive individuals) resemble smooth sacs with curved necks. SIZE: Stems 20 mm tall. BIOLOGY: Very common in shallow water; one of 11 species. RELATED SPECIES: **15.5d *Kirchenpaueria pinnata*** is very similar but lacks lateral nematothecae, and its gonothecae have ridges and a spiky apex. It occurs everywhere and is the most common intertidal hydroid in the SW Cape.

15.6 Toothed feather-hydroid *Aglaophenia pluma*

IDENTIFICATION: Upright, plumed, yellow stems with a root-like base. Stems usually unbranched. Hydrothecae (15.6d) with nine strong marginal teeth, a frontal nematotheca that projects as a short keel, and two lateral nematothecae. Reproductive individuals protected in a corbula resembling a tiny pine-cone. SIZE: 30 mm long, but very much longer in the branching variety. BIOLOGY: Very common; intertidal to 100 m.

15.7 Jointed hydroid *Salacia articulata*

IDENTIFICATION: Upright stems with side-branches that come off in pairs and jut out almost at right-angles, but all lie in one plane. Hydrothecae are carried on both sides of the stem and side-branches, and are distinctively sunken into them (15.7d). SIZE: Colony 50 mm. BIOLOGY: Lives in sheltered pools; common on the SW Cape coast.

15.8 Planar hydroid *Sertularella arbuscula*

IDENTIFICATION: Stem bright yellow and branched repeatedly in one plane so that the whole colony is flat. The hydrothecae alternate on either side of the branches, have four marginal teeth and three internal teeth, and are closed by a pyramid-shaped operculum (15.8d). SIZE: 50 mm. BIOLOGY: Grows on vertical subtidal rocks, aligned at right-angles to the water-flow; catches plankton. One of 21 species. RELATED SPECIES: **15.9d *Symplectoscyphus*** species are very similar, but their hydrothecae have only three marginal teeth and their opercula have three valves.

15.10 Wiry hydroid *Amphisbetia operculata*

IDENTIFICATION: Thin, wiry stems that divide repeatedly into pairs; these diverge at an acute angle, creating a thin, straggly bush. Hydrothecae arranged in opposite pairs on the stems and are long and thin, their lips extending as two sharp teeth, one often longer than the other. The gonothecae resemble elongate figs (15.10d). SIZE: Colonies 50–90 mm long. BIOLOGY: Intertidal to 100 m.

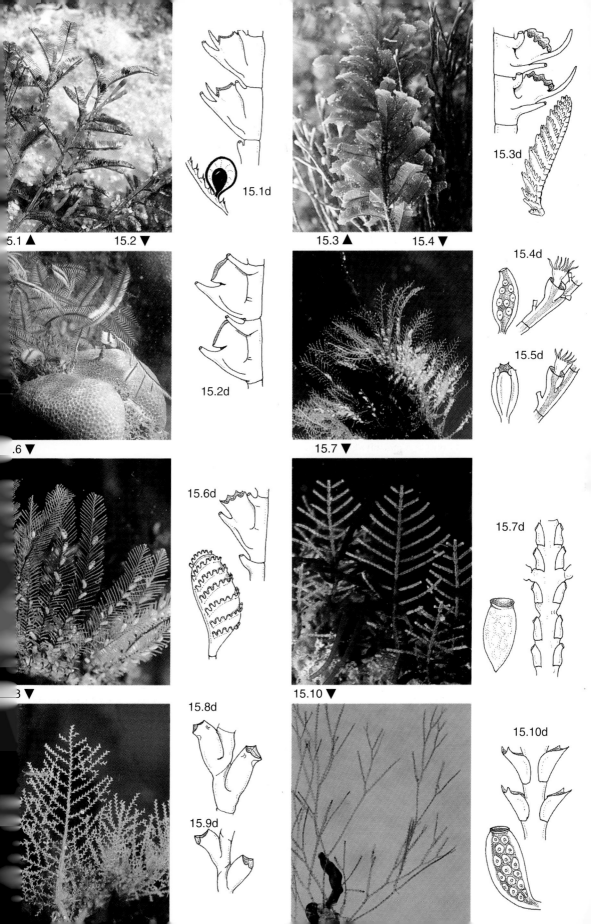

5.1 ▲   15.2 ▼

15.1d

15.3 ▲   15.4 ▼

15.3d

15.2d

15.4d

.6 ▼

15.7 ▼

15.5d

15.6d

3 ▼

15.10 ▼

15.7d

15.8d

15.9d

15.10d

Hydrozoa & Ctenophora : Bluebottles and Their Kin

The most unusual of the hydrozoans are floating forms that consist of colonies of highly specialised individuals. One individual is modified into a float and has a gas gland that inflates it with a mix of nitrogen and carbon monoxide. Gastrozooids (tubular individuals with a mouth) take in and digest food, which is then distributed to the rest of the colony. Dactylozooids consist of a single massive tentacle for defence and prey capture, while gonozooids have only one function: reproduction. Also shown on this page is the sea comb, an example of a completely different group, the phylum Ctenophora, which includes spherical, gelatinous, planktonic animals recognised by the rows of hair-like cilia that run down their bodies.

16.1 Bluebottle or Portuguese man-of-war *Physalia utriculus*

IDENTIFICATION: Unmistakable with its inflated blue-green float and long trailing tentacles. SIZE: Float 50 mm long, tentacles up to 10 m long, but contracting to about 30 cm. BIOLOGY: Floats on the surface of the sea and is vulnerable to being cast ashore by onshore winds. Because of the shapes of their floats, 'left-handed' and 'right-handed' individuals are blown to port or starboard respectively. On the West Coast of South Africa there tend to be more left-handed individuals because they are less likely to be stranded by the prevailing south-easterly winds. Bluebottles inflict painful stings, best treated with vinegar and ice. Severe cases are rare, but can lead to potentially lethal cardiovascular and respiratory collapse. Such cases should be treated by a medical practitioner with intravenous antihistamines and steroids. Despite their venomous stings, bluebottles are regularly eaten by the sea swallow (82.4) and by plough snails (78.4). There is some argument about the identity of the species that occurs in the Indo-Pacific and southern Africa: it may simply be a small variety of the larger, more virulent *Physalia physalia*, which occurs elsewhere.

16.2 Raft hydroid *Porpita pacifica*

IDENTIFICATION: The flat, circular, gas-filled float is distinctive, and is surrounded by a short, thin 'skirt'. Beneath the float hang a single central gastrozooid and rings of short tentacles dotted with minute spheres carrying stinging cells. The float is silvery because of the gas it contains, and the skirt and tentacles a translucent blue colour. SIZE: 30 mm diameter. BIOLOGY: Floats on the surface and feeds on planktonic animals. Its sting is mild and has no effect on humans.

16.3 By-the-wind sailor *Velella* sp.

IDENTIFICATION: Float oval and carrying a kinked vertical 'sail'. Beneath the float is a central mouth and a large number of short, simple tentacles. SIZE: 35 mm float length. BIOLOGY: Like *Porpita, Velella* is harmless to man, but can use its stinging cells to stun and capture tiny planktonic animals. It floats on the surface of the water and is driven by prevailing winds. Being a warm-water species, it is most common on the East Coast, but is regularly washed ashore further south during the summer months.

16.4 Comb jelly *Beroe* sp.

IDENTIFICATION: Shaped like a rugby ball with an opening at one end for the mouth. Transparent and bedecked with eight longitudinal bands of hair-like cilia which beat continually, passing rhythmic waves of movement down the bands to propel the animal, creating stunningly beautiful, flickering iridescent colours. SIZE: 30 mm. BIOLOGY: Planktonic, carnivorous, feeding on quite large shrimp-like prey. RELATED SPECIES: The sea gooseberry, *Pleurobrachia*, has a spherical body about 10 mm in diameter and two very long trailing tentacles which are feathery when fully expanded.

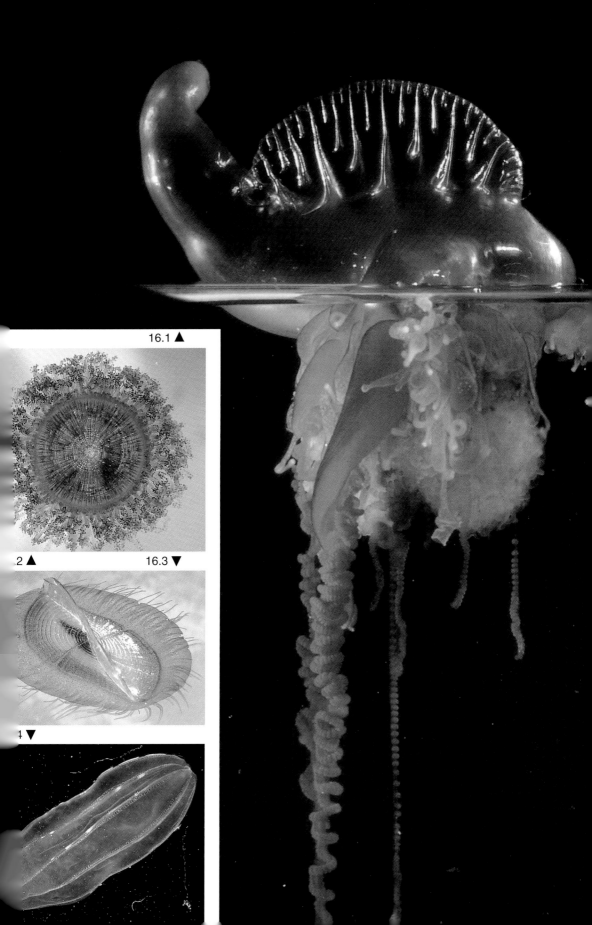

16.1 ▲

.2 ▲   16.3 ▼

⅄ ▼

Platyhelminthes : Flatworms

Only free-living flatworms are dealt with here, although the group also includes parasitic flukes and tapeworms. Flatworms have simple, flattened, leaf-like bodies, and glide along on a bed of fine hairs or by ripples of muscular contraction. A flexible tubular proboscis traps prey such as small crustaceans and molluscs. The gut ends blindly without any anus. Flatworms are hermaphroditic, but cross-fertilise; partners then lay strings of large yolky eggs. The southern African fauna is poorly known; there are many spectacular but unnamed species (e.g. 17.1), especially in the tropics.

17.2 Carpet flatworm   *Thysanozoon brocchii*

IDENTIFICATION: Body pink with a frilly margin, dorsal surface covered with pink to purple finger-like projections, which both camouflage the animal and increase the surface area available for oxygen exchange. SIZE: Normally 20–30 mm. BIOLOGY: Common under boulders in the intertidal zone.

17.3 Tentacled flatworm   *Planocera gilchristi*

IDENTIFICATION: Mottled brown with irregular black speckles. Margin frilly, but dorsal surface smooth except for a pair of short pointed tentacles towards the front end. SIZE: 20–30 mm. BIOLOGY: A predator of worms, small crustaceans and molluscs. Found under boulders near low tide.

17.4 Limpet flatworm   *Notoplana patellarum*

IDENTIFICATION: A flat, oval, grey-brown species without any dorsal processes, associated with large limpets. SIZE: 10 mm. BIOLOGY: Lives in the cavity between the side of the foot and the shell of large limpets, particularly *Cymbula oculus*. Feeds on small crus-taceans, including tiny commensal copepods, which share its specialised habitat.

Nemertea : Ribbon (or Proboscis) Worms

Thin, elongate, unsegmented worms with a soft elastic body that varies enormously in length as the worm stretches or contracts. Ribbon worms are carnivores, feeding mainly on small crustaceans and worms captured by a long flexible proboscis shot out from a special cavity in the head, unique to this group. Most South African species are undescribed (17.5 is a tropical example). One estuarine species, the 30-cm long *Polybrachiorhynchus dayi* (recognised by its branching proboscis), is collected for bait.

17.6 Collared ribbon worm   *Lineus ornatus*

IDENTIFICATION: Blue-grey with white lips and a broad white band across the back of the head. SIZE: Typically 30–100 mm. BIOLOGY: Poorly known. One of several purple-coloured *Lineus* species found under rocks or amongst shells and seaweeds. *L. olivaceus* has no markings on the head; *L. lacticapitatus* has a white tip to the head. *Cerebratulus fuscus* burrows on sandy beaches and is creamy-white or tinted pink.

Sipunculida : Peanut Worms

Tough, unsegmented worms with a short bulbous body and an elongate, tubular anterior process (the introvert), which can be squeezed out by muscular contraction, or retracted back within itself. The mouth at the tip of the introvert is surrounded by a frill of short tentacles, which gather detrital food particles.

17.7 Common peanut worm   *Golfingia capensis*

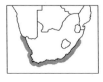

IDENTIFICATION: Body surface smooth and yellowish-brown. Introvert may exceed body length and has a smooth surface with a single ring of short, evenly-spaced tentacles around the tip. SIZE: 20–40mm. BIOLOGY: Usually found in clusters under stones on rocky shores, especially in areas where gravel accumulates. RELATED SPECIES: One of several *Golfingia* species distinguished by internal anatomy. *Phascolosoma* species are identified by small scaly conical papillae on the body surface, and *Siphonosoma* species by their short introverts and simple papillae.

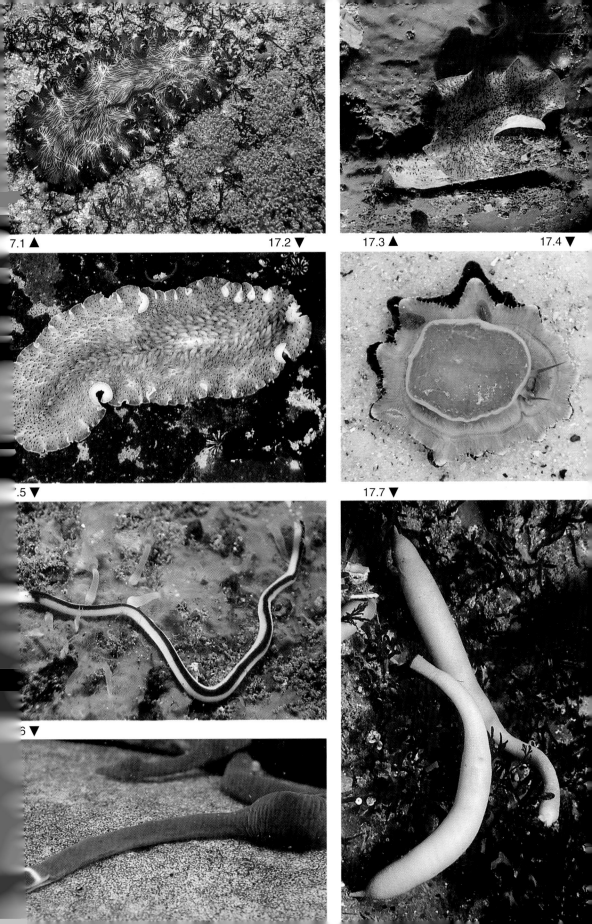

7.1 ▲ 17.2 ▼

17.3 ▲ 17.4 ▼

.5 ▼ 17.7 ▼

6 ▼

Polychaeta : Bristle-worms

Polychaetes are among the commonest and most diverse creatures on the shore, almost 800 species occurring in southern Africa. All have a head with a snout-like projection (prostomium) and a cylindrical front segment that surrounds the mouth (peristomium), but in some species the head is elaborately modified into beautiful fans or tufts of tentacles that are used for feeding. The body is divided into segments, each with a pair of lateral leg-like protuberances (parapodia) that carry one or two tufts of bristles (setae). Fertilisation is external, and the larvae are tiny and planktonic. Some species are not easy to identify from photographs, and drawings of their diagnostic features accompany the photographs.

18.1 **Common scaleworm** *Lepidonotus semitectus clava*

IDENTIFICATION: Upper surface of the body covered by 12 pairs of flat bean-shaped scales, each of which is roughened by blunt conical tubercles (18.1d). SIZE: 20 mm. BIOLOGY: Common in rocky-shore pools, rock crevices and beneath boulders, from the low-shore to the shallow subtidal. Carnivorous. RELATED SPECIES: There are about 58 species of scaleworm in South Africa.

18.2d *Lepidonotus durbanensis* (Mossel Bay to Moçambique) has 12 pairs of scales which are textured with low spherical tubercles.

18.3 **Milky scaleworm** *Antinoe lactea*

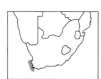

IDENTIFICATION: Body completely covered by 15 pairs of dorsal scales which are smooth, and coloured white with a grey crescent (18.3d). SIZE: 30 mm. BIOLOGY: Burrows in sandbanks in lagoons and estuaries. Often lives commensally in the burrows of the sandprawn *Callianassa kraussi* (39.1). RELATED SPECIES:

18.4d *Harmothoe aequiseta* (Walvis Bay to Durban) has 15 pairs of scales with dark spines.

18.5 **Plump bristle-worm** *Euphrosine capensis*

IDENTIFICATION: Body plump, oval and bright orange-red, with rows of short spines running transversely across each segment. SIZE: 20 mm. BIOLOGY: Found under stones on rocky shores. Often associated with sponges, upon which it may feed. Sluggish: scarcely moves when disturbed, apart from a tendency to curl up.

18.6 **Fireworm** *Eurythoe complanata*

IDENTIFICATION: Body long and thin, flattened; colour grey-green with lateral tufts of long white spines and short red gills. SIZE: 140 mm. BIOLOGY: Very common in pools on tropical and subtropical rocky shores and among coral. Spines sharp and break off easily, releasing a poison which causes fierce irritation. Should not be handled.

18.7 **Beadworms** *Syllis* spp.

IDENTIFICATION: Small, thin, thread-like worms with long tentacle-like cirri projecting from the sides of each segment of the body. As is the case in most members of the family Syllidae, the cirri are annulated (beaded in appearance). The head has two swollen palps that extend forwards, and three annulated tentacles. SIZE: 30 mm. BIOLOGY: Very common in tufts of seaweed. RELATED SPECIES: There are over 60 species in the family Syllidae, most being difficult to identify without microscopic examination of the teeth on the pharynx. *Syllis armillaris* (Namibia to Moçambique) has a long (35 mm), very thin body with short cirri that have only 8–12 annulations. *Syllis variegata* has 30 annulations and broken brown bars across the anterior segments.

18.8 **Bamboo worms** *Euclymene* spp.

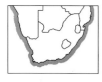

IDENTIFICATION: Cylindrical body with extremely long segments, resembling a bamboo stalk. Head with a flat plate and a raised margin (18.8d). Anus sunken into a frilly funnel. SIZE: 120 mm. BIOLOGY: Burrows head down in sand; feeds on organic particles. Forms long fragile sandy tubes. RELATED SPECIES:

18.9d *Nichomache* spp.: head with arched crest; build sandy tubes on boulders.

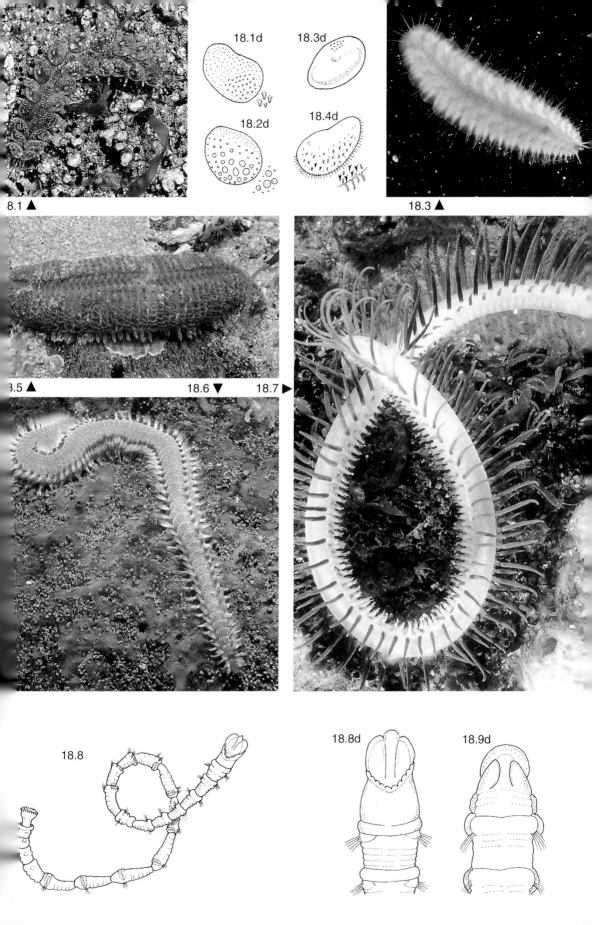

8.1 ▲

18.1d

18.3d

18.2d

18.4d

18.3 ▲

8.5 ▲   18.6 ▼   18.7 ▶

18.8

18.8d   18.9d

19.1 Mussel worm *Pseudonereis variegata*

IDENTIFICATION: Cylindrical, robust, dark khaki-green. Like all of the family Nereidae its prostomium has two short tentacles and a pair of swollen palps capped with tiny nipple-like segments. Four pairs of slender tentacle-like cirri flank the prostomium. Parapodia large and lobed, with two tufts of bristles. Identification of the 50 species of Nereidae requires close examination of the pharynx, which consists of two rings and is armed with a pair of large jaws and dotted with tiny teeth (paragnaths), the arrangement of which separates the species. Often the pharynx must be dissected or everted by gentle squeezing to reveal the paragnaths. *Pseudonereis variegata* has only two bar-like paragnaths on the lower ring of the pharynx (19.1d). SIZE: 150 mm. BIOLOGY: The commonest worm on rocky shores, living among seaweeds, reef-worms or mussels. Popular as bait, but its collection destroys large areas of mussel bed. Eats small animals and seaweed.

19.2 Estuarine nereid *Ceratonereis erythraeensis*

IDENTIFICATION: A slender, delicate worm which resembles *Pseudonereis variegata* but is smaller, white, and completely lacks paragnaths on the basal ring of the pharynx. SIZE: 30 mm. BIOLOGY: Burrows in sandbanks in lagoons and estuaries.

19.3 Comb-toothed nereid *Platynereis dumerilii*

IDENTIFICATION: Brown, of intermediate size; its paragnaths form characteristic comb-like rows of teeth. SIZE: 45 mm. BIOLOGY: Herbivorous; extremely common in seaweeds on rocky shores, forming mucous tubes that bind the seaweed. Swarms into the water in enormous numbers when spawning, at which time its body is modified to improve swimming ability, the parapodia becoming paddle-like and the eyes enlarged. This stage is referred to as a 'heteronereid'.

19.4 Bar-toothed nereid *Perinereis nuntia vallata*

IDENTIFICATION: Fairly large and slender; body dull brown. Distinguished by the single row of bar-like paragnaths stretching across the basal ring of the pharynx. SIZE: 70 mm. BIOLOGY: Occurs in sheltered sandbanks and under stones or among seaweeds on rocky shores. RELATED SPECIES:

1 9.5d *Nereis* spp.: several species, all with groups of tiny conical paragnaths on both rings of the pharynx, occur around the coast in rock pools; seldom abundant.

19.6 Gilled nereid *Dendronereis arborifera*

IDENTIFICATION: Distinguished from most other species of the family Nereidae by its pale colour and the red, feathery gills that lie on either side of the central part of the body. It also lacks any paragnaths on either of the rings of the pharynx. SIZE: 60 mm. BIOLOGY: Almost completely restricted to black mangrove muds, where it can be very common. Probably feeds on detritus, as its gut often contains mud particles.

19.7 Glycerine worm *Glycera tridactyla*

IDENTIFICATION: Elongate, slender and cylindrical, tapering at both ends. Pink when alive, becoming white if preserved. Prostomium narrow and acutely pointed. Proboscis cylindrical, very long, tipped with four sharp jaws. SIZE: 60 mm. BIOLOGY: Burrows in sandbanks; very active and lashes about when dug out. Proboscis can be shot out (as in photograph), extending almost half the body length when fully everted. Predatory: eats other worms and small crustaceans.

19.8 Nephthys' sand-worms *Nephtys* spp.

IDENTIFICATION: Body white, square in section. Parapodia with broad, flat lobes and two tufts of bristles with a curled gill between them (19.8d). Prostomium small, stubby, with four antennae. Proboscis eversible (though usually withdrawn in the body), lined with rows of soft prickles and ending in a ring of finger-like papillae. SIZE: 65 mm. BIOLOGY: Actively burrows in lagoonal sandbanks and sandy beaches. Named after the Egyptian goddess Nephthys.

19.1 ▲

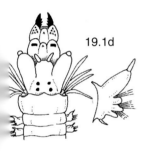

19.1d

19.2d

19.3d

19.4d

19.5d

19.6d

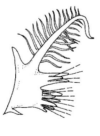

19.7 ▼

19.8 ▲

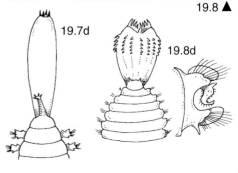

19.7d

19.8d

20.1 Wonder-worm *Eunice aphroditois*

IDENTIFICATION: Prostomium flanked by rounded palps and backed by five finger-like antennae. The next two segments lack parapodia but have two tentacular cirri that project upwards and distinguish all *Eunice* species from *Marphysa* species. Body long, cylindrical, mottled purple-brown with an iridescent blue-green sheen and a pale bar across segment 4. Feathery gills arise above the parapodia from about segment 8. SIZE: 35 cm, reaching up to 1 m! BIOLOGY: Common under boulders, especially where gravel allows them to burrow. Carnivorous; has large jaws that can inflict a painful bite. Used as bait. RELATED SPECIES:

20.2d *Eunice antennata* (East London–Moçambique) has beaded antennae. *Euniphysa tubifex* (Durban–Moçambique) has gills from segment 20; lives in branching tubes.

20.3 Case worm *Diopatra cuprea*

IDENTIFICATION: Head with two short, thin antennae and five much longer antennae arising from beaded, cylindrical bases (ceratophores). The central ceratophore has 9–12 rings. Large bushy gills sprout from segment 5 onwards (20.3d). Front of body uniformly brown, lacking obvious pattern. Lives in a robust mucous tube from which shell fragments jut. SIZE: 20 cm. BIOLOGY: Lives in sheltered sandy areas, its tubes embedded with just the top projecting. RELATED SPECIES: *D. neopolitana* (Namibia–Durban) is similar, but its front segments have a central oval spot (or series of 5 spots). *D. monroi* (Walvis–Cape Point) forms tubes of hardened mud without shell fragments, has dark bars across its anterior segments, and its central ceratophore has only 6–8 rings.

20.4 Estuarine wonder-worm *Marphysa sanguinea*

IDENTIFICATION: Prostomium with two swollen palps and five antennae, but the next two segments lack tentacular cirri (unlike *Eunice* spp.). Body distinctly flattened. All its bristles are slender and jointed (spinigerous setae). SIZE: 25 cm. BIOLOGY: Burrows in sheltered sandbanks. Used as bait by fishermen. RELATED SPECIES: Closely-related species are distinguished by the microscopic structure of their bristles.

20.5d *M. macintoshi* (Port Alfred–Moçambique, in muddy sandbanks) has a rounded body and spinigerous setae.

20.6d *M. corallina* (Transkei–Moçambique, under boulders and among coral) has jointed bristles with a more stocky, slightly hooked and flanged tip (falcigerous setae).

20.7d *M. depressa* (Saldanha–Durban, burrowing in sandbanks) has both falcigerous and spinigerous setae.

20.8 Three-antennaed worm *Lysidice natalensis*

IDENTIFICATION: Head with three finger-like antennae, which are about the same length as the two swollen palps (20.8d). No tentacular cirri on the first two segments of the body, and no gills. (Juveniles of *Marphysa* species sometimes also have three antennae, but possess gills.) Reddish-brown, speckled with white dots. SIZE: 75 mm. BIOLOGY: Lives on rocky shores among seaweed, in dead coral or in sand.

20.9 False earthworm *Lumbrineris tetraura*

IDENTIFICATION: Body earthworm-shaped, long, cylindrical, pale brown and lacking gills. Parapodia short, stubby, with a single tuft of a few bristles. Prostomium conical, lacking eyes or antennae (20.9d). SIZE: 60 mm. BIOLOGY: Burrows in sand-covered pools or sandbanks. RELATED SPECIES:

20.10d *L. coccinea* (Walvis–Moçambique): prostomium smoothly rounded; anterior parapodia with jointed bristles. Orange when alive. *L. cavifrons* (whole coast) also has rounded prostomium, but lacks jointed bristles.

20.11 Iridescent worm *Arabella iricolor*

IDENTIFICATION: Prostomium bluntly pointed, with four tiny eyes at its base (20.11d). Sometimes these are concealed if the prostomium is contracted. Body long, uniformly cylindrical, and iridescent bronze when the animal is alive. No gills; parapodia with a small number of bristles. SIZE: 80 mm. BIOLOGY: Burrows in sand among rocks.

20.1 ▲

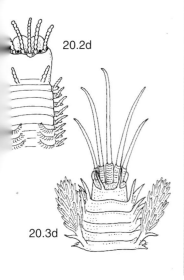

20.2d

20.3d

20.4 ▼

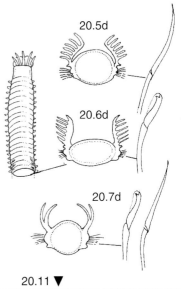

20.5d

20.6d

20.7d

20.11 ▼

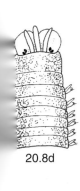

20.8d

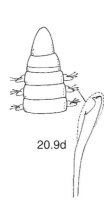

20.9d

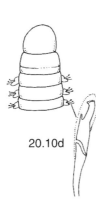

20.10d

20.11d

21.1 Shell-boring spionids *Polydora* spp.

IDENTIFICATION: Tiny worms which burrow in shells or limestone. The head has a pair of slender, grooved palps (characterising the family Spionidae). The 5th segment is swollen, with enlarged hook-like bristles (21.1d). Gills begin on segment 6 or 7. SIZE: 10 mm. BIOLOGY: Some are pests because they drill into oyster shells, creating unsightly 'mud-blisters', weakening or even killing the oyster. RELATED SPECIES:

21.2d *Boccardia polybranchia* (Namibia–Zululand) also has a modified 5th segment, but gills from the 2nd segment.

21.3d *Prionospio sexoculata* (Namibia–Zululand) has six eyes, feathery gills on segments 2 and 3, and a normal 5th segment.

21.4 *Scololepis squamata* (Lüderitz–Inhambane) has simple strap-shaped gills from the 2nd segment to the end of the body, and its 5th segment is not swollen.

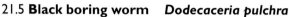

21.5 Black boring worm *Dodecaceria pulchra*

IDENTIFICATION: Small, pitch-black worms. Prostomium blunt and rounded, flanked by a pair of grooved palps which are followed by 4 pairs of long, thin, thread-like gills. SIZE: 10 mm. BIOLOGY: Occurs on intertidal rocky shores. Gregarious, large numbers living side by side and burrowing into encrusting corallines, their gills and palps projecting from burrow openings to obtain oxygen and to catch detrital particles for food.

21.6 Orange thread-gilled worm *Timarete capensis*

IDENTIFICATION: Bright orange. Prostomium blunt. Numerous tentacles on segments 1–3; long tangled thread-like gills over the rest of the body. SIZE: 90 mm. BIOLOGY: Lies buried in sediment or between mussels, only the long orange gills and tentacles visible. Previously known as *Cirriformia capensis*. RELATED SPECIES: *T. tentaculata* (Namibia–Moçambique) has a pointed prostomium, brown body and red gills. *T. punctata* (Natal–Moçambique) is brown, flecked with black. Its tentacles are striped.

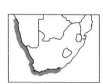

21.7 Woolly worm *Orbinia angrapequensis*

IDENTIFICATION: Slender, with a pointed prostomium. No appendages on the head. Body divided into two regions. The anterior region has 18–24 flattened segments in which the parapodia form lateral ridges; behind these lie rows of small conical fleshy projections (foot-papillae). In the posterior region the parapodia project dorsally, giving a 'woolly' appearance. SIZE: 50 mm. BIOLOGY: Abundant in sandbanks in sheltered lagoons and estuaries where they are frequently the predominant organisms. There are four species, but only *O. angrapequensis* is common. RELATED SPECIES:

21.8 *Naineris laevigata* (Namibia to Moçambique) has a rounded prostomium and an eversible proboscis with several finger-like lobes. *Scoloplos johnstonei* (NW Cape to Moçambique) co-occurs with *Orbinia* and is easily confused with its juveniles: it is more slender and lacks foot-papillae.

21.9 Bloodworm *Arenicola loveni*

IDENTIFICATION: Large, dark brown, with tufts of pale bristles and branched red gills in the centre of the body. Tail thin and lacking gills or obvious bristles. Proboscis balloon-like and can be inflated or withdrawn. SIZE: As thick as a thumb; 80 cm long. BIOLOGY: Occurs in estuaries and lagoons or off sandy beaches. Digs deep, U-shaped burrows, one end forming a funnelled depression. Water is drawn through the tube, oxygenating the sediment and encouraging bacterial growth. The 'farmed' microflora decomposes detritus and contributes to the worm's diet. A popular bait, but over-exploited in some estuaries. Contains haemoglobin and 'bleeds' red blood when damaged.

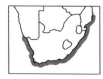

21.10 Club worm *Notomastus latericeus*

IDENTIFICATION: Resembles a miniature bloodworm. Body club-shaped, front end swollen; two pairs of bristle-tufts per segment (excluding the first). SIZE: 15 mm. BIOLOGY: Burrows in mud and sandflats. RELATED SPECIES: *Capitella capitata* (Cunene–East London) has bristles on all of the anterior segments. It is abundant in organically-rich sediments and a useful indicator of organic pollution.

21.1d 21.2d 21.3d

21.4 ▼

21.5 ▲   21.6 ▼

21.7 ▼

21.8 ▼

21.9 ▼

21.10 ▼

22.1 **Flabby bristle-worm** *Flabelligera affinis*

IDENTIFICATION: Body soft and floppy, covered with a pale-green mucous sheath. Thin, simple bristles project from the sides of the body, parapodia being poorly developed. Head surrounded with a cage of long annulated bristles. SIZE: 30 mm. BIOLOGY: Lives under boulders in silty or sandy pools.

22.2 **Cone-tube worm** *Pectinaria capensis*

IDENTIFICATION: Head with two comb-like arcs of 11–15 golden bristles. Lives in a tapering cylindrical tube made of sponge spicules arranged in beautiful brick-like rows. SIZE: 90 mm. BIOLOGY: Lives in sandbanks, head buried downwards to feed on organic matter. RELATED SPECIES: *P. neopolitana* (Lüderitz–East London) makes tubes of coarse, irregularly-arranged sand-grains.

22.3 **Cape reef-worm** *Gunnarea capensis*

IDENTIFICATION: Gregarious, forming massive intertidal reefs of sandy tubes. There is a double row of 40–50 stiff golden bristles on the head (22.3d), forming a 'door' (operculum) that blocks the tube when the worm withdraws. SIZE: Worms 50 mm; reefs up to 3 m in diameter and 50 cm high. BIOLOGY: Extends its iridescent black-blue tentacles from its tubes at high tide to feed on particles. Catches sand-grains and cements them together to build its tubes. The tubes are flanged and designed so that passing waves suck out faeces but concentrate food in the flanges. *Gunnarea* is an aggressive competitor for space and keeps its tubes clean by scraping off settling organisms with its operculum. Sometimes accused of 'invading' or 'taking over' the shore, but fluctuations in its numbers seem part of a natural cycle, storms periodically decimating old colonies. RELATED SPECIES:

22.4d *Idanthyrsus pennatus,* the Natal reef-worm (Durban to Moçambique), has feathery outer bristles in its operculum. It is often solitary or forms small aggregations.

22.5 **Tangleworms** *Thelepus* **spp.**

IDENTIFICATION: Soft, floppy body; head highly modified and completely obscured by a mass of thin, white, grooved tentacles resembling a tangled ball of cotton threads. Behind these lie shorter, red gills which are not branched. Bristles appear from the 3rd segment onwards (22.5d). SIZE: 50 mm. BIOLOGY: Like almost all members of the family Terebellidae, *Thelepus* lives in a mucous tube, which is decorated with sand or pieces of shell. The grooved tentacles extend considerably to capture food particles that deposit on them. RELATED SPECIES: There are 44 species in the family, all difficult to identify once preserved because the gills and tentacles tangle and drop off.

22.6d *Telothelepus capensis* (Saldanha to Port Elizabeth) has unbranched gills and a long tentacular lobe with a frilly margin.

22.7 *Nicolea macrobranchia* (Namibia to Durban) has two pairs of branched gills. The tips of its bristles are smooth, not spiny. It is common both under boulders and on sheltered sandbanks in lagoons.

22.8d *Terebella pterochaeta* (Walvis to Zululand) also has two pairs of branched gills, but has microscopic spiny tips to its bristles.

22.9d *Loimia medusa* (Cape Point to Moçambique) has three pairs of branched gills and makes sandy tubes in rock crevices. The tips of its bristles are smooth.

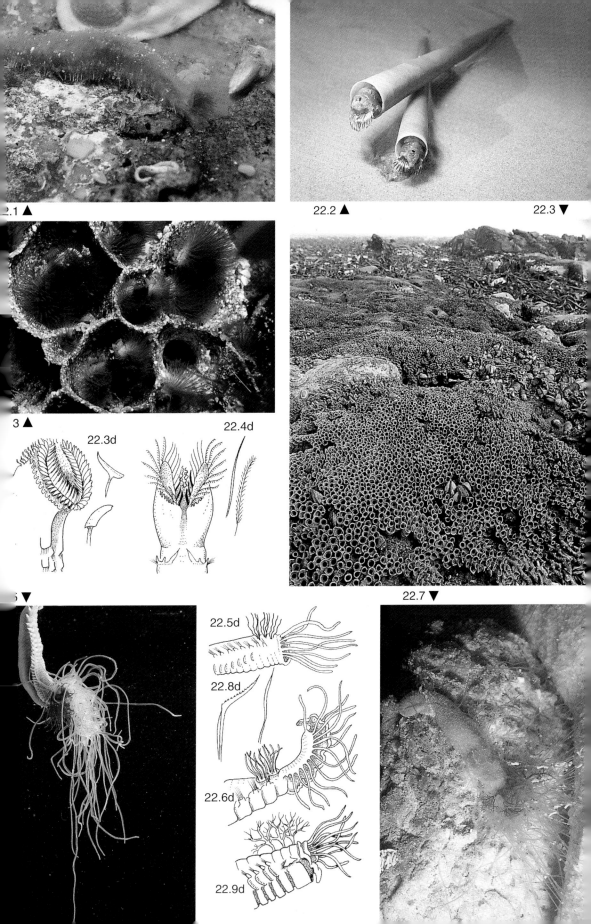

22.1 ▲

22.2 ▲

22.3 ▼

22.3d

22.4d

3 ▲

22.7 ▼

22.5d

22.8d

22.6d

22.9d

23.1 Feather-duster worm or giant fanworm *Sabellastarte longa*

IDENTIFICATION: Head crowned with two magnificent spiral whorls of feeding appendages. Commonly purple, white or orange. The absence of tiny external flaps (stylodes) from the branches of the crown helps distinguish this species. Body encased in a parchment-like mucous tube (which typifies the family Sabellidae). SIZE: 120 mm long. BIOLOGY: Solitary; lives in shady rock-crevices. Uses its fan to catch and sort food particles, eating only the smallest (and most nutritious). RELATED SPECIES:

23.2 *Sabellastarte sanctijosephi* (Durban–Maputo) also lacks stylodes; its crown

forms two semicircles. *Branchiomma* spp. all have stylodes; the crown forms two spirals in *B. natalensis* (Lüderitz–Cape Point, not Natal!), and two semicircles in *B. violacea* (Walvis–Durban). *Megalomma quadrioculatum* (Namibia–Moçambique) has a distinctive eye near the tip of most branches.

23.3 Gregarious fanworm *Pseudopotamilla reniformis*

IDENTIFICATION: Crown forms two arcs of feathery branches, with small eye dots on their mid-outer surfaces. SIZE: Body 25 mm, crown 10 mm. BIOLOGY: Lives in groups, forming mucous tubes, often between red-bait or sponges. The tip of each tube rolls up like a scroll when the animal withdraws, protecting it from predators.

23.4 Blue coral-worm *Pomatoleios kraussii*

IDENTIFICATION: Gregarious, building massive blue colonies of interwoven tubes which are calcareous (a feature of the family Serpulidae). Head with two rows of feathery branches and a stalked operculum that has two pointed 'wings' and a flat calcareous cap. SIZE: Body 15 mm, tubes 2 mm diameter. BIOLOGY: Abundant on moderately exposed shores at mid-tide, often fringing pools.

23.5 Estuarine tube-worm *Ficopomatus enigmaticus*

IDENTIFICATION: Builds large colonies of entwined calcareous tubes, each with a trumpet-shaped mouth and rings down its length. The head has a few feathery branches and a cone-like operculum that ends in short dark spines. SIZE: Body 10 mm, tube diameter 1 mm. BIOLOGY: Forms thick jagged growths on jetties in areas of low salinity in estuaries. Does no damage, but can be perilous for unwary bathers or boaters.

23.6 Spiral fanworms *'Spirorbis'* spp.

IDENTIFICATION: Minute coiled worms with spiral shells. Head with a small number of feathery filter-feeding branches and a stalked operculum that blocks the shell. SIZE: 2 mm. BIOLOGY: Abundant everywhere, dotting most rocks in the shallow subtidal and in pools. Several species, originally placed in the genus *Spirorbis,* but now reclassified: microscopic details distinguish them.

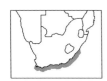

23.7 Red fanworm *Protula bispiralis*

IDENTIFICATION: Head with two bright orange or red spirals of feathery branches. Operculum absent. Front end of body with an obvious collar. Body protected by a tough white calcareous tube. SIZE: Body 65 mm, tube 10 mm diameter. BIOLOGY: Grows under boulders or in crevices. Catches tiny particles with its crown.

23.8 Operculate fanworm *Serpula vermicularis*

IDENTIFICATION: Head with two semicircles of feathery feeding appendages. Operculum funnel-shaped, with obvious longitudinal ridges that scallop the edge of the cone. Lives in a pinkish-white calcareous tube. Colour variable, usually red-pink but can be orange, purple or brown. SIZE: Body 15 mm, crown 10 mm. BIOLOGY: Solitary; attaches to sides of boulders.

23.9 Filigreed coral-worm *Filograna implexa*

IDENTIFICATION: Tiny, gregarious, forming delicate white lacy calcareous tubes. Head with eight orange feathery branches. SIZE: Body 4 mm, tubes 0.5 mm diameter. BIOLOGY: Occurs only subtidally, common in sheltered areas. Capable of reproducing asexually by splitting the body.

23.1 ▲

23.2 ▲

3 ▲

23.4 ▲

▲ 23.8 ▼

23.6 ▲

23.7 ▲

23.9 ▼

Arthropods

All arthropods have jointed limbs (*arthron,* joint; *podos,* foot) and a segmented body that is covered by a hard, jointed, external skeleton made of chitin, in some cases strengthened with calcium carbonate. Because the exoskeleton cannot expand, arthropods periodically shed the exoskeleton (a process called 'moulting') and can then rapidly expand. The cast-off skeleton is a remarkably faithful replica of the living animal. (The photograph in 42.2 depicts a crab's moulted exoskeleton, not a live animal.) Immediately after moulting, arthropods are soft and helpless until a new skeleton is deposited and becomes hard.

Arthropods include the insects and spiders, which are abundant on land but are scarcely represented in the sea, and the sea spiders (Pl. 24). The crustaceans (Pl. 25–48) are by far the most diverse group of arthropods in the sea, and include the crabs, shrimps and lobsters. Their bodies are normally made up of a fused head and thorax – the cephalothorax – plus a separate abdomen. They have numerous jointed appendages, which are variously adapted for walking, swimming, feeding, respiration and reproduction. Characteristically there are two pairs of antennae, and at least some of the limbs divide into two branches and are said to be 'biramous'. The exoskeleton may form a shield (the carapace) that covers and unites various segments of the body. Crustaceans have complex life cycles. Most have several larval stages that are planktonic and are dispersed widely by the currents before they metamorphose into miniature versions of the adult.

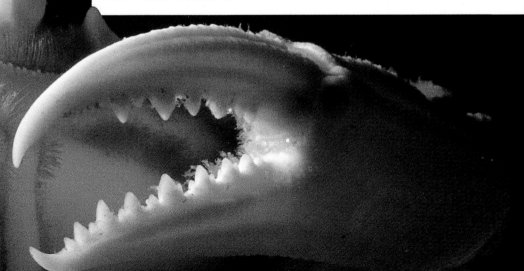

Pycnogonida : Sea Spiders

Although sea spiders have four pairs of long legs like true spiders, the two groups are unrelated. Sea spiders have a head armed with a tubular proboscis and often a pair of jointed, sensory palps and a pair of pincer-like feeding appendages (chelifers), a trunk of four segments, and a tiny conical abdomen. A pair of slender ovigerous legs usually hang below the head and are used by the males to carry the spherical egg masses. Most sea spiders are carnivorous, feeding on hydroids, sea anemones or bryozoans.

24.1 Compact sea spider   *Tanystylum brevipes*

IDENTIFICATION: Colour off-white. Body short and almost circular in outline, legs relatively short. Head with only one pair of appendages (the palps). Proboscis a large forward-pointing cylinder. Ovigerous legs only present in male. SIZE: About 10 mm across. BIOLOGY: The most abundant intertidal sea spider, found under boulders.

24.2 Scarlet sea spider   *Nymphon signatum*

IDENTIFICATION: Body bright pink, slender, with long spindly legs. Head with two pairs of appendages (palps and chelifers). Ovigerous legs present in both sexes. SIZE: 40–50 mm across. BIOLOGY: A striking species usually found amongst hydroids. One of several *Nymphon* species. RELATED SPECIES:
24.3 ***Queubus jamesanus*** (Saldanha–East London) lacks head-appendages, is bright yellow with white tips to the legs and has a long tapering proboscis.

Araneae : True Spiders

The bodies of true spiders consist of an anterior prosoma, equivalent to a fused head and thorax, and a posterior opisthosoma, or abdomen. The prosoma carries a pair of forward-directed poison fangs and a pair of leg-like feelers or pedipalps, which the males use for sperm transfer, as well as four pairs of elongate walking legs. The abdomen contains the lungs, reproductive organs and the silk glands so typical of this group. Although spiders are normally land animals, two species are common on rocky shores in the Cape.

24.4 Formidable shore spider   *Desis formidabilis*

IDENTIFICATION: Brown with a furry grey abdomen; fangs very large, about one-third body length. SIZE: 20 mm. BIOLOGY: Shore spiders trap bubbles of air in silk-lined crevices or shells in which they shelter during high tide. At night and when the tide is out, they emerge to hunt for isopods and amphipods. RELATED SPECIES: The chevron shore spider, *Amaurobioides africanus*, has a chevron pattern on its abdomen and smaller fangs than *Desis*.

Hexapoda : Insects

Insects have a distinct head, a thorax bearing three pairs of legs, and usually one or two pairs of wings, and a limbless, segmented abdomen. They abound in terrestrial and freshwater habitats but remarkably few occur in the sea. Some are associated with rotting seaweeds on sandy beaches, notably the kelp flies, *Fucellia capensis* and *Coelopa africana*. On rocky shores the long worm-like larvae of midges (*Telmatogeton* spp.) graze on seaweeds, and minute black water-beetles of the genus *Ochthebius* are found in salty pools around the high-water mark.

24.5 Marine springtail   *Anurida maritima*

IDENTIFICATION: Minute, mauve-coloured, six-legged, wingless creatures covered in short waxy hairs. SIZE: 2–3 mm. BIOLOGY: During high tides, shelters in air pockets under rocks and shells, or in crevices. Emerges during low tide to scavenge, sometimes congregating around dead or dying animals. Often seen floating in clusters on the surface of rock pools, supported by surface tension. Its waxy hairs repel water.

24.1 ▲ 　　　　　　　　　　24.2 ▼ 　　　　24.3 ▲

▼ 　　　　　　　　　　24.5 ▼

Smaller Crustacean Groups

The larger and better-known crustacean groups (including the barnacles, amphipods, isopods, lobsters, prawns, and crabs) are dealt with under separate headings on the following pages. Smaller-sized groups, for which space does not permit a fuller treatment, are introduced in an abbreviated form below.

25.1 Seed shrimps : *Ostracoda*

Small crustaceans in which the head and short, oval body are completely enclosed by a hard, bivalved carapace hinged along the centre of the back. Larger forms (1–4 mm) generally have a notch in the front of each shell, through which the hairy antennae project, while smaller ones are smoothly oval. Some ostracods are planktonic, but most crawl or plough through the surface layers of sand or mud, propelled by their antennae. The group includes carnivores, herbivores, scavengers and filter-feeders.

25.2 Copepods : *Copepoda*

A group of extremely abundant, very small crustaceans of variable body form. Most are free-living, but some are parasitic, particularly on fish. Many of these parasites have degenerate sac-like or worm-like bodies, with trailing egg-strings.

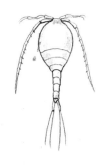

25.2 *Centropages brachiatus* is an example of the order Calanoida, which includes copepods with tapering, cylindrical bodies, very long antennae that project at right-angles to the head, and a single, central eye. They are usually the most conspicuous component of plankton samples and are the principal link in the food chain from phytoplankton to fish.

25.3 *Porcellidium* sp. is a representative of the bottom-dwelling copepods (order Harpacticoida) that have short antennae and include broad, flattened forms, which colonise the fronds of seaweeds, and elongated cylindrical ones, which live in the spaces between the sand-grains on coarse sandy beaches.

25.4 *Sapphirina* sp. is a stunningly beautiful planktonic member of the order Cyclopoidea, most of which are oval and flat. All have small antennae and are distinguished by the fact that none of their antennae divide into two branches.

25.5 Leaf shrimps : *Phyllocarida*

A small, primitive group of crustaceans in which the thorax and part of the abdomen are enclosed by a large, bivalved carapace. The eyes are stalked and separated by a characteristic, visor-like rostrum that projects from the front of the head. The abdomen ends in a pair of elongate, tapering appendages. The only South African representative of this unusual group of shrimps is *Nebalia capensis*, a detritus-feeder common under loose stones on rocky or mixed shores and often mistaken for an amphipod.

25.6 Cumaceans : *Cumacea*

Small, distinctively shaped crustaceans with an enlarged, swollen carapace and a very narrow, elongate abdomen ending in a pair of long slender uropods. Cumaceans usually burrow in sand or mud, where they either filter-feed or scrape organic matter from sand-grains, but they may emerge to swim in the water column. The species illustrated is *Iphinoe stebbingi*.

25.7 Tanaids : *Tanaidacea*

Small, cylindrical crustaceans with unstalked eyes. The anterior thoracic segments are fused to the head and covered by a short carapace, the other six thoracic segments remaining free. The first pair of legs bears strong nippers (a distinctive feature of the group), leaving six pairs of walking legs. Branched pleopods lie beneath some or all of the abdominal segments. The body ends in a rounded pleotelson with a pair of long, segmented uropods. Tanaids are generally tube-building or tunnel-dwelling animals, which either filter-feed or grasp pieces of detritus or small organisms from around the burrow entrance. Eggs are carried in a brood pouch under the body and develop directly into the adult form. The species depicted is *Anatanais gracilis*.

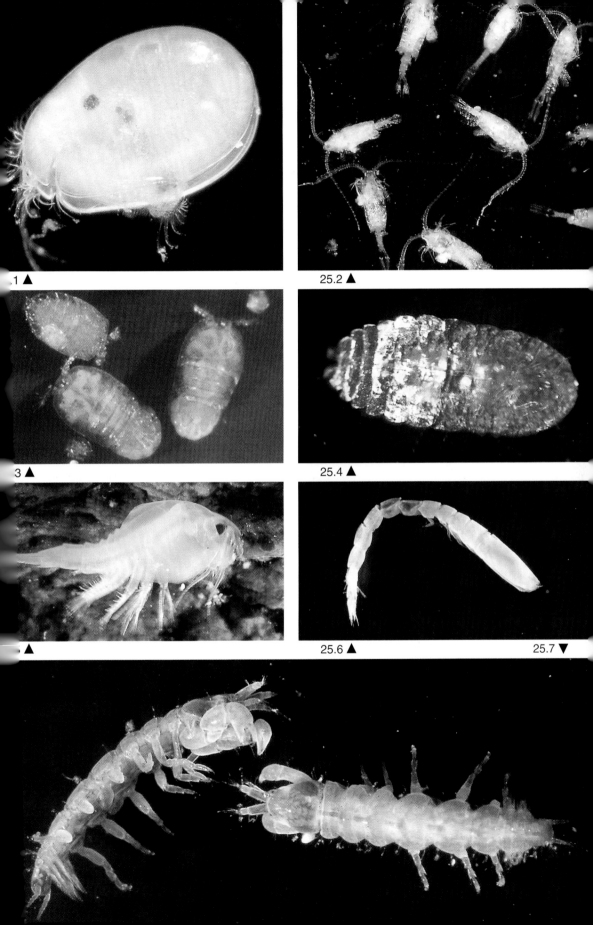

25.1 ▲

25.2 ▲

25.3 ▲

25.4 ▲

25.5 ▲

25.6 ▲

25.7 ▼

Cirripedia : Barnacles

Barnacles are highly-modified crustaceans, the adults of which are permanently attached to the substratum and encased in a shell. The legs have become long hairy cirri, which are extended through a hole at the apex of the shell and comb the water to filter out food particles. The legs can be withdrawn and the opening sealed by an operculum of four additional shell plates when the barnacles are threatened or exposed to air. Barnacles are hermaphroditic, but prefer to cross-fertilise. Being sessile, they can only accomplish this by having an extremely long penis – up to three times the length of the animal. The eggs are brooded and expelled as tiny, shrimp-like planktonic larvae.

26.1 Goose barnacle   *Lepas* sp.

IDENTIFICATION: Attached to floating objects by a long tough fleshy stalk. Body laterally flattened and enclosed by five shiny white shell plates. SIZE: 30 mm. BIOLOGY: Occurs in dense colonies on ships or floating objects, most commonly observed on driftwood cast ashore. The common name originates from the medieval myth that they grew into barnacle geese!

26.2 Volcano barnacle   *Tetraclita serrata*

IDENTIFICATION: Tall, dark-grey and volcano-shaped, with only four strongly-ribbed shell plates, the edges of which are difficult to distinguish. SIZE: 20 mm. BIOLOGY: One of the dominant invertebrates in the mid-intertidal zone of moderately sheltered shores. Slow-growing. RELATED SPECIES:

26.3 *Tetraclita squamosa rufotincta* is rosy pink and replaces *T. serrata* in northern KwaZulu-Natal.

26.4 Eight-shell barnacle   *Octomeris angulosa*

IDENTIFICATION: Moderately large with eight distinct, dirty-white shell plates. SIZE: 10–25 mm. BIOLOGY: A dominant species in the mid-to-low intertidal, forming extensive, often closely-packed sheets, particularly on wave-beaten rocks. Feeds by spreading its legs at right-angles to waves and passively holding them extended while the water passes through them.

26.5 Toothed barnacle   *Chthamalus dentatus*

IDENTIFICATION: A small, flat, dirty-white barnacle with a membranous (not calcareous) base. The projecting, finger-like ridges on the six shell plates produce the characteristic star-shaped outline. SIZE: 5–10 mm. BIOLOGY: Common in the upper intertidal, especially on the South and East coasts.

26.6 Giant barnacle   *Austromegabalanus cylindricus*

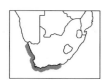

IDENTIFICATION: Very large with six tall, white-to-pink shell plates and a calcareous base. The tips of the opercular plates are modified into long 'fangs' which project up through the aperture. SIZE: Typically 30–40 mm, but can grow to a height of 150 mm. BIOLOGY: Usually subtidal, sometimes forming thick mats on floating structures. Tastes like lobster. A related species is a popular commercial crop in South America.

26.7 White dwarf barnacle   *Notomegabalanus algicola*

IDENTIFICATION: A small white species with six shell plates and a calcareous base. The flanges on the sides of each shell plate are perforated by transverse canals. SIZE: 5 mm. BIOLOGY: A short-lived and very fast-growing barnacle typically overgrowing mussel shells and low-shore rocks or encrusting floating structures.

26.8 Striped barnacle   *Balanus amphitrite*

IDENTIFICATION: Six smooth, white shell plates marked with vertical purple stripes. Base calcareous and flanges on sides of each shell plate solid. SIZE: 10–15 mm. BIOLOGY: Usually found in lagoons, estuaries and other similar wave-sheltered sites. RELATED SPECIES: Easily confused with *B. venustus* (Hermanus–Moçambique), which is smaller (5 mm), has pink stripes, and is usually found under low-shore boulders.

26.1 ▲

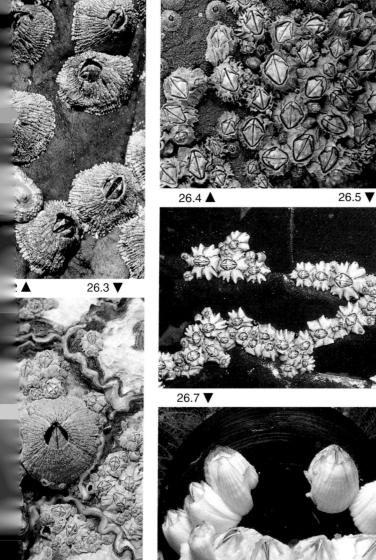

26.4 ▲   26.5 ▼

26.7 ▼

26.2 ▲   26.3 ▼

26.6 ▲   26.8 ▼

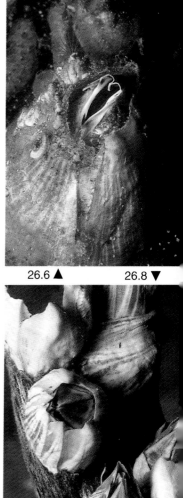

Isopoda : Isopods

Isopods are a diverse group of small crustaceans which occur in abundance in virtually all marine habitats from the intertidal to the deepest oceans. The body is usually flattened or depressed, although it can be cylindrical. The head has unstalked eyes and the thorax bears seven pairs of similar legs (hence the name *isos*, same, and *podos*, foot). The abdomen or pleon is extremely variable in form. Typically it consists of five segments, beneath each of which lies a pair of flap-like pleopods, used both for swimming and as gills. These are followed by a tail-fan made up of a central telson and a pair of uropods. Some or all of the pleon segments may fuse with the telson to form a so-called pleotelson. Alternatively the uropods may be folded beneath the telson to form a chamber to protect the delicate pleopods. Eggs are held in a brood pouch beneath the thoracic segments of the female and develop directly into the adult form. Over 270 species occur in southern Africa, only the most common being detailed here.

27.1 Keeled isopod *Glyptidotea lichtensteini*
IDENTIFICATION: Body elongate and often strikingly camouflaged by brown or pink blotches on a pale background. A pronounced keel runs along the centre of the back and ends in a spike on the front of the head. Pleon segments fused to telson. SIZE: Up to 40 mm. BIOLOGY: Found intertidally under boulders. Herbivorous.

27.2 Metallic isopod *Idotea metallica*
IDENTIFICATION: Body elongate and parallel-sided with an unusual shiny metallic-silver colour. Two of the pleon segments fully divided and a third partially divided from the rectangular telson. SIZE: Up to 30 mm. BIOLOGY: A cosmopolitan species usually found on drifting seaweeds. Sometimes large numbers are found swimming around in the shallows of sheltered bays.

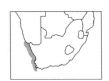

27.3 Reticulate kelp louse *Paridotea reticulata*
IDENTIFICATION: Very large, elongate isopods with net-like markings on a yellow or brown background. The pleon segments are fused to the rectangular telson, which has a thick grooved margin and ends in two sharp points (27.3d). Uropods flattened and folded under the telson to protect the pleopods. SIZE: Up to 55 mm. BIOLOGY: Grazes selectively on the spore-bearing parts of the fronds of kelps, which are heavily defended by anti-herbivore polyphenolic chemicals. Its gut contains surfactants and a chitinous lining that combat the polyphenols. RELATED SPECIES:

27.4d *Paridotea ungulata* (Walvis Bay–East London) grows to 55 mm. Usually green, never reticulated, common on *Ulva* or *Zostera*. Lateral margins of telson not thick and grooved. Corners of telson project as sharp points.

27.5d *Paridotea rubra* (Lüderitz–Cape Point) grows to 45 mm. Usually red-brown and found on the fronds of red algae. Corners of telson not pointed.

27.6d *Paridotea fucicola* (Lüderitz–Cape Point) is a smaller, narrow-bodied species. Telson slightly tapering, with a small notch at apex.

27.7 Fish louse *Anilocra capensis*
IDENTIFICATION: A smooth-bodied, slate-grey species parasitic on fish. Head triangular with short antennae. Legs ending in strongly-hooked claws that grip the host. Pleon has five segments; telson rounded. Uropods often extend well beyond body. SIZE: Up to 60 mm. BIOLOGY: An external parasite of fish, such as the hottentot (115.1). Usually found on the side of the head. Opens up a small wound and feeds on the blood and body juices of the fish.

7.1 ▲   27.2 ▼   27.3 ▶

27.3d   27.4d   27.5d   27.6d

27.7 ▼

28.1 **Slender checkered isopod**   *Mesanthura catenula*

IDENTIFICATION: Body very long and cylindrical, strikingly patterned by a wide black border around the margin of each segment. Antennae very short. Pleon segments fused. SIZE: Up to 20 mm. BIOLOGY: Found under rocks or amongst shell gravel. Habits poorly known. RELATED SPECIES: *Cyathura estuaria* (Saldanha–Moçambique) is uniform off-white in colour and common on estuarine sandbanks.

28.2 **Giant pill bug**   *Tylos granulatus*

IDENTIFICATION: Large, heavy-set, air-breathing isopods resembling giant wood-lice. Roll into a ball when disturbed. First antennae minute. Body surface granular, and pleon of five separate segments. SIZE: Up to 50 mm. BIOLOGY: A nocturnal species which spends the day buried up to 40 cm below the surface above the drift-line of sandy beaches. Emerges briefly during nocturnal low tides to feed on washed-up kelp. Entrance holes are covered by tiny 'molehills' of sand, but exit holes are open and flush with the surface. RELATED SPECIES: Replaced from Cape Point to Port Elizabeth by *Tylos capensis,* recognised by its smooth body surface and smaller size (up to 34 mm).

28.3 **Horned isopod**   *Deto echinata*

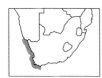

IDENTIFICATION: An air-breathing isopod easily recognised by the pair of long curved 'horns' arising from the back of each thoracic segment. These are much longer in males than in females. SIZE: Up to 30 mm. BIOLOGY: Associated with kelp and other drift algae washed up on rocky shores. Feeds mainly on these algae, but also takes carrion or live prey. Usually mixed with the similar-looking *Ligia* (28.4), to which it is only distantly related.

28.4 **Sea-slater**   *Ligia dilatata*

IDENTIFICATION: An air-breathing species with a broad, flattened, smooth body. First antennae greatly reduced, second antennae shorter than thorax. Pleon segments not fused, and end of telson rounded. Uropods rod-like and projecting well beyond body. SIZE: Up to 22 mm. BIOLOGY: Occurs in vast congregations amongst drift kelp on rocky shores in the Western and SW Cape. RELATED SPECIES: *Ligia glabrata* has a similar range and habits to *L. dilatata,* but its antennae are longer, reaching the end of the thorax. *L. natalensis* and *L. exotica* occur along the East Coast. The former has a rounded telson, while that of the latter has one central and two lateral points.

28.5 **Hairy isopod**   *Notasellus capensis*

IDENTIFICATION: Flattened, white isopods, with an elongate, oval body and long antennae and legs. Body and limbs covered with scattered spines. Pleon of one short separate segment, plus rounded pleotelson. Uropods elongate and cylindrical, projecting well beyond tip of pleotelson. SIZE: About 5 mm. BIOLOGY: Common under rocks in low-shore pools.

28.6 **Stebbing's isopod**   *Jaeropsis stebbingi*

IDENTIFICATION: Small, parallel-sided isopods with very short antennae. Abdominal segments all fused into a rounded pleotelson. Uropods minute. SIZE: Up to 5 mm. BIOLOGY: Under intertidal boulders, habits unknown. RELATED SPECIES: One of many similar-looking elongate, short-limbed species belonging to this and related genera.

28.1 ▲

28.2 ▲

28.3 ▲

28.5 ▼

28.6 ▼

28.4 ▲

29.1 Right-angle beach louse *Eurydice longicornis*

IDENTIFICATION: First segment of first antenna projects forward and rest of antenna bent sharply at right-angles to the side. Second antenna as long as thorax. Pleon of five free segments, telson rounded. Colour off-white with black markings. SIZE: Up to 9 mm. BIOLOGY: Common at mid-tide level of sandy beaches and below. A voracious carnivore which can rapidly strip the flesh from fish trapped in gillnets. Breeds throughout the year, but lives for only one year.

29.2 Wide-foot beach louse *Pontogeloides latipes*

IDENTIFICATION: Antennae not bent at right-angles; both pairs about one-third the length of thorax. Pleon of five separate segments, the first of which is covered laterally by last thoracic segment. Uropods and telson fringed with long hairs. SIZE: Up to 9 mm. BIOLOGY: A scavenger common on sandy beaches from the mid-intertidal to about 13 m depth.

29.3 Natal beach louse *Excirolana natalensis*

IDENTIFICATION: Antennae not bent at right-angle; first pair half the length of thorax, second pair two-thirds length of thorax. Pleon of five separate segments, the first overlapped on the sides by last thoracic segment. SIZE: Up to 9 mm. BIOLOGY: A scavenger found in the mid-to-high intertidal of sandy beaches.

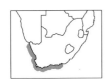

29.4 Tubetail isopod *Cymodocella magna*

IDENTIFICATION: Body smooth, pleon segments fused but lateral grooves may still mark their position. Sides of telson curved downwards and inwards to form a tube, which is directed slightly upwards at its tip. SIZE: 10 mm. BIOLOGY: Under intertidal boulders. RELATED SPECIES: This genus, which is characterised by the tubular telson, has six southern African representatives, the other five of which all have knobs or ridges on the telson.

29.5 Crimped cirolanid *Cirolana undulata*

IDENTIFICATION: Smooth off-white isopod with five separate abdominal segments. Posterior margins of thoracic segments finely crimped and those of abdominal segments finely toothed. Telson pointed, with a central ridge and crimp marks along margin (29.5d). SIZE: Up to 15mm. BIOLOGY: In pools and crevices on rocky shores. RELATED SPECIES: More than 25 *Cirolana* species occur in southern Africa, many of which are common. Amongst the most frequently seen are:

29.6d *Cirolana hirtipes:* a large (20 mm), smooth-bodied species with hairy legs. A groove runs across the head between the eyes. Found in sheltered sandbanks from Lüderitz to East London.

29.7d *Cirolana fluviatilis:* hind margin of last thoracic segment and of pleon segments 3–5 with rows of small tubercles; telson with two tubercles. Reaches 12 mm. Found from Knysna to Zululand, in estuaries.

29.8d *Cirolana venusticauda:* hind margins of last thoracic segment and pleon segments 3–5 are tuberculate, telson with dorsal projections. Grows to 12 mm. Found on rocky reefs from Lambert's Bay to East London.

29.9 Spike-back isopod *Parisocladus perforatus*

IDENTIFICATION: Male with a single distinctive curved 'horn' arising from last thoracic segment and arching back over the pleon. Telson with a keyhole-shaped notch at its tip (29.9d). The female has practically no 'horn' and lacks a notch in the telson, but its telson has two pairs of low knobs (29.9d). SIZE: 6 mm. BIOLOGY: Found intertidally under stones or in seaweeds. RELATED SPECIES:

29.10 *Parisocladus stimpsoni* (Lüderitz–East London): male with a much smaller process on the last thoracic segment and a narrow keyhole in the telson (29.10d); female with a very short process, and its telson has no notch (29.10d).

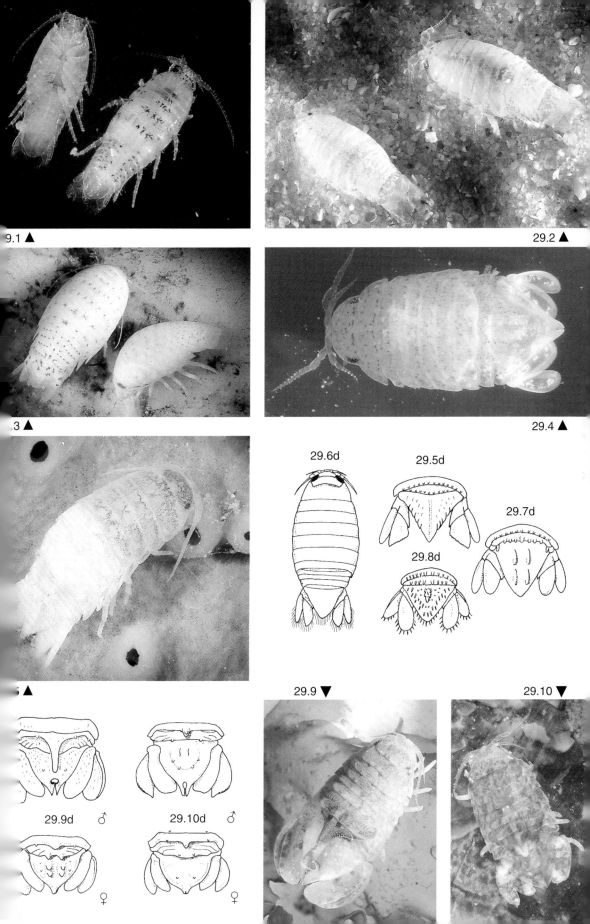

9.1 ▲

29.2 ▲

3 ▲

29.4 ▲

29.6d 29.5d

29.7d

29.8d

29.9 ▼

29.10 ▼

29.9d ♂

29.10d ♂

♀

♀

30.1 Variegated spherical isopod *Exosphaeroma varicolor*

IDENTIFICATION: Body flattened and mottled pink or black and white. Pleon formed of a single fused segment (with lateral grooves showing where the original segments are joined). Telson triangular, with two small ridges near base. SIZE: Up to 10 mm. BIOLOGY: Intertidal or shallow water, under stones or in weeds. RELATED SPECIES: The genus *Exosphaeroma* is characterised by one fused pleon segment and an unnotched telson, and includes 14 South African species distinguished mainly by the shape of the telson and uropods.

30.2d *Exosphaeroma hylecoetes* (Cape Point–East London). A purely estuarine form with a slightly furry body and hair-fringed uropods.

30.3d *Exosphaeroma planum* (Lüderitz–Port Alfred). A strongly flattened species with slight ridges along the thoracic segments. Telson large and triangular. Uropods broad, the outer branch oval and the inner squared off.

30.4d *Exosphaeroma porrectum* (Lüderitz–Port Elizabeth). Head, thoracic and pleon segments each with a row of 4–6 rounded knobs. Telson long and pointed, with several rows of knobs and a keel near the tip. On rocky shores.

30.5d *Exosphaeroma truncatitelson* (Namibia–Cape Agulhas). Body smooth. Telson broadly oval with squared tip. Uropods pointed. One of the most common isopods on both sandy shores and sand-inundated rocky shores.

30.6d *Exosphaeroma kraussii* (West Coast, on rocky shores). Body speckled. Tiny granules on hind margins of thoracic and pleon segments in male only. Telson triangular and pointed with small dorsal ridges; the uropods are broad and square-ended.

30.7d *Exosphaeroma laeviusculum* (Lüderitz–Cape Point). Very similar to *E. varicolor* but has a speckled body and a semicircle of tiny tubercles on the telson.

30.8 Hump-tailed isopod *Cymodoce valida*

IDENTIFICATION: A large, heavy-set and slow-moving isopod which is often attractively mottled in reds and browns. Telson with a pair of large dorsal humps. SIZE: 22 mm. BIOLOGY: One of the largest isopods found under boulders or on shallow reefs. RELATED SPECIES: More than 20 *Cymodoce* species occur in southern Africa, many of which have strongly sculptured telsons.

30.9 Button isopod *Sphaeramene polytylotos*

IDENTIFICATION: Unmistakable – head and body covered in flat-topped, button-like tubercles. Tip of telson with keyhole-like slit in the male, but not notched in female. SIZE: Up to 17 mm. BIOLOGY: Commonly found amongst mussels and barnacles on rocky shores and shallow reefs. Herbivorous; feeds on delicate filamentous algae and on simple green algae.

30.10 Roll-tail isopod *Dynamenella huttoni*

IDENTIFICATION: Body smooth and parallel-sided, sides of telson rolled down forming a gutter at tip. SIZE: Up to 18 mm, but usually much smaller. BIOLOGY: The most common isopod amongst tufts of algae in the intertidal. RELATED SPECIES: One of 10 *Dynamenella* species with similar habits. All are characterised by having the apex of the telson notched or with a narrow slit.

30.11d *Dynamenella ovalis* (Lüderitz–East London) resembles *D. huttoni*, but the body is flatter and oval, not parallel-sided.

30.12d *Dynamenella scabricula* (Lüderitz–Knysna). Slit at tip of telson widens anteriorly (like a keyhole). Body finely furry with transverse rows of tubercles. Two broad humps on the pleotelson.

30.13d *Dynamenella dioxus* (Lüderitz–Cape Point). Slit in telson widens anteriorly, roughly heart-shaped. The body is smooth except for the last pereon segment of the male, which has two large, sharp triangular processes.

30.14d *Dynamenella australis* (Lüderitz–Cape Agulhas). Notch in telson with a knob at its base; inner branch of uropod slender and curved. Often shelters under limpets.

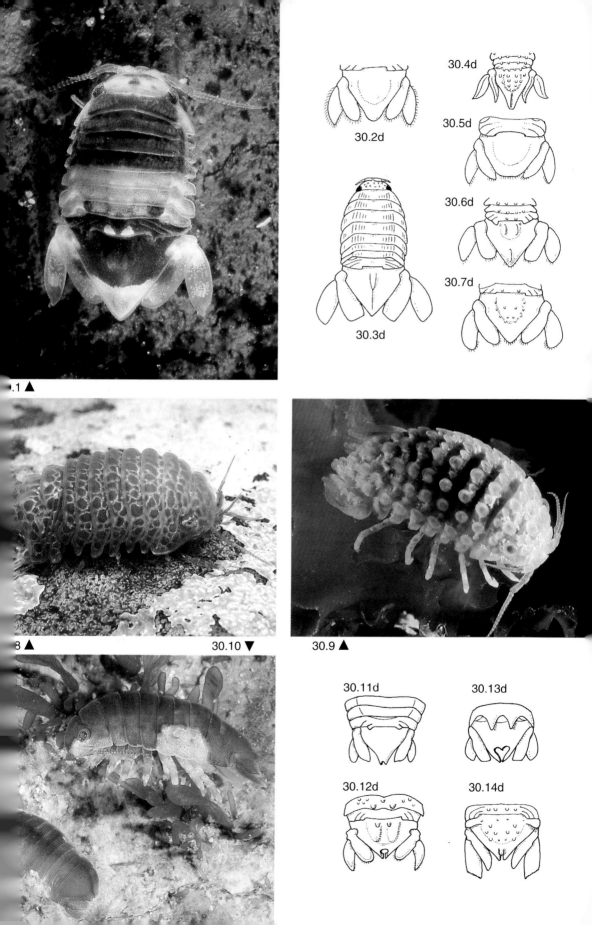

30.4d

30.2d

30.5d

30.3d

30.6d

30.7d

.1 ▲

8 ▲   30.10 ▼

30.9 ▲

30.11d   30.13d

30.12d   30.14d

Amphipoda : Amphipods

Although small and easily overlooked, amphipods are a diverse group of crustaceans found in profusion in most habitats. The body is compressed from the side (as opposed to that of isopods, which is generally flattened), and is often protected by large side-plates. The head bears unstalked eyes. The first two pairs of legs generally form nippers, while the remaining five end in a simple claw (hence the derivation: *amphi*, both, and *podos*, foot). Each abdominal segment typically carries two branched appendages, the first three being called pleopods and the remaining three uropods. Most amphipods crawl or burrow amongst seaweeds, or in sediment, feeding on detritus or carrion, but many build tubes and filter food particles from the water. Eggs are carried in a conspicuous brood pouch below the body, and hatch directly into the adult form. Over 300 species of amphipod occur in southern African waters and only a few of the most common or conspicuous of these are dealt with here.

31.1 Skeleton shrimp   *Caprella equilibra*

IDENTIFICATION: Highly modified with an extremely elongate, cylindrical body. The first body segment and pair of legs are fused to the head, while the second pair forms large, powerful nippers. The third and fourth pairs of legs have disappeared, while the last three are strongly clawed to grasp the substratum. The abdomen is reduced to a tiny stub. SIZE: 10–20 mm. BIOLOGY: Clings to hydroids or algae, reaching out to grasp plankton. Occasionally moves, inchworm-fashion, to a new perch. One of eight closely-related species in southern African waters.

31.2 Pocket amphipod   *Amaryllis macrophthalma*

IDENTIFICATION: A sturdy, compact amphipod with enlarged side-plates. The eye is elongate and the antennae stout and equal in length. Second pair of legs without nippers. The side-plate of the third abdominal segment has a characteristic upturned shape with a small 'pocket' above the corner. Colour brown with blue side-plates. SIZE: About 10 mm. BIOLOGY: A scavenger common in the holdfasts of seaweeds or under loose boulders.

31.3 Compact amphipod   *Lysianassa ceratina*

IDENTIFICATION: Body sturdy and compact, the limbs protected by long side-plates. Eyes elongate and antennae short and stout. Second pair of legs lacking nippers. Colour white to brown. SIZE: 10 mm. BIOLOGY: Scavenges under stones or amongst algae on rocky shores, or in sand or gravel. RELATED SPECIES: The most common of over 30 easily confused local species in the family Lysianassidae, most of which have a compact body form, short antennae and enlarged side-plates.

31.4 Big-eye amphipod   *Paramoera capensis*

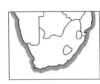

IDENTIFICATION: Delicate, shrimp-like amphipods with long, equal antennae. Eyes unusually large, almost meeting on top of the head. First and second legs both with small nippers. Uropods uniformly elongate. Colour very variable, most often pale with a red 'saddle' across back, sometimes mauve. SIZE: About 10 mm. BIOLOGY: Frequently the most abundant amphipod amongst seaweeds, also common swimming in the plankton, particularly at night.

31.5 Red striped amphipod   *Ceradocus rubromaculatus*

IDENTIFICATION: One of the larger amphipods, easily recognised by its distinctive pink-to-red colour pattern. First antennae much longer than second. First two pairs of legs with nippers, the second pair enlarged in adult males. Posterior edges of first three abdominal segments, and of their side-plates, are cut into a characteristic series of teeth. Last pair of uropods with two long equal branches that project well beyond the first two pairs. SIZE: 20 mm. BIOLOGY: Usually found under stones lying on coarse sand or gravel.

1.1 ▲   31.2 ▲   ▼ 31.3   31.4 ▼

.5 ▼

32.1 Seaweed amphipod *Hyale grandicornis*

IDENTIFICATION: First pair of antennae much shorter than second. Second pair of legs with large nipper in males, weak in females. Last pair of uropods very short and with only a single branch. Colour green or brown, usually with one or more circular white dots on each thoracic segment. Preserved specimens become bright orange. SIZE: 10 mm. BIOLOGY: The most abundant amphipod on rocky shores. Survives high in the intertidal by nestling amongst the seaweeds on which they graze. RELATED SPECIES: The most common of seven *Hyale* species found in southern Africa.

32.2 Louse amphipod *Temnophlias capensis*

IDENTIFICATION: Highly unusual amphipods in that the body is flattened and isopod-like with reduced side-plates. The legs lack nippers and are splayed to the sides, and the uropods are greatly reduced. Colour green to chocolate brown with characteristic rows of fluorescent blue spots along the back. SIZE: 4–7 mm. BIOLOGY: Common clinging to sponges and ascidians on the underside of boulders in intertidal pools and gullies, usually in tight groups.

32.3 Sponge amphipod *Leucothoe spinicarpa*

IDENTIFICATION: Colour uniform pale-pink to mauve. Nippers extremely large, the first pair of a characteristic form in which the 'finger' is formed by the last two segments, which close against a long, spine-like projection arising from the preceding segment. SIZE: About 10 mm. BIOLOGY: Lives commensally within the body cavities of sponges and ascidians. Believed to filter particles from the stream of water pumped through the body of the host.

32.4 Brack-water amphipod *Melita zeylanica*

IDENTIFICATION: Colour uniform pale-green or yellow. Second pair of nippers large and unusual in that finger closes against the inner surface, rather than the edge of the hand. Third uropods long, reaching well beyond first two pairs, and of characteristic form, the inner branch being only one-fifth the length of outer. SIZE: About 10 mm. BIOLOGY: Abundant in estuarine weed-beds.

32.5 Burrowing amphipod *Urothoe grimaldii*

IDENTIFICATION: Very short, broad-bodied amphipods with wide, setose legs. Colour off-white. SIZE: 4–6 mm. BIOLOGY: Superbly adapted for burrowing. The wide legs and broad body create a channel through which water and sand are driven by the powerful pleopods, driving the animal forwards. At the same time the mouthparts search the sand for food particles. Males have greatly enlarged eyes and elongate second antennae. One of eight *Urothoe* species found in southern Africa and identified by detailed structure of the legs.

32.6 Ornate amphipod *Cyproidea ornata*

IDENTIFICATION: A small but common and conspicuous species, best identified by its brilliant yellow and black colour pattern. The back is humped and the side-plates of segments 3 and 4 enormously enlarged, shielding the limbs. When the abdomen is tucked under the body, the animal has an almost spherical appearance. SIZE: 3–4 mm. BIOLOGY: Particularly common under the sea urchin *Parechinus* (93.6), but also moves around openly on rocky reefs, suggesting that the bright colours may warn predators that it is distasteful.

32.7 Ridgeback amphipod *Ochlesis lenticulosus*

IDENTIFICATION: Unmistakable by virtue of its spectacular mauve and yellow colour pattern. The body is rounded and compact, with a distinct 'keel' or ridge running down the centre of the back. This is cut into a triangular tooth on each of the first three abdominal segments. SIZE: 4–8 mm. BIOLOGY: A conspicuous species found on bryozoans and sponges on rocky shores and reefs.

2.1 ▲

32.2 ▲   32.5 ▼

.3 ▲   32.4 ▼

32.6 ▼   32.7 ▼

33.1 **Hunchback amphipod** *Iphimedia gibba*

IDENTIFICATION: Extraordinary, hunchbacked amphipods in which the top of the first body segment has become elongated, so that the head is directed downwards. First four side-plates elongate and pointed. Colour brown with numerous fluorescent blue spots. A bright-yellow stripe runs down the centre of the back and across the hind margin of each body segment. SIZE: 5 mm. BIOLOGY: Found in coarse sediments and moving openly on rocky reefs. Has been seen feeding on the toxic bryozoan *Alcyonidium nodosum* (48.5) and may obtain protection from predators as a result.

33.2 **Beach hopper** *Talorchestia capensis*

IDENTIFICATION: Body off-white. First pair of antennae much shorter than second. Second pair of legs of male with large nippers. Last three pairs of legs elongate and splayed to the sides to support animal upright. SIZE: 10–20 mm. BIOLOGY: A familiar, air-breathing scavenger found in vast numbers at or above the drift-line of sandy beaches. Often concealed under drift material, exploding into activity if disturbed. At night, hops clumsily about the beach in search of freshly deposited seaweeds. Shortly before dawn, returns up the beach and burrows into the dry sand. RELATED SPECIES: In *Talorchestia quadrispinosa* (False Bay–Namibia) the males have a prominent pair of spines on the back of each of the first two abdominal segments.

33.3 **Four-eyed amphipod** *Ampelisca palmata*

IDENTIFICATION: Head with two pairs of small red eyes, each bearing a bead-like lens. Antennae long and hairy, legs without nippers. SIZE: About 5 mm. BIOLOGY: Constructs fragile tubes of fine sand or mud, and extends its antennae to filter food particles from the water. RELATED SPECIES: One of thirteen *Ampelisca* species found in southern African waters and identified mainly by differences in the structure of the last pair of legs.

33.4 **Hitchhiker amphipods** *Jassa* spp.

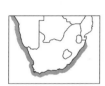

IDENTIFICATION: Adult males with very large second nippers that have an enormous 'thumb' projecting from the palm of the hand. Outer branch of third uropod with characteristic upturned tip and two triangular teeth on upper surface. SIZE: About 5 mm. BIOLOGY: Build open-ended mud or silt tubes attached to rocks and other solid objects in areas of high water flow. Common fouling organisms on piers, buoys and ships' hulls. Filter-feeders and predators of small crustaceans. Three closely related species occur in southern Africa.

33.5 **Stout-antenna amphipod** *Corophium acherusicum*

IDENTIFICATION: A cylindrical species with the last pair of legs much longer than the others. The second antennae of males are greatly inflated and toothed on the lower margin. First pair of legs with a weak claw, second pair very hairy but not clawed. Last three abdominal segments fused. SIZE: Typically about 4 mm. BIOLOGY: Abundant in sheltered, muddy areas, including estuaries. A burrow is excavated by the anterior legs and the spoil ejected by rapid beating of the pleopods. The animal then propels water through the burrow with the pleopods and filters particles from the current with its hairy legs.

33.6 **Nesting amphipod** *Cymadusa filosa*

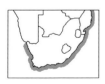

IDENTIFICATION: Antennae elongate, eyes white, body mottled brown with large side-plates. Last pair of uropods with characteristic short, pad-like branches, the outer with two strong curved hooks. SIZE: 10 mm. BIOLOGY: Weaves tubular nests out of fragments of algae or seagrass leaves. Herbivorous.

33.1 ▲   33.2 ▼   33.3 ▼   33.4 ▼

33.5 ▼   33.6 ▼

Stomatopoda : Mantis Shrimps

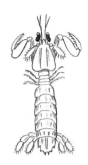

Highly specialised predators easily recognised by their massive raptorial second thoracic limbs, which resemble those of praying mantids. The eyes are large and stalked and the anterior half of the thorax is covered by a short carapace. The flattened abdomen is equipped with powerful, paddle-like swimming pleopods and ends in an armoured telson and a pair of large uropods. Mantis shrimps live in burrows or in rock and coral crevices. Many are brightly coloured and defend their territories, threatening intruders. Two functional groups occur. 'Spearers' impale soft-bodied prey, such as shrimps and fish, with a rapid upward thrust of the barbed finger. 'Smashers' seldom have barbed fingers, but strike and disable hard-shelled prey, including crabs and molluscs, with the reinforced heel of the raptorial limb. Such strikes have a force approaching that of a small-calibre bullet and can easily crack the glass of an aquarium. Twenty-five species of stomatopods occur in southern Africa, most in tropical waters.

34.1 Cape mantis shrimp   *Pterygosquilla armata capensis*

IDENTIFICATION: A 'spearer' with 6–8 teeth on the finger of the raptorial claw (34.1a & 34.1d). Telson keeled, with six large teeth and a notch at the tip (34.1d). Colour pale pink with metallic eyes (34.1b). SIZE: Up to 200 mm. BIOLOGY: The only West Coast stomatopod. Burrows into soft sediments; may congregate in large swarms near the surface where it is preyed on by seals, hake and other fish. RELATED SPECIES:
34.2d *Lysiosquilla capensis* (Cape Point–Port Elizabeth) has 15–17 teeth on the finger of the raptorial claw.
34.3d *Harpiosquilla harpax* is a tropical sand-burrower with a barred abdomen and 8–9 teeth on the raptorial claw. Its telson lacks a central notch.

34.4 Sickle mantis shrimp   *Gonodactylus falcatus*

IDENTIFICATION: A 'smasher' with an untoothed sickle-like finger to the raptorial appendage. Usually bright green, but dark, almost black animals also found. Five rounded bumps on telson (34.4d). SIZE: About 60 mm. BIOLOGY: Found under stones on tropical shores. RELATED SPECIES:
34.5d *Gonodactylus chiragra* has three humps on the dorsal surface of the telson.
34.6d *Gonodactylus lanchesteri* has numerous small tubercles and spines dotted over the telson in addition to 3–5 larger bumps.

Mysidacea : Mysids or Opossum Shrimps

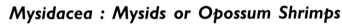

Small, shrimp-like crustaceans with stalked eyes. The thorax is covered by a carapace but (unlike true shrimps or euphausids) this is not fused to the last four thoracic segments. There are six or seven pairs of thoracic limbs, none of which bears nippers. The abdomen is usually slender, with reduced pleopods and ends in a tail-fan formed of a central telson and one pair of large, branched uropods. A round statocyst, or balancing organ, is visible within the base of each uropod, and is characteristic of the group. Mysids are detritus feeders and scavengers which often form large swarms. The eggs are brooded in a large marsupium or brood pouch, hence the common name.

34.7 Surf mysid   *Gastrosaccus psammodytes*

IDENTIFICATION: Telson is deeply notched at the tip with 6–7 conspicuous spines along each margin (**34.7d**). SIZE: About 12 mm. BIOLOGY: Abundant in the surf zone of exposed sandy beaches. Burrows during the day, but emerges into the water column by night. Populations are often segregated, with the brooding females closest inshore and males and juveniles further offshore. RELATED SPECIES: Replaced in Namibia by *G. namibensis* (**34.8d**) and in northern Natal by *G. bispinosa* (**34.9d**) and *G. longifissura* (**34.10d**). *Mesopodopsis wooldridgei* (**34.11d**) forms dense swarms beyond the breaker line and *Mesopodopsis africanus* (**34.12d**) is estuarine. These species are all distinguished mainly by the shape and spination of the telson. *Mysidopsis major* is familiar to Western Cape divers, forming dense swarms in kelp beds.

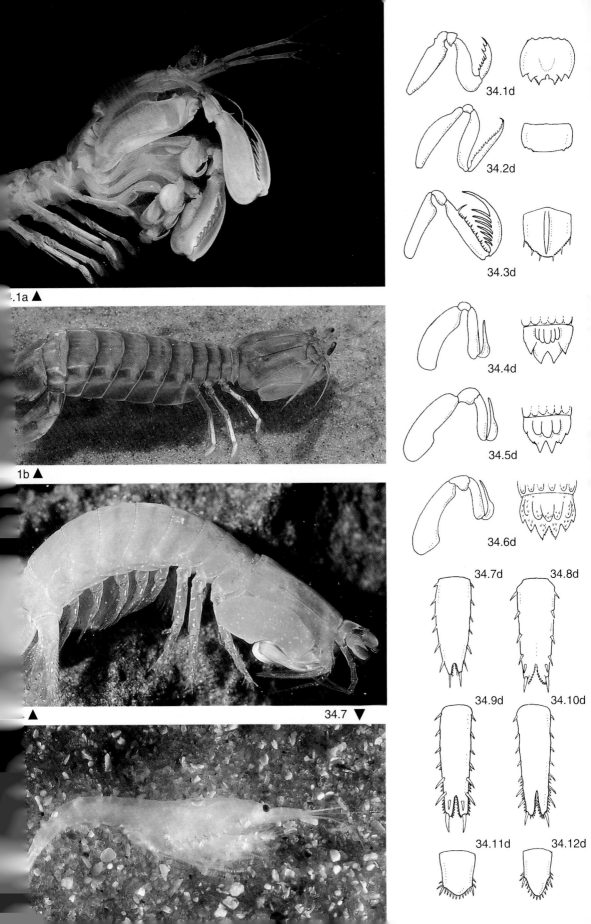

34.1a ▲

1b ▲

▲   34.7 ▼

34.1d

34.2d

34.3d

34.4d

34.5d

34.6d

34.7d   34.8d

34.9d   34.10d

34.11d   34.12d

Euphausiacea : Krill

Euphausids, or krill, are shrimp-like, with stalked eyes and a carapace covering the thorax. They swim almost continually, using their hairy abdominal appendages (pleopods) for propulsion. The slender thoracic limbs are fringed by stiff hairs and together form a funnel-shaped basket, through which water is sieved. Many species form huge swarms, and are an important food for fish, whales and other predators.

35.1 Light euphausid *Euphausia lucens*

IDENTIFICATION: First six pairs of thoracic limbs similar in size and shape, seventh and eighth pairs greatly reduced. Carapace has a single tooth mid-way on its lower margin. No dorsal teeth on abdomen. A conspicuous triangular tooth projects from the base of the first antenna. SIZE: 15 mm. BIOLOGY: Forms dense swarms near the surface during daylight along the South and West coasts.

Macrura : Rock Lobsters

Rock lobsters are robust, large crustaceans with a long tail ending in a well-developed tail-fan. The thorax and head are covered with a single shield, or carapace. None of the walking legs ends in nippers. Popularly called crayfish, they are correctly termed 'spiny lobsters' or 'rock lobsters' to avoid confusion with the clawed freshwater crayfish (confusion that might cost a good deal – freshwater crayfish hold less attraction for gourmets). Rock lobsters have an elaborate life cycle involving thirteen larval stages and many months floating in the sea before metamorphosis into the adult stage. How the larvae return to the parental grounds after so long at sea remains a mystery.

35.2 West Coast rock lobster *Jasus lalandii*

IDENTIFICATION: Orange-brown body; tail-fan orange, blue and green. The thorax is covered with spines, and the front of the carapace has two large spines and a smaller central projection (rostrum) between the eyes. SIZE: 30 cm. BIOLOGY: The most important commercial rock lobster in southern Africa, 2200 tonnes being harvested in 1992. In past years the industry took up to 10 000 tonnes, but steadily eroded the supply of animals of legal size (then 89 mm carapace length). *Jasus* grows slowly. Males of 80 mm are 5–7 years old; females about 12 years old. Stunted growth in 1990–92 severely reduced catches. To spread the fishing effort over a wider size-range, the size-limit was reduced to 80 mm in 1992 and (for commercial fisheries) to 75 mm in 1993. *Jasus* is intolerant of low oxygen levels and, on the northern West Coast, regularly becomes concentrated inshore to avoid water that has little oxygen. Occasionally hundreds of thousands are stranded on the shore. *Jasus* feeds on mussels, urchins and even barnacles, and is even capable of eliminating many species in areas where rock lobsters are abundant. Rock lobsters are eaten by octopus, dogsharks and seals, but their natural mortality rate is low, making them susceptible to over-fishing.

35.3 South Coast rock lobster *Palinurus gilchristi*

IDENTIFICATION: Body orange, with alternating pinky-orange and white bands on the legs and antennae. A broad plate between the eyes is flanked by two outer horns and has about 11 shorter spines along its front edge. Abdominal segments 2–5 have anterior and posterior transverse grooves, which are filled with fine hairs and linked together near the mid-line by two longitudinal grooves, creating an 'H' pattern. The outer surface of the fourth-last joint of the legs is flat and hairy. SIZE: 30 cm. BIOLOGY: The second-most important commercial species in South Africa, *P. gilchristi* is caught in lobster pots in deep rocky areas (50–170 m). RELATED SPECIES: *Palinurus delagoae*, the Natal deep-sea lobster, is very similar. It is reddish-mauve with irregular ivory-white patches, and its legs and antennae are banded. The fourth-last joint of each leg is almost cylindrical and lacks hairs. Abdominal segments 2–5 have only a trace of an anterior groove, which is not linked by medial grooves to the posterior groove. It prefers open areas of mud and rubble at depths of 180–300 m, and is caught by trawling.

5.1 ▲

2 ▲   35.3 ▼

36.1 **East Coast rock lobster** *Panulirus homarus*

IDENTIFICATION: Two sharp horns project forwards between the eyes. The abdomen has a transverse scalloped groove on each of abdominal segments 2-5. Colour brown to brick-red (or olive-green in one variety) with orange spines and blue and green markings on the head. SIZE: 25 cm. BIOLOGY: Occurs on rocky reefs in the surf zone at depths of 1-36 m, coinciding with brown mussels, their major food. Emerges at night to feed, hiding in holes by day. As in other southern African rock lobsters (with the exception of *Jasus*), the ventral surface of the 3rd segment of the antenna has a rasp (the stridulating organ) which is rubbed against the skeleton to produce a squeaky or rattling sound when the animal is threatened. *P. homarus* readily sheds some of its legs (by a process called 'autotomy') to distract predators. These legs can be regrown later. Moray eels often share holes with *P. homarus*, and attack one of its main predators, the octopus. Male lobsters plaster the under-surface of females with a sperm jacket (spermatophore) which hardens on contact with seawater, to be broken open later by the female when she extrudes her eggs. The fertilised eggs are then attached to the pleopods on her abdomen, remaining there until the larvae hatch. RELATED SPECIES: The long-legged spiny lobster, *Panulirus longipes* (central KZN to Sodwana), has conspicuous white spots all over its abdomen, and the horns between its eyes are flattened and oval in section. Grooves run across the abdominal segments, but they are not scalloped. Orange stripes run down its legs, although some individuals have spotted legs. The penicillate spiny lobster, *Panulirus penicillatus* (Zululand and Moçambique), is dark brown to olive-black, with orange spines and thin white lines down the legs. Its horns are rounded in section.

36.2 **Painted rock lobster** *Panulirus versicolor*

IDENTIFICATION: Strikingly coloured, the carapace green with a black-and-white pattern on the sides, and the abdomen dull green with a characteristic white bar across each segment. The legs are black with longitudinal blue or white stripes. Juveniles have striking white antennae. Some individuals are blue instead of black. SIZE: 30 cm. BIOLOGY: Adults are rarely found on the South African coast but juveniles are common, hiding in holes with only their extraordinarily long bright white antennae projecting. RELATED SPECIES: The ornate spiny lobster, *Panulirus ornatus* (Durban to East Africa), is blue-green with orange spines, white antennae and transverse bands on its legs. There are no grooves or white bands across the abdominal segments. The spurs that extend downwards on the sides of each abdominal segment have a conspicuous diagnostic white spot. A shallow-water species (1–25 m), *P. ornatus* is the most abundant species in East Africa, where it is commercially fished.

36.3 **Shoveller crayfish** *Scyllarides elizabethae*

IDENTIFICATION: Unmistakable, with its short, broad, flattened antennae. The body is dull brown, with a rough texture and orange pattern. The legs are strikingly banded vermilion. SIZE: 25 cm. BIOLOGY: Occurs most commonly in relatively deep water (about 150 m) on gravel, but sometimes found close inshore in the shallows. Shovels its way through the surface layers of sediment, feeding on worms and molluscs. Also known as the slipper crayfish or shovel-nosed crayfish.

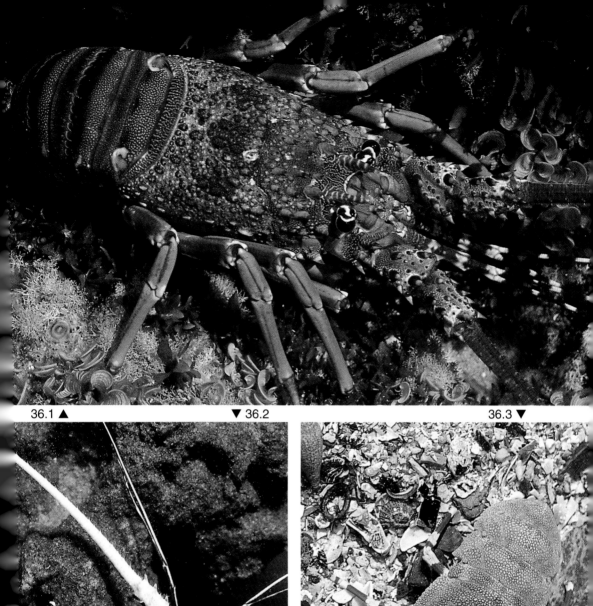

36.1 ▲   ▼ 36.2   36.3 ▼

Swimming Prawns & Cleaner Shrimps

Swimming prawns (Penaeidea) are distinguished from other prawns because the sides of the first abdominal segment overlap the sides of the second segment, the last abdominal segment has a dorsal keel, and the first three pairs of walking legs end in nippers. The various species are recognised by the shape of the long rostrum that projects forward between the eyes, and the number of teeth on the rostrum, as indicated by a 'rostral formula'. For example, 9–11/1 indicates 9–11 teeth on the upper edge of the rostrum and 1 on the lower edge. Commercially-fished deep-water prawns are seldom seen alive, and are not described here. Swimming prawns breed at sea, shedding their eggs into the water, but their juveniles develop in nursery areas in estuaries. The cleaner shrimp belongs to another group, the Stenopodidea, similar to swimming prawns except for the possession of large nippers on the third pair of walking legs.

37.1 Tiger prawn *Penaeus monodon*

IDENTIFICATION: Rostrum moderately long, about twice as long as the eyestalks. Upper surface of rostrum with 7 teeth, lower surface with 3. Grooves on either side of the rostrum extend back for about three-quarters of the carapace length. Top of carapace ridged behind the rostral teeth, but this ridge is not grooved. Abdomen barred when the animal is fresh. SIZE: 90 mm. BIOLOGY: Commonly netted in estuaries and shallow water. With *Metapanaeus monoceros* it makes up 75–95% of the commercial catch in Moçambique and Natal. RELATED SPECIES:

37.2d *Fenneropenaeus indicus*, the white prawn (Transkei northwards), is plain white. Rostral formula 7–8/4–5. Dorsal ridge on the carapace is not grooved.

37.3d *Penaeus semisulcatus*, the zebra prawn (Durban to central Moçambique), has a rostral formula of 6–7/3 and the ridge on the top of the carapace behind the rostral teeth is characteristically grooved. The fifth pair of walking legs has a small outer appendage (an exopod) which helps distinguish this species from *P. monodon*.

37.4d *Penaeus japonicus*, the bamboo prawn, occurs further south than most swimming prawns, from False Bay to Natal. Its rostral formula is 9–11/1, its tail-fan mottled yellow and brown, and the central portion of its tail-fan (the telson) has three spines on either side amongst the marginal hairs.

37.5d *Penaeus canaliculatus*, the striped prawn (East London northwards), is very similar to *P. japonicus*, with a rostral formula 9–11/1, but its body is greenish-brown with faint bands, and its telson has no spines on its margins.

37.6d *Metapenaeus monoceros*, the brown prawn (East London northwards), has a rostrum that is about twice the length of the eyestalks and projects well beyond the eyes; 9–10 teeth on the upper surface of the rostrum, but none on the lower surface. It is one of the commonest estuarine and shallow-water prawns, contributing substantially to the commercial catch in Moçambique and on the Tugela Bank in Natal.

37.7 Surf shrimp *Macropetasma africana*

IDENTIFICATION: More slender and active than most swimming prawns. Distinguished by the possession of a small outer appendage (an exopod) on the first pair of walking legs. Rostrum same length as eyestalks, upper edge strongly convex. Rostral formula 9–10/0, with a distinctive gap between the last two rostral teeth. SIZE: 50 mm. BIOLOGY: Found almost exclusively in or just behind the surf zone of exposed sandy beaches.

37.8 Cleaner shrimp *Stenopus hispidus*

IDENTIFICATION: Strikingly coloured, with red and white bands on the body and legs, tinges of blue, and long white antennae. The third pair of walking legs is obviously larger than the other walking legs, spiny, and ends in nippers. SIZE: 50 mm. BIOLOGY: Occurs on tropical coral reefs. Removes parasites and infected bacterial growths from the surfaces of fish, which will queue for the attentions of cleaner shrimps. Cleaner fish fulfil a similar function (see *Labroides dimidiatus* 127.7).

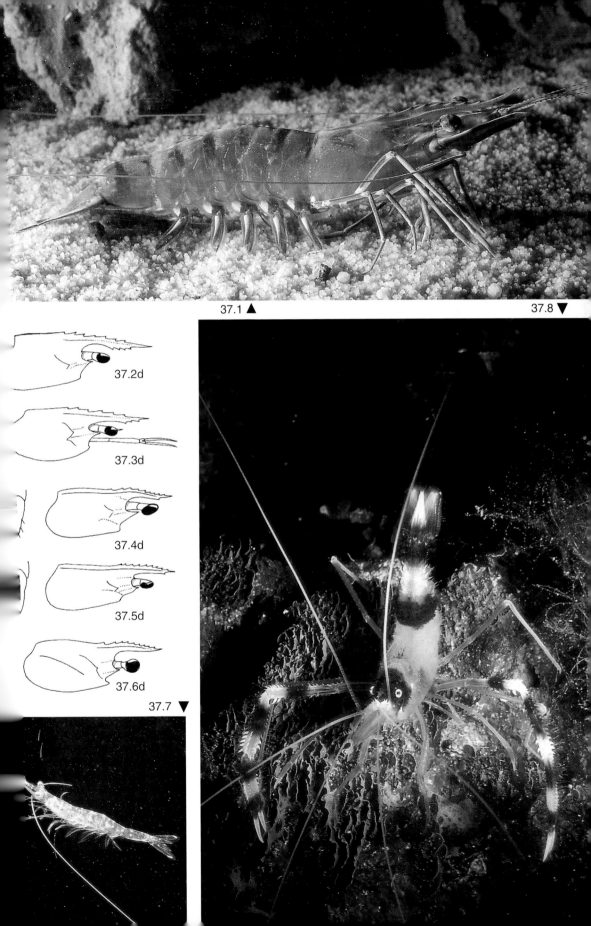

37.1 ▲

37.8 ▼

37.2d

37.3d

37.4d

37.5d

37.6d

37.7 ▼

Caridea : Benthic Prawns

Bottom-dwelling (benthic) prawns, ill-suited to swimming, fall in the group Caridea, distinguished by the fact that the sides of the second abdominal segment overlap those of the first and third segments, the dorsal surface of the last abdominal segment is rounded, not keeled, and the third walking legs lack nippers. Females attach their eggs to the swimming appendages of the abdomen. Some species occur in estuaries, but none depends on estuaries as nursery grounds, and none is of commercial importance.

38.1 Commensal shrimp   *Betaeus jucundus*

IDENTIFICATION: Small, equal-sized nippers on first walking legs. Body almost transparent. Carapace lacks a rostrum, but extends forwards to cover the eyes (38.1d). SIZE: 15 mm. BIOLOGY: Burrows in sheltered sandbanks in estuaries and lagoons. Often lives commensally in the burrows of the sandprawn *Callianassa kraussi* (39.1), but the nature of the interaction is unknown.

38.2 Cracker shrimp   *Alpheus crassimanus*

IDENTIFICATION: First pair of walking legs with strong nippers, one much larger than the other. Carapace keeled dorsally. The central rostrum is either minute or absent and there are no horns on the front edge of the carapace (38.2d). SIZE: 30 mm. BIOLOGY: Lives in kelp holdfasts or burrows among stones or in coral rubble. Its larger nipper emits a sharp click when snapped closed (familiar to divers in kelp-beds). In the process, a spurt of water is jetted outwards. Both the sound and the jet of water play a role in territorial defence, communicating the size and strength of a territory owner. RELATED SPECIES: There are several species of *Alpheus* that live buried in estuarine weed-beds or among coral branches. Often they share burrows with gobies, aiding in burrow construction and benefiting from the gobies' alert reactions to predators.
38.3d *Synalpheus anisocheir* (Knysna to Inhambane) is similar but its carapace extends forwards over the eyes like a hood, and has a short but distinct central rostrum with a projecting horn on either side of it.

38.4 Sand shrimp   *Palaemon peringueyi*

IDENTIFICATION: Transparent, with vertical bars across the body; opaque when preserved. Front legs with slender nippers and yellow spots at the joints. Rostrum well developed, toothed and almost saw-like. SIZE: 25 mm. BIOLOGY: Lives in intertidal pools and shallow areas with a sandy bottom. Scavenges fragments of dead animals.

38.5 Broken-backed shrimp   *Hippolyte kraussiana*

IDENTIFICATION: Abdomen bent near the middle, giving a hunchback appearance. Rostrum long, tipped with three tiny projections; rostral formula 1–2/4–5. SIZE: 15 mm. BIOLOGY: Lives among estuarine weed-beds, particularly in seagrass. Colour varies continually with that of the background, but is commonly yellow, dull green or almost black.

38.6 Oriental shrimp   *Alope orientalis*

IDENTIFICATION: Body pale green, mottled brown; rostrum short, rostral formula 4–5/0. A large spine immediately above and behind each eye; second pair of walking legs with a small nipper, and the 'wrist' divided into about seven tiny joints. SIZE: 25 mm. BIOLOGY: Occurs among weed-covered rocks in intertidal pools; very common in warmer waters.

38.7 Zebra shrimp   *Gnathophyllum fasciolatum*

IDENTIFICATION: Stubby, barrel-shaped body with striking black and white stripes. Legs white with yellow and black bars. Rostrum short and pointed, lacking teeth. SIZE: 12 mm. BIOLOGY: Found on tropical shores in the shallow subtidal zone, hiding under stones. Slow-moving but territorial, chasing away other individuals of the same species and sex.

1 ▲

38.2 ▲

38.1d   38.2d   38.3d

▲   38.6 ▼

38.5 ▲   38.7 ▼

Sandprawns, Mudprawns, Mole & Porcelain Crabs

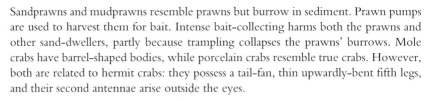

Sandprawns and mudprawns resemble prawns but burrow in sediment. Prawn pumps are used to harvest them for bait. Intense bait-collecting harms both the prawns and other sand-dwellers, partly because trampling collapses the prawns' burrows. Mole crabs have barrel-shaped bodies, while porcelain crabs resemble true crabs. However, both are related to hermit crabs: they possess a tail-fan, thin upwardly-bent fifth legs, and their second antennae arise outside the eyes.

39.1 Common sandprawn *Callianassa kraussi*

IDENTIFICATION: Pink, fragile and translucent; internal organs visible through the skeleton. Nippers well developed; one much larger than the other (particularly in males). The 'wrist' of the nipper (third joint from the tip) is longer than broad; telson (central part of the tail-fan) wider than long. SIZE: 60 mm. BIOLOGY: Abundant in estuaries, even in those closed to the sea for years. Builds deep burrows and sifts the sediment for food, chucking particles from the burrow entrance to create miniature volcanoes, oxygenating and turning over tons of sediment. It profoundly affects other organisms, promoting bacteria but burying diatoms and reducing the meiofauna (animals <1 mm long). Females carry eggs on the abdomen; offspring dig tiny side-tubes off the parent's burrow. RELATED SPECIES: The round-tailed sandprawn, *Callianassa rotundicaudata* (Saldanha Bay to Durban), has a wrist that is longer than broad and its telson is as wide as long. It burrows in sand under stones on the open coast.

39.2 Estuarine mudprawn *Upogebia africana*

IDENTIFICATION: Green-brown and robust. Nippers equal-sized and subchelate ('thumb' almost absent, 'finger' bending back to meet it). The first two walking legs lack a spine at the base, pointing inwards. SIZE: 40 mm. BIOLOGY: Lives intertidally in muddy sand in estuaries; absent from closed estuaries. Pumps water through its U-shaped burrows and filters out nutritious particles on its hairy mouthparts. Its gut is laden with symbiotic bacteria which may supplement its food. RELATED SPECIES: The coastal mudprawn, *Upogebia capensis,* lives in mud under stones on the open coast (Lüderitz to Mossel Bay). The first two walking legs have a spine at the base.

39.3 Mole crab *Emerita austroafricana*

IDENTIFICATION: Barrel-shaped body; abdomen tightly tucked under the thorax, and ending in a tail-fan. Legs stubby and spade-like. Antennae as long as the body but usually rolled up and hidden. SIZE: 25 mm. BIOLOGY: Buries itself on exposed sandy beaches, extending its hairy antennae to filter particles from the waves. Moves up and down the shore with the tides. Collected as bait, and often called 'sea lice', a poor name because they are not lice and there are too many other animals with this name. RELATED SPECIES: *Hippa adactyla* (Transkei to Zanzibar) lives in the same habitat. Its body is flatter and broader, and its antennae only a third of the body length.

39.4 Lamarck's porcelain crab *Petrolisthes lamarckii*

IDENTIFICATION: Crab-like, body flat and circular; nippers very large. Antennae long and thin, but often concealed. Brown or mottled green. Front margin with three low projections between the eyes; 'wrist' of nipper longer than broad. SIZE: 25 mm. BIOLOGY: Hides under stones high on rocky shores. Voluntarily sheds its nippers when threatened – the abandoned limb merrily snaps away to distract predators.

39.5d *Pachycheles natalensis* (south of Durban to Moçambique): margin between eyes smoothly rounded: 'wrist' about as broad as long. Pink to red.

39.6d *Porcellana dehaanii* (East London to Maputo): front margin between eyes with three obvious teeth, all of which are smooth. Body pink.

39.7d *Porcellana streptocheles* (Namaqualand to Port Elizabeth): three teeth on margin between the eyes; central tooth finely serrated. White or pink.

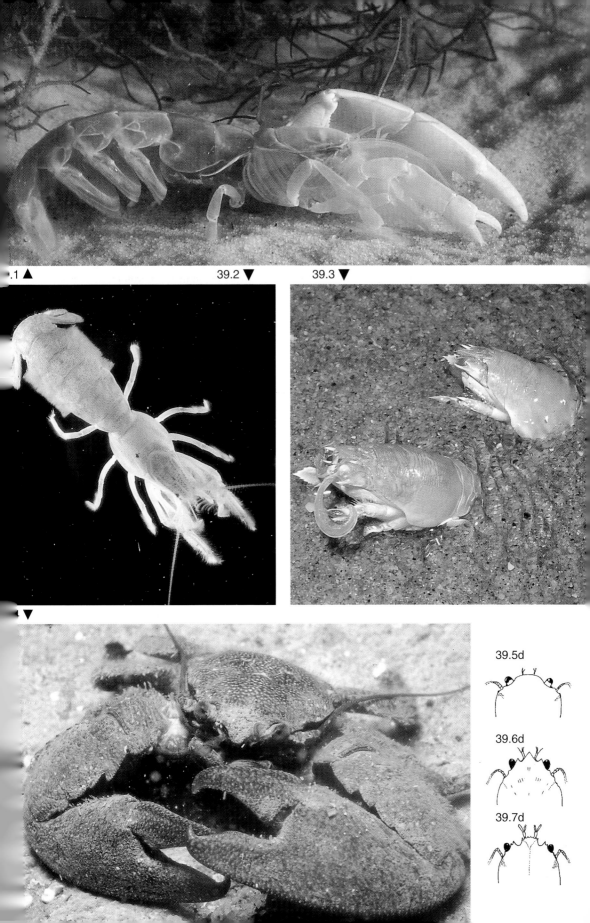

39.1 ▲   39.2 ▼   39.3 ▼

39.5d

39.6d

39.7d

Anomura : Hermit Crabs

Hermit crabs are well known for their curious habit of colonising empty gastropod shells to protect their soft-skinned abdomens. To fit the spiral of their adopted homes, the abdomen has become twisted and the appendages on the inside of the curve reduced in number or lost. The first pair of legs bears nippers, the left one often being enlarged to seal the opening of the shell when the crab withdraws. The second and third pairs of legs are used for walking, while the last two pairs are reduced to grip the shell. As they grow, hermit crabs have to move into progressively larger homes, and they often fight for the best shells. Most species are scavengers, but some filter food particles from the water using their antennae. About 45 species occur around South Africa.

40.1 Yellow-banded hermit *Clibanarius virescens*

IDENTIFICATION: Body olive-green, nippers about equal in size, spotted with yellow, the fingers yellow with black tips. Walking legs banded with yellow. SIZE: 10–20 mm. BIOLOGY: Abundant hermit in mid-to-high shore pools in Natal. RELATED SPECIES: **40.2 Clibanarius longitarsus** has blue stripes along the length of the legs, and is found from Transkei northwards, particularly in estuaries and calm lagoons.

40.3 Blue-eyed hermit *Calcinus laevimanus*

IDENTIFICATION: Nippers dark brown with white fingers, that of the left side greatly enlarged to block the shell opening when the crab withdraws. Eyes blue, eyestalks and first antennae banded brilliant blue and orange. Legs pale brown ending with white bands. SIZE: 10–30 mm. BIOLOGY: A beautiful species common in tide-pools on the Natal coast. Prefers low-spired shells, especially those of *Turbo* and *Nerita* (68.2, 68.4).

40.4 Land hermit *Coenobita cavipes*

IDENTIFICATION: Colour cream to brown, the left nipper being the larger. Readily recognised by its unusual terrestrial habits. SIZE: 30–40 mm. BIOLOGY: An air-breathing hermit which buries above the high-water mark by day, emerging at night to scavenge. Returns to the sea only to wet the body and in order to spawn. RELATED SPECIES: *C. rugosus* has an oblique ridge of knobs across the outer surface of the left hand, which it grates to produce sound.

40.5 Pink hermit *Paguristes gamianus*

IDENTIFICATION: Body pink, limbs hairy and spiny, left and right nippers equal in size. Fourth leg ends in a simple spine (unlike the minute claw found in all other southern African genera). SIZE: 10–20 mm. BIOLOGY: The most common hermit on rocky shores on the Atlantic seaboard, often clustered under boulders in low tide-pools. RELATED SPECIES: *P. barnardi* lacks spines on the last joint of the second walking leg, and occurs only on the Southern Cape Coast.

40.6 Common sand hermit *Diogenes brevirostris*

IDENTIFICATION: Body dirty-white with darker speckles and spines, and brown stripes on last segments of walking legs. Left nipper thorny and much larger than the right. SIZE: 30 mm. BIOLOGY: Extremely common on sheltered sandbanks and in sandy pools. Buries in sand when exposed by the tide, emerging when submerged to scavenge. RELATED SPECIES: The commonest of four closely-related southern African *Diogenes* species.

40.7 Giant red hermit *Dardanus megistos*

IDENTIFICATION: A large red hermit with equal-sized nippers. Surface of nippers and walking legs covered with comb-like rows of red and brown bristles. SIZE: Up to 10 cm. BIOLOGY: An uncommon but spectacular species found on shallow reefs and sandbanks.

.1 ▲

40.2 ▲   40.3 ▼

4 ▲

▲   40.6 ▼

40.7 ▼

Brachyura : True Crabs

Crabs are the most specialised crustaceans. The abdomen no longer forms a tail, but is tucked beneath the thorax. The abdominal limbs have lost their original swimming function and serve only to hold the eggs in the female or transfer sperm in the male. The tail-fan has disappeared altogether. The first pair of walking legs bears nippers (chelipeds); the remaining four pairs are used for walking. Crabs scuttle sideways, thus lengthening their stride without entangling their legs. The head and thorax are covered by a shield-like carapace which houses gills on either side of the body. The eggs hatch into planktonic larvae (recognised by a long spine on the back), which settle and are transformed into miniature adults. About 300 species inhabit South Africa.

41.1 Red-clawed mangrove crab *Sesarma meinerti*

IDENTIFICATION: Carapace almost square, with a series of 4–5 short lateral grooves and a strong tooth at the front corners. The eyes fold outwards and can be housed in depressions running across the front of the carapace. 'Chest' strongly granular. Body dark, almost black. Nippers granular and approximately the same size; bright red, shading to yellow. SIZE: 50 mm. BIOLOGY: Lives in estuaries, particularly those with mangroves. Near the high-tide mark it builds deep holes with 'hooded' openings. Scavenges, but is particularly fond of mangrove leaves, racing from its burrows to capture them as they fall. RELATED SPECIES: *Sesarma guttatum* (Transkei–Moçambique) has uniformly bright-red nippers and a distinctive sharp tooth on the fourth joint of the nipper-bearing legs.

41.2 Marsh crab *Sesarma catenata*

IDENTIFICATION: Carapace square, with lateral grooves, yellow to brown. Nippers yellow-brown, moderate and equal-sized, and with a characteristic furry lining around the hinge. SIZE: 25 mm. BIOLOGY: Very common; burrows in estuarine sand and mudflats near the high-water mark. Feeds by sucking mud particles to remove organic matter and micro-organisms; scavenges on dead animals. RELATED SPECIES: Two very similar species lack the hairy lining to the nippers. *Sesarma eulimene* is abundant in estuaries between the Bashee and Inhambane; it has dull blue-yellow nippers. *Sesarma ortmani* is more scarce, occurs in tropical estuaries, and has brilliant orange nippers.

41.3 Columbus' crab *Planes minutus*

IDENTIFICATION: Unmistakable, with a soft, smooth body that is uniformly blue. SIZE: 30 mm. BIOLOGY: Found almost world-wide; often washed ashore with bluebottles, and forms part of a 'blue community' of organisms that float or swim near the surface of the sea. Named Columbus' crab because it was apparently first noted by Columbus while he was on voyage to the West Indies. RELATED SPECIES: *Planes cyaneus* replaces it in Natal and Moçambique.

41.4 Masked crab *Mursia cristimanus*

IDENTIFICATION: Carapace roughly oval, pale pink with red tubercles. About eight short, blunt marginal teeth and a single, much larger tooth projecting outwards from the side of the carapace. Cheliped broad and liberally supplied with spikes. SIZE: 50 mm. BIOLOGY: Hides beneath boulders in the subtidal zone. The nippers are characteristically held close to the 'face', hence the name 'masked crab'. RELATED SPECIES: 41.5 *Calappa hepatica,* the box crab (Durban northwards), also has a mask-like arrangement of the nippers but is easily recognised by expansions on the sides of the carapace that cover and hide the legs.

41.6 Lunar box crab *Matuta lunaris*

IDENTIFICATION: Carapace almost round, with a sharp protruding mid-lateral spine. Cream, with tiny red dots; legs yellow, last pair with paddle-shaped tips. Nippers spiny, and folding neatly to mask the 'face'. SIZE: 40 mm. BIOLOGY: Burrows in sheltered sandbanks; feeds on worms.

1 ▲

41.2 ▲

▲   41.5 ▼

41.4 ▲   41.6 ▼

42.1 Cape rock crab *Plagusia chabrus*

IDENTIFICATION: Carapace smooth and velvety in texture, with two notches between the eyes that house the first antennae, and three marginal teeth. Colour red-brown, with orange-yellow ridges on the legs, and rows of yellow tubercles on the nippers. SIZE: 50 mm. BIOLOGY: Occurs in pools and on shallow reefs. Frequently observed cropping low-growing seaweeds with its nippers, but also feeds on small animals caught up with the seaweed. RELATED SPECIES:

42.2 *Plagusia depressa tuberculata,* the tuberculate crab, occurs on exposed rocky shores in Natal and Moçambique, and is very similar except for its pale-brown colour and the numerous tubercles on its back.

42.3 Green rock crab *Grapsus fourmanoiri*

IDENTIFICATION: Very similar to the Natal rock crab (42.4) in structure, but dull khaki-green tinged with yellow and with pale nippers. The sides of its carapace are almost straight, not obviously convex as in the Natal rock crab. SIZE: 35 mm. BIOLOGY: Lives in the intertidal zone, hiding in rock crevices and emerging to feed during low tide, particularly at night. Feeding habits unknown. Juveniles of the Natal rock crab can easily be confused with this species because they are also drab, but the shape of the sides of the carapace is a reliable distinguishing feature.

42.4 Natal rock crab *Grapsus grapsus tenuicrustatus*

IDENTIFICATION: Carapace black, usually speckled with green; sides obviously convex. The posterior part of the carapace has about eight fine grooves that run outwards to the edge. Legs red-brown and yellow; nippers dark red; 'hand' weakly ridged. SIZE: 50 mm. BIOLOGY: Abundant in Natal and southern Moçambique, scurrying around in groups on rocks above the water level, plucking seaweeds and picking up small animals. Males particularly gaudy when mature; the females more drab and often a dull mottled green. A very similar subspecies, with the boringly repetitive name of *Grapsus grapsus grapsus,* occurs in northern Namibia and Angola.

42.5 Shore crab *Cyclograpsus punctatus*

IDENTIFICATION: Body very smooth except for a granular 'chest'. Sides of carapace convex, lacking marginal teeth, although a single low tooth lies just outside each eye. Hands smooth. Body black or dark brown near the front, blending into a grey-green network towards the back. Legs orange to brown, flecked red. SIZE: 30 mm. BIOLOGY: Lives high on the shore, often aggregating densely under boulders. Scavenges by night during low tide, feeding mostly on drift seaweeds, but also animal matter.

42.6 Estuarine rock crab *Metopograpsus thukuhar*

IDENTIFICATION: Distance between the base of the eyes more than half the width of the carapace; no lateral teeth on the carapace except for one that flanks each eye. Sides of carapace very straight but delicately notched with about six short grooves. The legs have stiff, spiky hairs. 'Hand' of nipper with a single ridge. SIZE: 35 mm. BIOLOGY: Confined to estuaries, where it is most common on rocky banks; often climbs mangrove trees. RELATED SPECIES: *Metopograpsus messor* is very similar, but distributed from East London to Moçambique.

42.7 Flat-bodied crab *Percnon planissimum*

IDENTIFICATION: Body flat, with two notches on the front of the carapace and three teeth on each side; mottled green-brown and with bright-green narrow bands. At least the lower half of eyes vivid red. Front edge of the largest leg-joints strongly serrated. Nippers of the male swollen, resembling tiny balloons. SIZE: Body 35 mm. BIOLOGY: Very common in Natal, scuttling secretively around boulders in pools and in shallow water.

.1 ▲

42.2 ▲

3 ▲

42.4 ▲   42.7 ▼

▲   42.6 ▼

43.1 European shore-crab *Carcinus maenas*

IDENTIFICATION: Carapace oval, mottled khaki-green, with five marginal teeth on each side. Legs robust, with flattened but pointed tips. Nippers strong, outer surface smooth-textured, finger and thumb with about 12 teeth. SIZE: 50 mm. BIOLOGY: Alien to South Africa, probably introduced on oil rigs. First recorded in Table Bay docks in 1983, but by 1990 had spread 120 km to Saldanha Bay. Confined to reefs or bays protected from direct wave action. A voracious predator, *C. maenas* poses a threat to many local molluscs, including some used in aquaculture. When it invaded the East Coast of America, it caused millions of dollars of damage to the shellfish industry.

43.2 Three-spot swimming crab *Ovalipes trimaculatus*

IDENTIFICATION: Carapace triangular but with rounded corners; five marginal teeth on either side. Sandy-coloured, flecked with red-brown dots and three spots. Like most swimming crabs (family Portunidae), its last pair of legs ends in an oval, paddle-shaped joint. Nippers ridged and with strong cutting teeth. SIZE: 40 mm. BIOLOGY: An aggressive predator on bivalves and gastropods (particularly *Bullia*) and lives on surf-beaten sandy beaches. It can use its paddles to swim, but more often it scuttles over the sand, digging backwards if threatened. Previously known as *Ovalipes punctatus*.

43.3 Blue swimming crab *Portunus pelagicus*

IDENTIFICATION: Carapace broad, with long drawn-out spines mid-laterally and eight smaller marginal teeth. Nippers ridged and usually blue. Body a dull variegated brown, acquires a spectacular pink-brown mottling when the crab matures. SIZE: 120 mm. BIOLOGY: Lives in sandy bays. Scavenges on dead animals and preys on molluscs and crustaceans. Fights viciously if cornered; can inflict painful wounds. Sold in markets in Moçambique. RELATED SPECIES:

43.4 *Portunus sanguinolentus*, the blood-spotted swimming crab, is similar, but its carapace has three characteristic red-brown, white-ringed spots.

43.5 Heller's swimming crab *Charybdis helleri*

IDENTIFICATION: Carapace transversely ridged, and much wider than the distance between the front pair of marginal teeth. Six blunt teeth between the eyes; five marginal teeth on either side. Nippers ridged and spiny. SIZE: 50 mm. BIOLOGY: Occurs in rock pools and bays. Eats small molluscs, crustaceans and worms. RELATED SPECIES: In *Thalamita* species (Durban northwards) the carapace is scarcely wider than the distance between the first marginal teeth. *T. admete* has two broad lobes between the eyes and a serrated base to the second antenna; *T. woodmasoni* has four lobes and a smooth base to the second antenna. Both genera have several other tropical species.

43.6 Smith's swimming crab *Charybdis smithii*

IDENTIFICATION: Carapace smooth, front edge with four pairs of short teeth, sides with four broad incisor-like marginal teeth and a single pointed tooth. Outstretched nippers easily four times the carapace width. Nippers with 5–6 longitudinal rows of tubercles. SIZE: 120 mm. BIOLOGY: Predatory. Normally tropical and uncommon. First described from False Bay in 1838 but not seen again until 1978. In 1983 enormous numbers were recorded in False Bay during unusually warm sea temperatures. It then disappeared, only to appear again in inexplicably large numbers in 1993.

43.7 Mud crab *Scylla serrata*

IDENTIFICATION: Carapace oval, with nine pairs of equal-sized marginal teeth. Nippers massive, but with smooth hands. Colour green-brown, but limbs (particularly the paddle-shaped last pair) with a net-like pattern. SIZE: The giant of the swimming crabs: 30 cm. BIOLOGY: Adults burrow in estuarine mud, but migrate to sea to reproduce. Predatory, but feeds on surprisingly tiny prey, including small molluscs. At least one unwary captor has lost a finger to its powerful nippers. Often inappropriately called the Knysna crab, it occurs in many tropical parts of the world. Commercially fished in Australasia, although its slow growth makes it susceptible to over-exploitation.

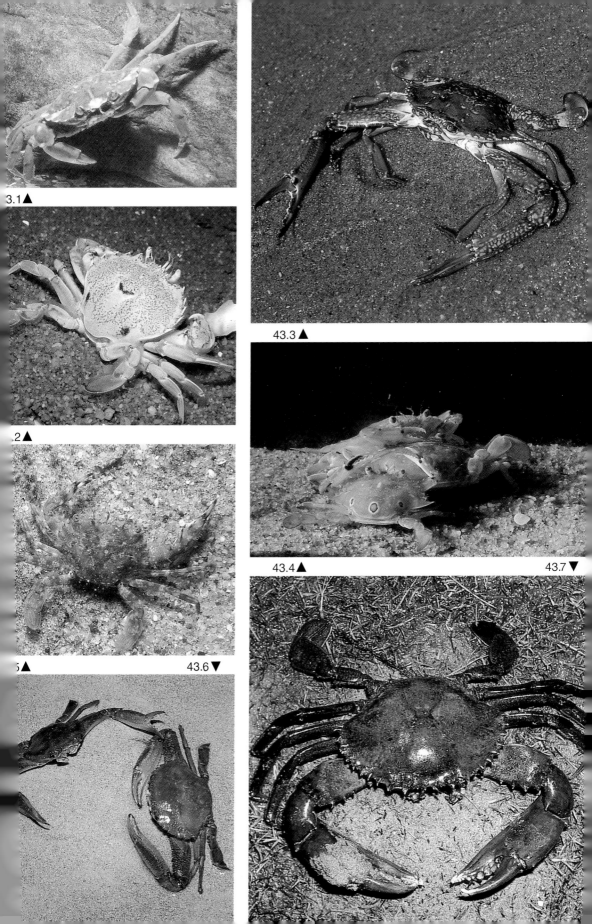

3.1 ▲

.2 ▲

43.3 ▲

43.4 ▲

43.7 ▼

5 ▲

43.6 ▼

44.1 **Horn-eyed ghost crab** *Ocypode ceratophthalmus*

IDENTIFICATION: Easily recognised by the pointed 'horns' that cap its eyes (although these are absent in juveniles). Body robust and grey-green, legs long and pointed. One nipper larger than the other, and with a wide, hairy ridge across the palm. SIZE: 40 mm. BIOLOGY: Digs holes up to 1 m deep, high on sheltered sandy beaches; emerges during low tide at night to scavenge, and feeds on a range of animals including newly-hatched turtles. A rubbing motion of the hairy ridge on the nipper (stridulation) is used to produce a rasping sound during courtship or aggression. RELATED SPECIES: *Ocypode cursor* (Northern Namibia) is easily recognised by a tuft of hair-like setae extending from the tips of the eyes.

44.2 **Pink ghost crab** *Ocypode ryderi*

IDENTIFICATION: Body pale pink, with distinctive mauve joints to the legs; similar in shape to the previous species, with a square carapace and long, robust legs. Eyes on long stalks but lacking extended 'horns'. The larger of the two nippers has a granular stridulating organ on the palm which consists of a single row of granules. SIZE: 35 mm. BIOLOGY: Abundant on tropical beaches that are directly exposed to the sea. Burrows deeply by day, emerging by night to feed on deposited carrion and small animals. RELATED SPECIES: *Ocypode madagascariensis* (Durban to Moçambique) is almost identical but is sandy-coloured and its legs do not have mauve joints.

44.3 **Urville's fiddler crab** *Uca urvillei*

IDENTIFICATION: The males of all fiddler crabs have one greatly-enlarged nipper which, in this case, has a granular outer surface and is bright yellow to orange, sometimes grading to pale blue. Both the finger and thumb have a series of low teeth, but there is usually a larger tooth near the mid-point of each. The carapace is blue-black, and may have pale blue spots, although mature males can be a pure royal-blue colour. Females have two small nippers. In both sexes, the eyestalks are very long and slender and arise close together, the distance between their bases being only slightly greater than the width of the eyestalks. Legs blue-black. SIZE: 25 mm. BIOLOGY: Common on sandbanks in estuaries and lagoons, emerging during low tide to feed on surface sediments, rolling the sediment into small, neat balls (pseudofaeces) as it extracts organic particles and micro-organisms. Males aggressively defend their holes against others, and wave their nippers in a vertical up-and-down gesticulation to attract females. RELATED SPECIES: *Uca vocans* (Transkei to Moçambique) has similar closely-spaced eyestalks. It differs in having a brownish-yellow carapace and, in the male, a golden nipper with an obvious tooth near the tip of the thumb; a spine on the fourth joint of the nipper is also unique to the species. Previously called *Uca marionis*.

44.4 **Pink-clawed fiddler crab** *Uca lactea annulipes*

IDENTIFICATION: Males have a black-to-grey carapace with a white network; nipper bright salmon-pink, outer surface scarcely textured, curving smoothly up to its upper margin. The inner surface has two marked oblique ridges. Several equal-sized weak teeth on the thumb and finger. Bases of eyestalks widely-spaced, the distance between them being about three times the eyestalk diameter. Legs black. SIZE: 20 mm. BIOLOGY: Abundant: forms huge aggregations on open sandflats in estuaries and lagoons. Thousands emerge almost simultaneously at low tide to feed on surface sediments, darting into their holes if alarmed. During courtship the males extend the enlarged nipper laterally, swinging it inwards in a 'come-hither' gesture. RELATED SPECIES:

44.5 *Uca chlorophthalmus* (Transkei to Moçambique) also has widely-spaced eyestalks, but red legs. Male's nipper red; the upper edge has a fine ridge and a groove, and the inner surface has only one faint ridge. Previously called *Uca gaimardi*. Another species, *Uca inversa* (Moçambique only), is larger and has two white patches on the male's carapace. The male's large nipper is red, with a prominent tooth near the tip of the movable finger. Legs red in both sexes.

44.1 ▲

44.2 ▼

44.3 ▼

44.4 ▼

45.1 Army crab   *Dotilla fenestrata*

IDENTIFICATION: Body sandy-coloured, the shape and size of a pea, with grooves on the dorsal surface. Nippers small and equal-sized. Walking legs with a distinct oval 'window' on the sides of the fourth joint. SIZE: 10 mm. BIOLOGY: Lives in dense aggregations on sheltered sandbanks, burrowing shallowly and emerging in countless numbers to feed during low tide, sucking organic material from the sediment and depositing tiny pellets of processed sand. Migrates up and down the shore in battalions, keeping ahead of the advancing tide. Most abundant in the tropics north of Durban, but small numbers extend south to Breede River.

45.2 Long-eyed crab   *Macrophthalmus grandidieri*

IDENTIFICATION: Body width almost twice the length; the back of the carapace has two parallel shallow grooves running from near the mid-line to the sides. Eyestalks very long: much longer than the distance between their bases. SIZE: 25 mm. BIOLOGY: Burrows in moist sandy mud near the mid-tide level in sheltered lagoons and estuaries. Retreats sideways down its burrow if disturbed, and sits at the mouth of the burrow with just one eye peering out to survey the world. Feeds on fine particles of detritus. RELATED SPECIES: *Macrophthalmus boscii* (Port Alfred to Inhambane; estuarine) is more squat, its carapace only slightly wider than long; the length of its eyestalks is about equal to the distance between their bases. *Macrophthalmus depressus* (Moçambique) also has a squat body but can be recognised by two parallel rows of granules in the sides of the carapace. It lives in mangrove mud.

45.3 Sandflat crab   *Cleistostoma edwardsii*

IDENTIFICATION: Small, with a sandy-coloured, flat, almost circular body, the outline being broken by a single tooth near the front corners of the carapace. Eyestalks smooth. SIZE: 8 mm. BIOLOGY: Scuffles under the surface of the sand in waterlogged areas in lagoonal and estuarine sandbanks. Feeds on fine particles of detritus. RELATED SPECIES: *Cleistostoma algoense* is almost identical and has a similar distribution and habitats, but differs in its hairy eyestalks.

45.4 Three-legged crab   *Thaumastoplax spiralis*

IDENTIFICATION: Unmistakable because it has only three pairs of walking legs (in addition to the nipper-bearing chelipeds). All other crabs have four pairs of walking legs. (They may shed legs when attacked, but if they do so, the stumps of the discarded legs are clearly evident.) The body is smoothly oval in outline, flat and mottled grey-brown. SIZE: 8 mm. BIOLOGY: Lives in temporary burrows low on the shore on sheltered sandbanks. Often found in the burrows of the sandprawn *Callianassa kraussi* (39.1), and may be commensal with it.

45.5 Pea crab   *Pinnotheres dofleini*

IDENTIFICATION: Pea-shaped, soft-bodied, and straw-coloured or yellow. Eyes stunted and minute. SIZE: 10–20 mm. BIOLOGY: Females are always found inside the shells of bivalves, parasitising their hosts by stealing food from the mucous strings on their gills. The tiny eyes and flimsy legs are a reflection of the crab's sheltered, parasitic mode of life. The males are minute and rarely seen, but can move between hosts in search of females. There is probably more than one species in southern Africa; a large form is found in the horse mussel *Atrina* (53.10) and a small version lives inside the black mussel *Choromytilus meridionalis* (52.2). (Occasionally they horrify diners who discover them in their meal of mussels, although any seafood gourmet should delight in the unintended supplement to the meal.)

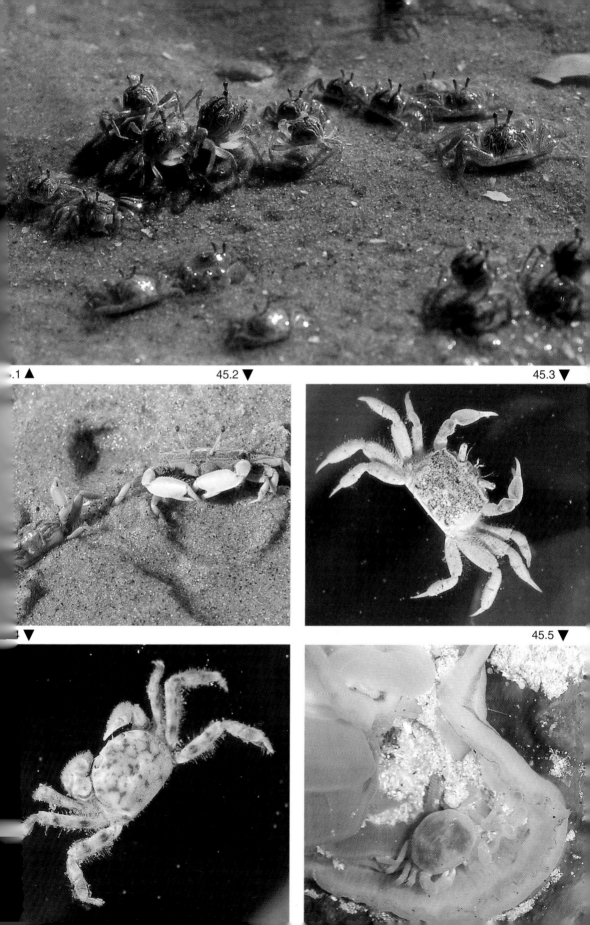

45.1 ▲   45.2 ▼   45.3 ▼

45.4 ▼   45.5 ▼

46.1 Smith's xanthid crab *Eriphia smithii*

IDENTIFICATION: Carapace oval, brown; front half dotted with tubercles but not hairy; at least five marginal teeth. Nippers large and robust, the smaller of the two with spiny tubercles whereas the larger is scarcely tuberculate. Eyes orange. SIZE: 50 mm. BIOLOGY: Like all species in the family Xanthidae, *Eriphia smithii* is slow-moving and sluggish, but has massive, powerful nippers with which it crushes the shells of molluscs. It is the most abundant xanthid crab on rocky shores in Natal, hiding in holes from which it is difficult to extract.

46.2 Red-eyed xanthid crab *Eriphia laevimanus*

IDENTIFICATION: Carapace brown, often mottled with grey on the sides, dotted with small knobs and armed with at least five small lateral teeth on each side. Nippers very strong but smooth, not tuberculate. Eyes bright red. SIZE: 40 mm. BIOLOGY: Less common than *Eriphia smithii*, but with similar habits. RELATED SPECIES: *Eriphia scabricula* (Durban to tropical Indo-Pacific) has a similar shape but is smaller and has a sparse coating of short hairs on its carapace and nippers.

46.3 Kelp crab *Pilumnoides rubus*

IDENTIFICATION: Carapace oval, orange to purple-orange; the whole of its upper surface is densely tuberculate, and the margins lined with both small and large teeth. The nippers are of similar size, heavily dotted with knobs, and end in a black finger and thumb. SIZE: 30 mm. BIOLOGY: Lives under stones, among mussels or in kelp holdfasts. Feeds on small molluscs. This crab was previously identified as *P. perlatus*, a South American form thought to have been introduced to South Africa. The South African form, however, has now been recognised as a separate endemic species. RELATED SPECIES: Several xanthid crabs have carapaces that are not knobbly but are hairy. *Pilumnus hirsutus* (Saldanha Bay to Durban) has three marginal teeth and a fine fur of short hair through which longer bristles project. *Parapilumnus pisifer* (St Helena Bay to Cape Vidal) is similar but has four large tubercles on the second-last joints of its walking legs.

46.4 Chocolate crab *Atergatis roseus*

IDENTIFICATION: Carapace chocolate-coloured above and pink below, smoothly oval when viewed from above, completely lacking marginal teeth. Eyes small and close-set. Nippers smooth except for a faint ridge; finger and thumb black, armed with white knob-like teeth. SIZE: 60 mm. BIOLOGY: A relatively scarce but handsome crab that hides beneath boulders. Certainly predatory, although its exact diet is unknown.

46.5 Coral crab *Tetralia glaberrima*

IDENTIFICATION: Carapace shiny, roughly oval, with a single marginal tooth. Eyes extend outwards to reach almost to the outer edge of the carapace. Nippers large, the outer surface of one of them being faintly woolly. Colour usually bright orange. SIZE: 20 mm. BIOLOGY: Lives in coral heads. Aggressively attacks other crustaceans that encroach on its host; also defends the host against other corals that might compete with it. RELATED SPECIES: Several other trapesiid crabs live among branching corals. All *Trapezia* species have a toothed frontal margin. *Trapezia cymodoce* is red or orange with a black finger and thumb. *T. guttatus* is pink with tiny red dots and *T. rufupunctata* pink with large red spots.

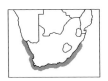

47.1 Cryptic sponge crab *Cryptodromiopsis spongiosa*

IDENTIFICATION: Body soft, slightly furry, yellow or cream; carapace broader than long; fifth legs smaller than the others and bent upwards to lie over the back. Tips of legs knobbly. SIZE: 25 mm. BIOLOGY: Found under stones in the shallow subtidal. Like all members of the family Dromiidae, the cryptic sponge crab carries a cloak of sponge or ascidian on its back; these organisms are unpalatable, thus protecting the crab. Pieces of sponge or ascidian are cut off the rocks by the crab and held in place over their backs by the modified fifth legs, which have tiny nippers at their tips. The 'cloak' then grows to cover the crab, sometimes concealing it almost completely. RELATED SPECIES: **47.2 *Pseudodromia latens*** (Saldanha Bay to East London) is easily recognised because its body is longer than wide, and its fifth leg much longer than the fourth. A close yellowish fur covers the body. From above, only two teeth are visible on the rostrum between the eyes. Almost completely enclosed in a ball-shaped ascidian cloak, this species is familiar to divers because it often climbs on sea fans, which it may consume.

47.3 Shaggy sponge crab *Dromidia hirsutissima*

IDENTIFICATION: Body densely covered with a magnificent shaggy coat of long, stiff, brown hairs. Carapace broader than long, with two or three marginal teeth; fifth legs about equal in size to fourth legs, but twisted upwards over the carapace. SIZE: 40 mm. BIOLOGY: Carries a cloak of sponge, ascidian or seaweed on its back, but is seldom completely hidden by it. RELATED SPECIES: *Dromidia unidentata* (Moçambique) has a short fur and a single tooth near the middle of the carapace margin.

47.4 Masked crab *Nautilocorystes ocellata*

IDENTIFICATION: Unmistakable because of its extraordinarily long antennae, which are almost as long as the body. Carapace longer than broad, with 4 marginal teeth and 4 grey blotches ('ocelli'). Tips of legs markedly flattened. SIZE: 35 mm. BIOLOGY: Digs backwards into sheltered sandbanks, extending only its long antennae, which are lined with hairs that interlock to form a tube down which water is drawn.

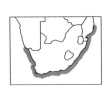

47.5 Crown crab *Hymenosoma orbiculare*

IDENTIFICATION: Body round and flat, with a small triangular projection between the eyes (the rostrum). The hinge of the nippers is smooth, not hairy, and the tips of the finger and thumb sharply pointed. Pale, mottled with brown, but often overgrown by microalgae. SIZE: 15 mm. BIOLOGY: Lives in estuaries and lagoons, shuffling into the sand and feeding on small crustaceans. RELATED SPECIES: *Rhynchoplax bovis* occurs in tropical and subtropical estuaries, is smaller (5 mm), and its nippers have a furry hinge and spoon-shaped tips.

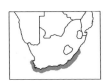

47.6 Toothed decorator crab *Dehaanius dentatus*

IDENTIFICATION: Carapace longer than broad, with two horns projecting between the eyes. On each side there are three marginal teeth, the first and third being very strong, with a weaker tooth between them. Tufts of hooked hairs dot the front of the carapace. SIZE: 20 mm. BIOLOGY: Lives among seaweeds and can change its colour to match the alga it lives in. Cuts fragments of seaweeds and attaches them to the hooked hairs on its back, where they continue to grow, increasingly decorating the crab so that it becomes almost impossible to see. RELATED SPECIES: *Dehaanius quattuordentatus* (East London to Moçambique) is smaller, has four (or sometimes three) equal-sized marginal teeth, and hooked hairs over the whole of the carapace. Often it becomes so smothered with decorations that it is almost impossible to recognise it as a crab.

47.7 Cape long-legged spider crab *Macropodia falcifera*

IDENTIFICATION: Body small in relation to the bizarre long, thin legs; nippers thicker than legs, spiny and covered with short, spiky hairs. Rostrum very long: longer than antennae. Carapace longer than broad, with two pairs of long, erect spines. Fourth joints of walking legs end in a long spine. SIZE: Body 10 mm long, legs 35 mm. BIOLOGY: Slow-moving; occurs in deep, calm waters.

7.1 ▲

47.2 ▲   47.3 ▼

47.4 ▼

47.5 ▼

47.6 ▼

47.7 ▼

Bryozoa : Moss or Lace Animals

Abundant and often exquisitely beautiful, bryozoans are seldom recognised, resembling many other organisms including seaweeds, hydroids and corals. They build colonies of minute (1 mm long) individuals (zooids) which lie side by side but are individually enclosed in tiny coffin-like skeletons (zooecia) made of chitin or lime. Through an opening in each zooecium (the orifice), a crown of filter-feeding tentacles extends. Some zooids may be specialised into beaked structures (called avicularia, resembling a bird's head), which snap at creatures attempting to settle on the colony. Classification depends on microscopic details of the zooecia, but based on their body form bryozoans can be divided artificially into four groups: (1) flat forms that encrust rocks; (2) forms that encrust other organisms; (3) upright, branching colonies that are bushy or cactus-like; (4) species with heavy lime skeletons, resembling miniature corals.

48.1 Subovoid bryozoan   *Watersipora subovoidea*

IDENTIFICATION: Forms concentric rings of zooecia, creating flat, roughly round crusts. Dark brown with reddish edges, or orange-brown with black dots at the orifices. Zooecia urn-shaped, pitted with large pores; orifice with a collar that is notched on its rear margin (48.1d). SIZE: Colony 40 mm wide. BIOLOGY: The commonest of several species forming crusts beneath boulders. SIMILAR SPECIES:

48.2d *Cryptotheca nivea* (Cape Point to East London): pale pink; its zooecia are straight-sided and their orifices are flanked by a pair of small round avicularia.

48.3d *Escharoides contorta* (Cape Town to Cape Point) is pale yellow-brown and its orifices also have a pair of avicularia. Young zooecia have four spines behind each orifice, although these are lost when they age.

48.4d *Steganoporella buskii* (Cape Point to Durban) is also pale but has a thin membrane over the front of each zooecium, beneath which is a lime sheet peppered with pores. A few zooecia have an inverted 'V' marked on the operculum.

48.5 Nodular bryozoan   *Alcyonidium nodosum*

IDENTIFICATION: Lives exclusively on the whelk *Burnupena papyracea,* forming an orange or purple cloak over the shell and throwing the surface into regularly-spaced lumps. SIZE: Covers the shell. BIOLOGY: *Alcyonidium* protects its host against predators because it contains highly toxic chemicals. Rock lobsters avoid whelks coated with *Alcyonidium*. It is not known what benefit the bryozoan obtains from the association.

48.6 Membranous lace animal   *Membranipora tuberculata*

IDENTIFICATION: Encrusts flat-bladed algae, forming lace-like white sheets. The zooecia are hexagonal, with a pair of rounded tubercles on their anterior corners (48.6d). SIZE: Colonies 100 mm. BIOLOGY: Grows on seaweeds, sometimes covering extensive areas but apparently doing no harm. The flat nudibranch *Corambe* feeds on *Membranipora* and is camouflaged to resemble it. RELATED SPECIES:

48.7 *Membranipora membranacea* (Saldanha–Durban) is abundant on kelp. Its rectangular zooecia have tiny knobs at the corners.

48.8 Verticellate lace animal   *Electra verticillata*

IDENTIFICATION: Forms long, narrow colonies with oblique or transverse rows of zooecia. Long spines project from the surface of the colony, giving it a hairy appearance. The zooecia (48.8d) are narrow and have an obvious, round, dark orifice armed with one long and four short spines. SIZE: Colonies 50 mm long, 5 mm wide. BIOLOGY: Invariably grows on narrow-bladed algae.

48.9 Magellanic lace animal   *Beania magellanica*

IDENTIFICATION: Delicate lacy flat colonies. Individual zooecia are joined to one another by six thin tubes. Two tiny avicularia guard the front end of each zooecium (48.9d). SIZE: Colony 30 mm. BIOLOGY: Encrusts algae, rocks or other bryozoans. RELATED SPECIES: **48.10d *Beania vanhoeffeni*** (Cape Peninsula only) is similar but its zooecia have about 10 pairs of short spines that project inwards.

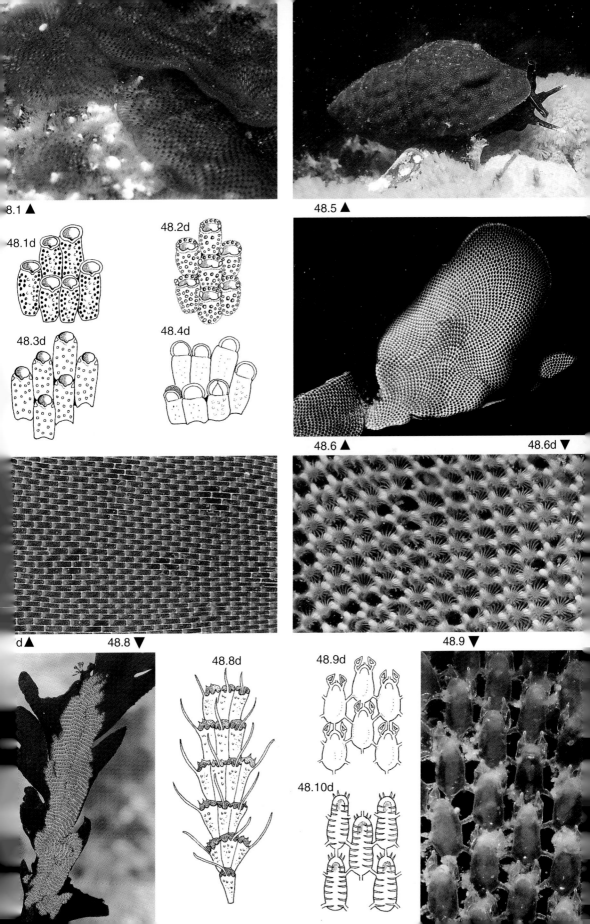

48.1 ▲

48.1d

48.2d

48.3d

48.4d

48.5 ▲

48.6 ▲   48.6d ▼

d▲   48.8 ▼

48.9 ▼

48.8d

48.9d

48.10d

49.1 Dentate moss animal *Bugula dentata*

IDENTIFICATION: Forms bushy tufts that hang downwards; grey but becoming deep blue when sunlit. Zooecia rounded at their upper ends and armed with about four long spines. Avicularia commonly flank each zooecium and resemble birds' heads (49.1d). SIZE: Colony 50 mm in length. BIOLOGY: Extremely common in the warmer waters of the East Coast, hanging down from the vertical walls of subtidal gulleys. RELATED SPECIES:

49.2d *Bugula avicularia* (Cape Point to Durban) has a much more sparsely-branched straw-coloured colony; its zooecia have a sharply-pointed upper corner and most are armed with a bird's-head avicularium. Mature colonies are dotted with pearly, round, reproductive bodies (ovicells).

49.3 Neritic moss animal *Bugula neritina*

IDENTIFICATION: Upright, bushy colonies in which the branches curve inwards. Orange in colour. Zooecia almost rectangular, with sharp outer corners; avicularia absent, but rounded reproductive bodies (ovicells) are frequent (49.3d). SIZE: Colony 30 mm tall. BIOLOGY: Grows on rocks or attached to seaweeds, and is often abundant on wharfs and underwater pylons. Can seriously foul the hulls of ships.

49.4 Eyelash moss animal *Bicellariella ciliata*

IDENTIFICATION: Colony resembling a miniature willow-tree, with a central upright trunk and much-divided drooping branches that are very pale, often white. Zooecia (49.4d) rounded, projecting outwards from the branches, and ending in 4–7 very long, curved spines (the 'eyelashes'). A single bird's-head avicularium lies at the base of each zooecium. SIZE: Colony 60 mm tall. BIOLOGY: Grows beneath rocky overhangs low on the shore and in the subtidal.

49.5 Spiral moss animal *Menipea triseriata*

IDENTIFICATION: Colony upright, bushy, with flat branches that divide regularly and are arranged in a spiral manner (although the spiral is not always obvious). Straw-coloured, or orange if growing in dark caves. Zooecia almost rectangular, arranged in pairs and with four stubby spines on the upper end near the orifice and a single minute avicularium at the base (49.5d). SIZE: Colony 40 mm tall. BIOLOGY: Often grows among sponges or larger bryozoans, deriving protection against predators from them. Abundant in deeper water, but extending into intertidal pools. RELATED SPECIES:

49.6 *Menipea crispa* ranges from Lambert's Bay to Durban and forms upright tree-like colonies with in-curving branches; zooecia oval, with a pair of long spines that curve at their tips and a single tubular avicularium at the base. Occasional larger triangular avicularia occur on the edges of the branches (49.6d).

49.7 Busk's moss animal *Onchoporella buskii*

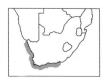

IDENTIFICATION: Colony forming flat leafy blades, white or pale blue to green. The zooecia are only lightly strengthened with lime, and form a network on the surface of the blades. Each zooecium is shaped like an elongate hexagon, and has five pores below the orifice. SIZE: Colony 20 mm. BIOLOGY: Usually grows on algae or attached to other bryozoans.

49.8 Cactus-bush bryozoan *Margaretta triplex*

IDENTIFICATION: Colony low-growing, with short, cylindrical, upright branches that send off side-stems, somewhat like a miniature cactus. The zooecia resemble Grecian urns in shape, and their surfaces are covered with a delicate net-like pattern (49.8d). SIZE: Colony 10 mm tall. BIOLOGY: Occurs subtidally in sites sheltered from strong wave action, mixed with low-growing seaweeds.

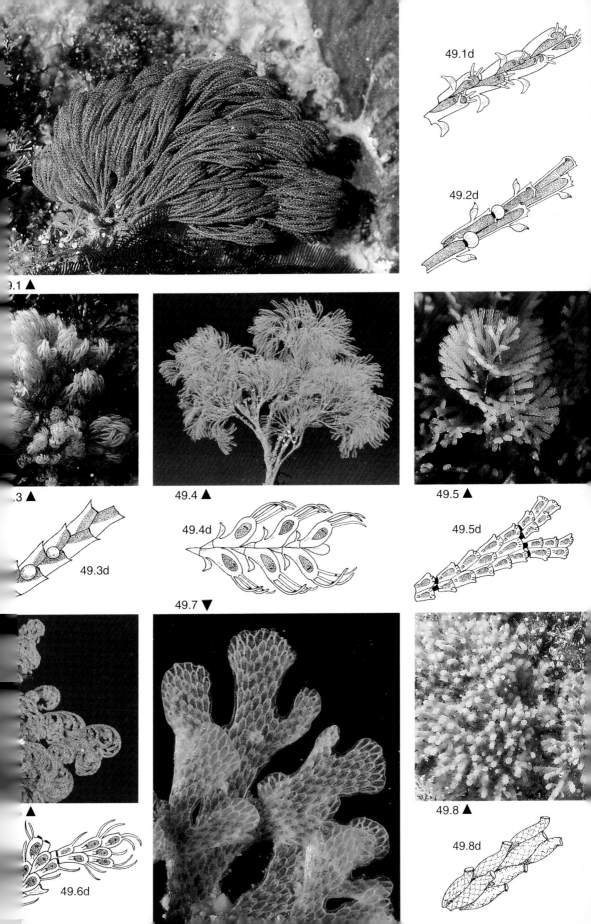

49.1d

49.2d

49.1 ▲

49.3 ▲

49.4 ▲

49.5 ▲

49.3d

49.4d

49.5d

49.7 ▼

▲

49.6d

49.8 ▲

49.8d

50.1 Scrolled false corals *Chaperia* spp.

IDENTIFICATION: Bright-yellow to orange colonies, which form projecting scroll-like twisted plates. The zooecia are diamond-shaped with a round, soft central membrane, a single basal avicularium and two or more spines (50.1d). These spines are visible even with the naked eye, giving the colony a bristly texture like the chin of a man who has not shaved for two days. SIZE: Colonies up to 25 cm wide, 30 mm tall. BIOLOGY: There are several species in the genus. All grow beneath or on the sides of rocky overhangs. Often associated with the sea fan *Acabaria rubra,* which defends it against predators with its stinging cells and benefits by sheltering in the rigid skeleton built by *Chaperia.*

50.2 Pore-plated false coral *Laminopora bimunita*

IDENTIFICATION: Colony formed from twisted plates that are united in an irregular manner. Colour usually purple-brown, less often orange with paler orange edges. Zooecia diamond-shaped, completely lacking in spines but with a pair of tiny triangular avicularia that flank the orifice (50.2d). SIZE: Colony 100 mm in diameter. BIOLOGY: Grows on vertical rock-faces where it is free from silt, and extends from a depth of 2 m down to 50 m. SIMILAR SPECIES:

50.3d *Adeonellopsis meandrina* (East London to Durban) has colonies with an almost identical appearance, but they are dark purple-brown and have hexagonal zooecia with one or two tiny triangular avicularia and a central pore.

50.4 Staghorn false coral *Gigantopora polymorpha*

IDENTIFICATION: Upright, much-branched coral-like colonies with stubby, flattened, dull orange branches. Zooecia rectangular, the front surface dotted with coarse pores, and the orifice covered with a narrow arch carrying two tiny triangular avicularia (50.4d). SIZE: Colony 60 mm in height. BIOLOGY: The colonies are robust and heavily laden with calcium carbonate, and grow very low on the shore or in the subtidal zone down to 20 m, usually on vertical rock surfaces.

50.5 Lacy false coral *Schizoretepora tesselata*

IDENTIFICATION: The colony consists of twisted upright plates that are loosely connected and punctured by regularly-spaced holes, giving a lacy appearance. Zooecia oval, with four long spines around the orifice, although these spines are usually evident only in young zooecia. Dotted around on the surface of the colony are isolated avicularia, which have triangular 'beaks' and resemble tiny barnacles (50.5d). SIZE: Colony 50 mm in diameter. BIOLOGY: Forms delicate, brittle colonies on the sides of rocky reefs, from the low-tide mark down to about 20 m.

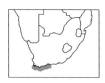

50.6 Pore-tubed bryozoan *Tubulipora pulcherrima*

IDENTIFICATION: Exquisite, delicately-branched colonies with tubular zooecia that curve away from the branches and have a simple, round, terminal orifice. SIZE: Colonies 10 mm across the crown. BIOLOGY: Grows only in sheltered places, often beneath boulders, and forms beautiful colonies that resemble miniature staghorn corals. Its specific name refers to its beauty (*pulcherrima* being Latin for 'most beautiful').

50.7 Cylindrical false coral *Cellepora cylindriformis*

IDENTIFICATION: Forms salmon-pink, irregular, finger-like colonies. The surface is dotted with knobbly zooecia which are irregularly oriented. SIZE: Colonies 20–50 mm tall. BIOLOGY: Grows under large boulders in the lower intertidal and shallow subtidal zones. RELATED SPECIES: *Celleporaria capensis* (Cape Peninsula to Port Elizabeth) forms thick orange encrustations but never has upright columns; a pair of short spines flanks the orifices of young zooecia.

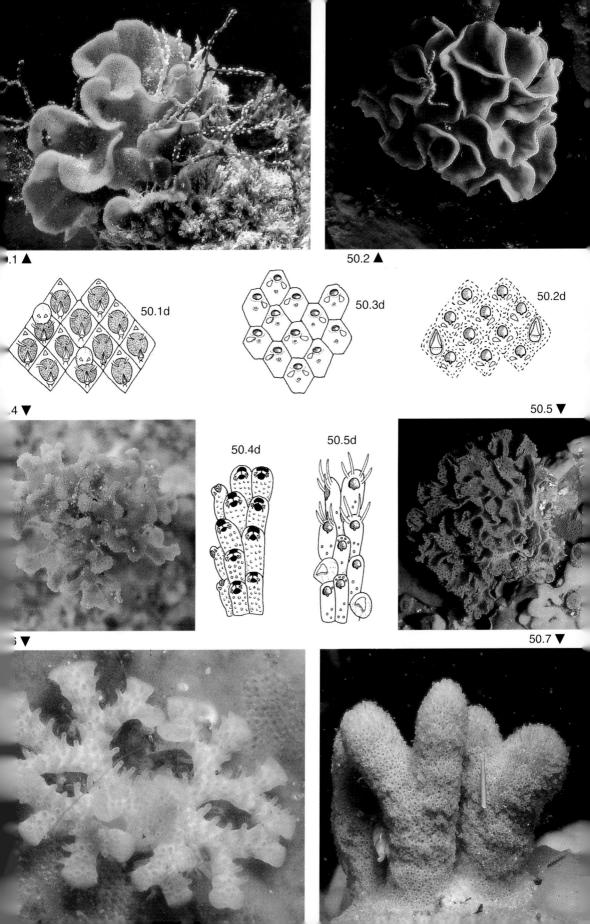

50.1 ▲

50.2 ▲

50.1d

50.3d

50.2d

50.4 ▼

50.5 ▼

50.4d

50.5d

50.6 ▼

50.7 ▼

Molluscs & Lamp Shells

The phylum Mollusca is one of the largest marine groups, with more than 5000 species in southern Africa. All have an unsegmented body that is divided into a head, a visceral mass with the digestive and reproductive systems, and a foot. Most have a ribbon-like rasping 'tongue' (a radula), which processes the food and is unique to molluscs. Nearly all secrete a lime shell that covers the body.

Members of another phylum, Brachiopoda, also have a calcium carbonate shell, but are only distantly related to the Mollusca. Lamp shells have two plates (valves) to their shell, one ventrally beneath the body and another dorsally above it. Most are attached by a short stalk (peduncle). The shape of the shells and this protruding 'wick' give these animals the name of 'lamp shells'. Brachiopods once dominated the early seas, but are now represented by a handful of species.

The phylum Mollusca is divided into seven classes, five of which include oft-encountered species. Tusk shells, class Scaphopoda, are little known but easy to distinguish by their tusk-shaped shells. Members of class Bivalvia, including mussels, clams and oysters, have a shell that consists of two valves but, unlike the lamp shells, the valves are hinged together dorsally and extend down laterally on either side of the body to encase it. Most bivalves have large gills, used both for respiration and to filter out tiny food particles. Because of this diet, bivalves lack a radula. Many burrow in sand and have a large, wedge-shaped foot for this purpose. Mussels attach to rocks by a beard-like byssus, while oysters and their kin cement one valve to the rock-face.

The class Polyplacophora ('many plate bearers') contains the chitons, easily recognised because their shells are split into eight dorsal plates that cover the centre of the body. Surrounding the plates is a girdle, and beneath the body lies a broad foot that provides tenacious attachment to rocks. Chitons are herbivores and have strongly toothed radulae.

Class Gastropoda is the largest in the phylum and incorporates the snails, winkles, whelks and sea slugs. Most species belong to the subclass Prosobranchia and have a spiral shell, a well-developed head with tentacles and a radula, and a large, flat foot used for locomotion. Primitive members are herbivores, rasping seaweeds and micro-algae. More advanced forms are predators, and have a long proboscis and a cylindrical siphon. To house the latter, the shell has an anterior canal, or groove, betraying the animal's predatory habits.

Most gastropods are shelled, but many of the subclass Opisthobranchia (sea slugs, sea hares and nudibranchs) have forsaken their shells and lost their original gills. The reason for these losses is unknown, but it has been speculated that their ancestors were sand-burrowers, for whom a heavy shell and external gills would have been a hindrance. Lacking a shell, modern forms protect themselves in different ways. Many produce toxins that make them poisonous. Others consume anemones or bluebottles and then build the stinging cells of their prey into their own tissues. To advertise their unpalatable nature, most are vividly patterned and exquisitely coloured, so that it is a travesty to call them by their mundane common name – 'sea slugs'. Although there are several groups within the subclass, pride of place goes to the order Nudibranchia ('naked gills'), probably the most stunningly beautiful creatures in the sea.

The final large class, the Cephalopoda, includes the octopus, squid and cuttlefish, among the most highly evolved of all invertebrates. The 'foot' is modified into eight or ten long tentacles armed with suckers. Most cephalopods are active, predatory swimmers, including the deep-sea giant squids 20 m long, inspiration for mythical stories about sea monsters. Cephalopods have eyes as complex as those of humans, and a greater capacity for learning than any other invertebrate.

Brachiopoda : Lamp Shells

Sessile, filter-feeding animals with calcareous, bivalved shells superficially resembling those of bivalve molluscs, but in fact enclosing the body dorsally and ventrally instead of laterally. The ventral valve is normally the larger and is pierced posteriorly by a hole through which a short stalk or pedicle emerges to anchor the animal to the substratum. Internally the major organ is a large, feathery, horseshoe-shaped filter-feeding organ called the lophophore. This may be supported by a delicate internal skeleton (the brachidium). Brachiopods are relatively uncommon today, but thousands of species occur in the fossil record, indicating that they were once a dominant marine group.

51.1 Ruby lamp shell   *Kraussina rubra*

IDENTIFICATION: Shells pink with 20–30 prominent ridges radiating from the posterior opening. Internal skeleton present and resembling a delicate pair of horns. SIZE: About 20 mm across. BIOLOGY: The commonest brachiopod on the shore, usually attached in small groups to the undersurface of loose rocks. RELATED SPECIES:

51.2 *Kraussina crassicostata* (Cape Point–Mossel Bay) has only 10–12 very coarse ribs.

51.3 *Terebratulina abyssicola* (Port Alfred–East London) has an internal skeleton which forms a complete loop. The shell is longer than broad, rather angular and only very faintly ribbed.

51.4 *Terebratulina meridionalis* (Cape Peninsula) has a more smoothly rounded shell with fine branching ridges on its surface.

51.5 Disc lamp shell   *Discinisca tenuis*

IDENTIFICATION: Shell consists of a pair of flat, semitransparent, brown horny discs with concentric growth ridges. Lophophore visible through shell. Internal skeleton absent. SIZE: Average diameter 20 mm. BIOLOGY: Attached one on top of another to form rafts of shells that are frequently washed up along the driftline.

Scaphopoda : Tusk Shells

A small group of molluscs easily recognised by their long, tapering, tubular shells, which are open at both ends. Scaphopods live buried upright in sand or mud. The narrow end projects from the sediment surface and is used for water exchange. Long sticky tentacles spread from the lower opening into a feeding chamber and trap fine food particles.

51.6 Tusk shell   *Dentalium regulare*

IDENTIFICATION: White to pink with 20–30 longitudinal ribs which may become eroded in older specimens. SIZE: Up to 32 mm. BIOLOGY: Rarely found alive but shells fairly commonly cast ashore. RELATED SPECIES:

51.7 *Dentalium strigatum* (Cape Point–East London) is white with 13–14 strong longitudinal ribs.

51.8 *Dentalium plurifissuratum* is a larger (60 mm) species with about 18 primary longitudinal ribs interspersed with finer ridges.

51.9 *Dentalium salpinx* has a large (70 mm) solid shell with a broad aperture. The apex of the shell is finely ridged, becoming smooth towards the aperture. Found in deep water off the Western Cape.

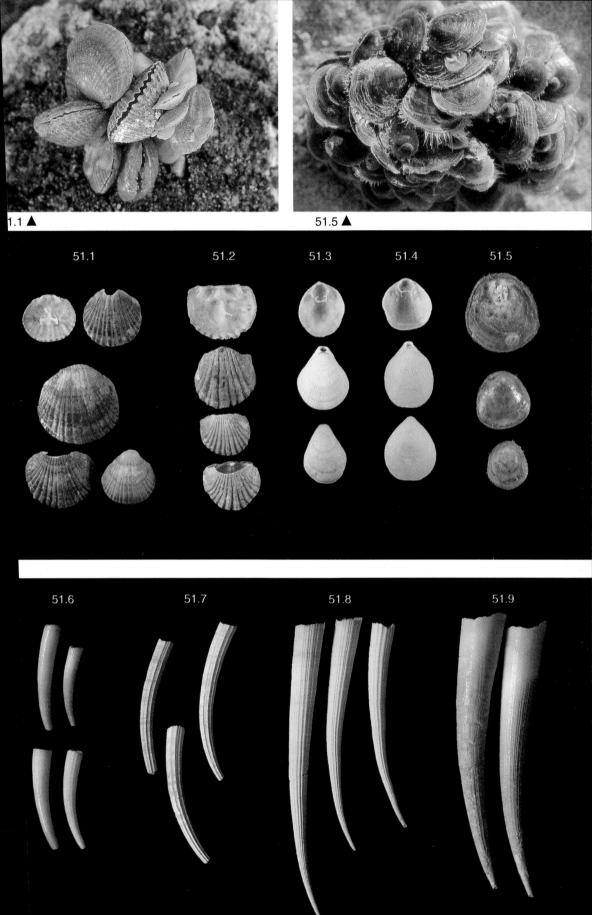

1.1 ▲

51.5 ▲

51.1　51.2　51.3　51.4　51.5

51.6　51.7　51.8　51.9

Bivalvia : Bivalve Molluscs

As the name implies, bivalves are enclosed and protected by a pair of shell valves, which are hinged together along the back by an elastic ligament. This stretches when the animal clamps the valves together and springs them apart when the animal relaxes. Bivalves have limited mobility and most either cement one valve to the substratum, fasten themselves down with byssal threads, or burrow into sand, mud, wood or even rock. Only a few can swim. Nearly all bivalves are filter-feeders, sucking water in through an inhalant siphon, sieving it through enlarged, sheet-like gills, and then expelling the waste water through an exhalant siphon. Much of the body consists of gonad, and in most species enormous numbers of sperm and eggs are shed into the water, where they develop into planktonic larvae. A few species brood their eggs.

52.1 Ribbed mussel   *Aulacomya ater*

IDENTIFICATION: Easily recognised by the strong wavy ridges running the length of the shell, which is brown in juveniles, but becomes blue-black in larger individuals. Eroded areas are white. SIZE: Up to 90 mm. BIOLOGY: Forms extensive beds low in the intertidal and on rocky reefs to about 40 m depth, particularly in the colder waters of the West Coast. Slow-growing, taking at least ten years to reach its maximum size. An important source of food for the rock lobster *Jasus lalandii* (35.2).

52.2 Black mussel   *Choromytilus meridionalis*

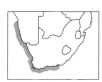

IDENTIFICATION: Shell smooth and shiny, black eroding to blue. Distinguished from *Mytilus galloprovincialis* (52.3) by its narrower cross-section and absence of pits in the resilial ridge (the narrow white band alongside the hinge ligament on the inner surface of the shell). SIZE: Up to 150 mm. BIOLOGY: Common on low-shore rocks and on shallow, flat reefs in the Cape, particularly in areas subject to sand cover or abrasion. The flesh of females is coloured dark chocolate by the gonads, which permeate through the body, while that of males is pale yellow. Grows rapidly, reaching 60 mm within a year. Popular with seafood lovers, but like other bivalves can become poisonous following toxic red tides (which are normally restricted to the West Coast).

52.3 Mediterranean mussel   *Mytilus galloprovincialis*

IDENTIFICATION: Shell smooth, typically black or blue (52.3b), shading to brown on lower surface, rarely light brown. Broad in cross-section and usually widest at base. Distinguished from *Choromytilus* (52.2) by pits in the resilial ridge (see above). SIZE: Up to 140 mm. BIOLOGY: Thought to be a recent introduction from Europe, but now the dominant intertidal mussel throughout the West Coast, often forming a dense band in the low intertidal (52.3a). Rare below low water. The flesh of females is orange and of males off-white. A fast-growing mussel raised commercially.

52.4 Brown mussel   *Perna perna*

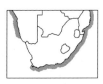

IDENTIFICATION: Shell smooth and yellow-brown, sometimes tinged with green or with a chevron pattern. Specimens from crowded intertidal beds often very elongate and almost rectangular in cross-section; submerged, fast-growing specimens are taller and narrower. Resilial ridge pitted (see 52.2 above). SIZE: Up to 125 mm. BIOLOGY: The dominant mussel on the South and East coasts, forming dense beds from the mid-intertidal to a few metres depth. The flesh of females is bright orange and of males off-white.

52.5 Bisexual mussel   *Semimytilus algosus*

IDENTIFICATION: A small, elongate species with a delicate, smooth brown shell that bulges below the apex. Resilial ridge narrow, not pitted. SIZE: Up to 50 mm. BIOLOGY: A possible introduction to southern Africa, previously being known only from South America. Most specimens are hermaphroditic with male gonad on one side, female on the other.

52.1 ▲

52.4 ▼

52.3a ▲

52.3b ▼

52.1

52.2

52.3

52.4

52.5

53.1 Brack-water mussel *Brachidontes virgiliae*

IDENTIFICATION: Shells small, brown; both the inner and outer surfaces are ribbed. SIZE: 25 mm. BIOLOGY: Confined to the upper reaches of estuaries. RELATED SPECIES: **53.2 *Brachidontes semistriatus,*** the semistriated mussel (Port Elizabeth northwards), is also small and brown, but smooth or ribbed only externally. It forms close-packed groups in intertidal rock–crevices, especially in Natal.

53.3 Ledge mussel *Septifer bilocularis*

IDENTIFICATION: Shell thick and either brown, green or orange; external surface covered by radiating ridges. An interior shelly ledge bridges the anterior (pointed) end of each shell valve. SIZE: Up to 43 mm. BIOLOGY: Found singly or in small groups under rock ledges from mid–tide downwards.

53.4 Estuarine mussel *Arcuatula capensis*

IDENTIFICATION: An attractive mussel with a thin, glossy, greenish shell which is smooth but marked with radiating reddish-brown lines. Inner surface of shell smooth. SIZE: Up to 76 mm. BIOLOGY: Restricted to estuaries, where it may bury in mud amongst seagrass beds or attach to rocks, seaweeds or wood.

53.5 Half-hairy mussel *Gregariella petagnae*

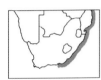

IDENTIFICATION: A small, reddish-brown mussel in which the anterior end is swollen and the posterior half of the shell is covered by fibrous hairs. SIZE: Up to 18 mm. BIOLOGY: Occurs singly in rock crevices and clumps of seaweed.

53.6 Ear mussel *Modiolus auriculatus*

IDENTIFICATION: Shell broad with dorsal margin raised and narrowed to form a low ear-like hump. Colour generally reddish brown, juveniles with a tufted outer layer posteriorly. SIZE: Usually less than 50 mm. BIOLOGY: Fairly common in low-shore pools and crevices along the East Coast; abundant in Moçambique where it can grow to over 80 mm.

53.7 Oblique ark shell *Barbatia obliquata*

IDENTIFICATION: Lower margin of shell concave. Shells sculptured with fine radiating ridges, and covered with a black velvety fur, which is often eroded away except along the edges. Eroded areas white. Inner surface shows brown markings posteriorly. SIZE: Up to 58 mm. BIOLOGY: Attached by byssus in crevices or under rocks in sandy pools. RELATED SPECIES:

53.8 *Striarca symmetrica* (Port Elizabeth eastwards) is small, squat and parallel-sided with a velvety surface. The external ligament is diamond-shaped.

53.9 *Arca avellana* (Port Alfred eastwards) has a very rectangular shape and long straight hinge. The exterior surface is fibrous.

53.10 Horse mussel *Atrina squamifera*

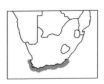

IDENTIFICATION: Very large, with a fragile translucent-brown shell that does not meet at the wide posterior end. A series of 6–12 ribs runs the length of each valve and bears conspicuous scale-like cusps. A bunch of long byssal threads projects from the ventral margin. SIZE: Up to 390 mm. BIOLOGY: Lives buried vertically in sheltered mud or sand, with the wide posterior end projecting above the surface. The commensal crab *Pinnotheres dofleini* (45.5) is often found in the body cavity. RELATED SPECIES: *Pinna muricata* (Port Elizabeth–Moçambique) is almost square-cut posteriorly, has smaller cusps externally, and a longitudinal furrow running down the middle of the inner surface of each valve.

53.1

53.2

53.3

53.4

53.5

53.6

53.7

53.8

53.9

▼

53.10 ▼

54.1 **Cape pearl oyster** *Pinctada capensis*

IDENTIFICATION: A large squarish oyster which attaches itself to the substratum with byssal threads and has equal valves banded with dark radiating lines. The outer surface is decorated with concentric, projecting, wavy growth ridges. SIZE: Up to 130 mm. BIOLOGY: Attached to rocks in pools or on shallow reefs. Species of this genus are well known as a source of both pearls and mother-of-pearl, but few pearls of value have been found in the Cape species.

54.2 **Cape rock oyster** *Striostrea margaritacea*

IDENTIFICATION: A large, heavy oyster with a deep, multi-layered, cup-shaped lower valve, which is cemented to the substratum, and a thin, flat upper valve often bearing fine radial threads. SIZE: Up to 180 mm. BIOLOGY: Common on rocky reefs from low water to about 5 m depth. Excellent eating and exploited throughout its range. About 500 000 are commercially collected in Natal each year. Natal specimens take 33 months to reach marketable size of 60 mm length. Previously known as *Crassostrea margaritacea*. RELATED SPECIES:

54.3 *Ostrea atherstonei* (Saldanha Bay–Natal S. Coast) is a large flat oyster with a shallow lower valve that is not hollowed below the hinge. It occurs on shallow reefs, where it is often overgrown by fouling organisms and difficult to see. The eggs are brooded in the gill chamber. *Hyotissa numisma* (Transkei eastwards) is common under rocks and reaches only 55 mm. The oldest portion of the shell often has black dots, and worn areas reveal a bubble-like crystalline structure.

54.4 **Natal rock oyster** *Saccostrea cuccullata*

IDENTIFICATION: Lower valve deeply hollowed below the hinge and cemented to the rock; upper valve relatively flat. Margins of both valves form a series of neatly interlocking undulating folds, which are often tinted mauve. Previously known as *Crassostrea cuccullata*. SIZE: Up to 70 mm. BIOLOGY: The dominant oyster in Natal, forming a distinct band in the mid-to-upper intertidal.

54.5 **Saddle oyster** *Anomia achaeus*

IDENTIFICATION: Small, transparent oyster-like shells which resemble fish scales. Easily recognised by the hole in the lower valve, through which a calcified byssal plug extends to cement the bivalve to the substratum. SIZE: Usually under 20 mm. BIOLOGY: Under stones, on algal fronds and on shells of larger molluscs.

54.6 **Dwarf fan shell** *Talochlamys multistriata*

IDENTIFICATION: This small scallop has a single 'ear' and comes in a wide variety of colours including orange and violet. Shell valves equal in size, with 50–70 fine prickly radiating ridges. SIZE: Up to 37 mm. BIOLOGY: An attractive bivalve found under rocks or in algal holdfasts. Swims readily if disturbed.

54.7 **Scallop** *Pecten sulcicostatus*

IDENTIFICATION: Shells large, right (lower) valve convex, left one flattened; ears equal in size; 12–15 ridges on each shell. Colour pink to brown, darker above. SIZE: Up to 106 mm. BIOLOGY: Lies on the surface of clean sand or mud, but can swim by clapping the valves together. The edges of the mantle have a series of brightly-coloured simple eyes.

54.8 **File shell** *Limaria tuberculata*

IDENTIFICATION: Shell valves equal in size and translucent white, with small 'ears' and radiating ribs. Animal pink with a prominent fringe of mobile, segmented tentacles. SIZE: Up to 42 mm. BIOLOGY: Lives unattached under boulders or in algal holdfasts, and can swim clumsily by clapping the shells together. The long tentacles secrete a sticky, repugnant mucus and can be shed at will to distract attackers.

54.4 ▲

54.6 ▲

54.8 ▲

54.1

54.2

54.3

54.4

54.5

54.6

54.7

54.8

55.1 Cockle *Trachycardium flavum*

IDENTIFICATION: Shell taller (hinge to aperture) than long (side to side), with 26–29 radiating ribs; anterior ribs with semicircular scales, posterior ones with erect scales. Colour brown. SIZE: Up to 64 mm. BIOLOGY: Buries shallowly in muddy sand using its large muscular foot. RELATED SPECIES:

55.2 *T. rubicundum* (Moçambique–E. Transkei) has a shell which is flecked with pink or brown and has a red margin internally.

55.3 Smooth trough shell *Mactra glabrata*

IDENTIFICATION: Large fat oval clams with smooth glossy shells. Colour cream, juveniles often with radiating lines. Siphons fused. SIZE: Up to 114 mm. BIOLOGY: Burrows just below the surface in fine sand, sometimes forming dense beds. Good eating, but collectors are legally limited to eight clams per day.

55.4 Angular surf clam *Scissodesma spengleri*

IDENTIFICATION: A large, almost equilateral clam in which the posterior margin is formed into a broad, angular, flattened surface. Shells smooth and off-white with a thin brown surface 'skin' (periostracum). SIZE: Up to 113 mm. BIOLOGY: Burrows shallowly in the surf zone of sandy beaches. Particularly common in False Bay where shells are frequently washed up after storms.

55.5 Otter shell *Lutraria lutraria*

IDENTIFICATION: Shell large and elongate, valves gaping widely at both ends. Hinge with a prominent spoon-shaped depression internally. Colour white to dirty yellow with brown 'skin'. SIZE: Up to 137 mm. BIOLOGY: Immobile, buried up to 30 cm deep in muddy sands and extending elongate, non-retractile siphons to the surface.

55.6 Pencil bait *Solen capensis*

IDENTIFICATION: Very elongate, almost cylindrical shells with widely gaping ends. Anterior (upper) end bent outwards to form a lip. Siphons short and fused, leaving a keyhole-shaped opening on the sand surface when retracted. SIZE: Up to 160 mm long. BIOLOGY: Burrows deeply in firm clean sand in estuaries and lagoons. Good eating, and prized as bait. Collectors usually thrust a hooked wire down the burrow, then twist it to snag and withdraw the animal. RELATED SPECIES: *S. cylindraceus* lacks any lip around the anterior end and replaces *S. capensis* from the Transkei northwards.

55.7 Rectangular false cockle *Cardita variegata*

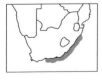

IDENTIFICATION: Small, almost rectangular, with 20–25 coarse ribs radiating from the hinge, which is displaced to the anterior corner. Colour off-white. SIZE: Up to 35 mm. BIOLOGY: Lives attached to the underside of rocks on reefs. RELATED SPECIES: **55.8 *Thecalia concamerata,*** 'dead man's hands' (Port Nolloth–Transkei), has a similar external appearance, but the ventral margin of the shell is folded in to form a brood chamber within which the young are incubated.

55.9 Rough false cockle *Carditella rugosa*

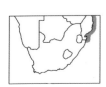

IDENTIFICATION: Small triangular cockles with 16–18 bumpy ribs radiating from the hinge. Colour off-white flecked with brown. SIZE: Up to 10 mm. BIOLOGY: Common on sheltered sandbanks. RELATED SPECIES: *Carditella capensis* occurs on the west coast and is almost equal-sided. It is found, along with *Tellimya trigona*, a tiny smooth triangular species, on west coast lagoonal sandbanks.

55.10 Giant clam *Tridacna squamosa*

IDENTIFICATION: Easily recognised by its large size and heavy wavy aperture through which the beautifully-coloured mantle lobes project. Outer surface of the shell white with pronounced leaf-like projections (scutes). SIZE: 40 cm. BIOLOGY: Both filter-feeds and 'farms' microscopic algae housed in the brightly-coloured mantle lobes; hence confined to shallow, clear waters, mainly on coral reefs. Rare in South Africa; *Tridacna maxima* has smaller scutes and is the species commonly seen in Zululand.

55.6 ▲

55.10 ▲

55.1

55.2

55.3

55.6

5.4

55.5

55.7

55.8

55.9

56.1 Smooth platter shell   *Loripes clausus*

IDENTIFICATION: Shell valves thick and flat with an almost perfectly circular outline. Surface smooth and white with faint growth rings. Ligament internal and cuts obliquely across the hinge plate. SIZE: Up to 34 mm. BIOLOGY: An estuarine form which burrows deeply in muddy sand near low tide. Feeding unusual in that water flows through the animal anteriorly–posteriorly, entering through a mucus-lined tube secreted by the foot (instead of through the inhalant siphon), then out through an elongate exhalant siphon.

56.2 Toothless platter shell   *Anodonta edentula*

IDENTIFICATION: Shell valves thin and deeply concave, with a 'beak' bulging beyond the hinge, which is very thin and toothless. Surface white and smooth, with faint growth rings. SIZE: Up to about 50 mm. BIOLOGY: Estuarine, burrowing just below the surface of soft muds.

56.3 Dwarf rusty clam   *Lasaea adansoni turtoni*

IDENTIFICATION: A tiny, globular clam with smooth valves, often stained red-brown around the hinge, but otherwise plain white or pink. Formerly misidentified as *Kellya rubra*. SIZE: 4 mm or less. BIOLOGY: Lives on rocky shores, usually nestling amongst other filter-feeders. Particularly abundant amongst the byssal threads of mussels or amongst barnacles in the mid-to-high intertidal. Hermaphroditic, starting life as a male and changing to a female. The eggs and larvae are brooded in the shell and emerge fully developed.

56.4 Port Alfred tellin   *Tellina alfredensis*

IDENTIFICATION: Valves smooth and laterally compressed, posterior end tapering and bent slightly to the right. Colour a lovely pink, vivid on the inner surface, especially when alive. SIZE: Up to 85 mm long. BIOLOGY: Buries left side down in shallow water off clean sandy beaches. The long, mobile inhalant siphon is moved around on the sand surface, 'vacuuming' up light detrital particles.

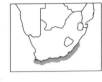

56.5 Gilchrist's tellin   *Tellina gilchristi*

IDENTIFICATION: The valves of the shell are slightly extended posteriorly, but not bent to the side. Colour white or pink, usually with distinctive pink rays which distinguish it from other small species of tellin in the Cape. SIZE: Up to 28 mm. BIOLOGY: Burrows in sand in the subtidal zone on sandy beaches, but can also be found intertidally in sheltered bays and lagoons. RELATED SPECIES:

56.6 *Tellina trilatera* (Orange River–Transkei) has delicate, smooth, glossy-white shells up to 49 mm long. The anterior end is broadly rounded and much shorter than the tapering posterior end.

56.7 *Tellina capsoides* (Durban northwards) has relatively thick, chalky-white shells sculptured with numerous concentric growth ridges. The posterior end is angular with a long, straight hinge line. Grows to 58 mm.

56.8 Ridged tellin   *Gastrana matadoa*

IDENTIFICATION: Valves white to fawn and etched with strong, sharp concentric growth ridges. Hinge of left valve with one wedge-shaped tooth, that of the right valve with two diverging teeth. SIZE: Up to 44 mm. BIOLOGY: Burrows in sheltered sandbanks and estuaries.

56.9 Littoral tellin   *Macoma litoralis*

IDENTIFICATION: A small species with delicate, flattened, cream-to-white shells. Surface smooth and silky. SIZE: Up to 33 mm. BIOLOGY: Burrows a few centimetres below the surface of muddy sandbanks in estuaries.

.5 ▲

56.1 56.2 56.3 56.4

56.5

56.6

56.7

56.8 56.9

57.1 Shipworm   *Bankia carinata*

IDENTIFICATION: Extremely elongate, cylindrical bivalves which occupy burrows cut deep into submerged wood by the reduced, rasp-like shells. The tubes are lined with chalky shell material, and the open end can be closed off by a unique, feather-like, segmented pallet. SIZE: Up to 20 cm body-length. BIOLOGY: Notorious pests, which can riddle and destroy wooden structures. Shipworms retain their filter-feeding capacity but are also able to digest wood with the aid of symbiotic bacteria in the gut.

57.2 White mussel or wedge shell   *Donax serra*

IDENTIFICATION: Wedge-shaped, with coarse wavy ridges running across the truncated posterior end. Inner margin of shells wrinkled and inner surface usually purple. Ligament large and external, about one-third from hind end. SIZE: Up to 88 mm long. BIOLOGY: Abundant on wave-exposed beaches and extensively exploited for bait and eating (legal limit 50 per day). In the Southern and Eastern Cape, adults concentrate about mid-tide, with juveniles around low water, but on the Atlantic Coast the adults are mainly subtidal. Minimum legal size of 35 mm is reached after about 20 months.

57.3 Ridged wedge shell   *Donax madagascariensis*

IDENTIFICATION: A triangular shell easily recognised by the parallel ridges which run diagonally across the entire outer surface. SIZE: Up to 26 mm. BIOLOGY: Common on sandy beaches in Natal. The ridges on the shell are thought to facilitate burrowing. RELATED SPECIES:

57.4 *Donax sordidus* (Cape Point–Transkei) is very similar in shape and size but the shell is smooth, apart from a few fine ridges on the truncated posterior end. Abundant in the E. Cape, where it migrates up and down the beach with the tide.

57.5 Slippery wedge shell   *Donax lubricus*

IDENTIFICATION: Anterior end long, posterior end short and truncated. Surface smooth and white or flesh-coloured, sometimes with purple rays. SIZE: Up to 35 mm. BIOLOGY: Seldom found alive but shells often washed ashore. RELATED SPECIES:

57.6 *Donax bipartitus* (East London northwards) can be recognised by the violet coloration of the inner surface of the truncated posterior end of each shell valve.

57.7 Round-ended wedge shell   *Donax burnupi*

IDENTIFICATION: Valves elongate, quite fragile and compressed, the posterior ends tapering and rounded. Surface smooth and glossy, pink or cream to brown, often with darker rays. Interior with a violet mark behind the hinge. SIZE: Up to 36 mm. BIOLOGY: Burrows in subtidal sands. Its shells often wash up in large numbers.

57.8 Sunset clam   *Hiatula lunulata*

IDENTIFICATION: Elongate oval shells which gape posteriorly. Ligament external, attached to a distinct projecting platform. Hinge near the middle of the shell valves. Colour pale violet with darker rays. SIZE: Up to 37 mm. BIOLOGY: Estuarine, burrowing in fine clean sand in shallow water. May achieve very high densities.

57.9 Sand tellin   *Psammotellina capensis*

IDENTIFICATION: Valves of shell almost oval, although the hinge is displaced into the anterior half of the shell. Outer surface of valves yellow with violet rays, or violet with yellow rays. SIZE: Grows to 21 mm. BIOLOGY: Burrows shallowly in sandbanks in estuaries and sheltered lagoons. Most common just inside the mouths of estuaries but is tolerant of low salinities.

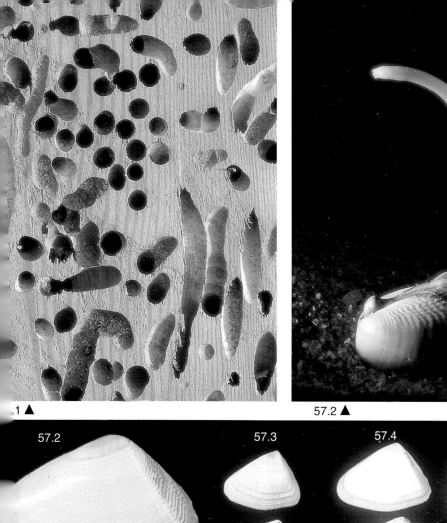

57.1 ▲   57.2 ▲

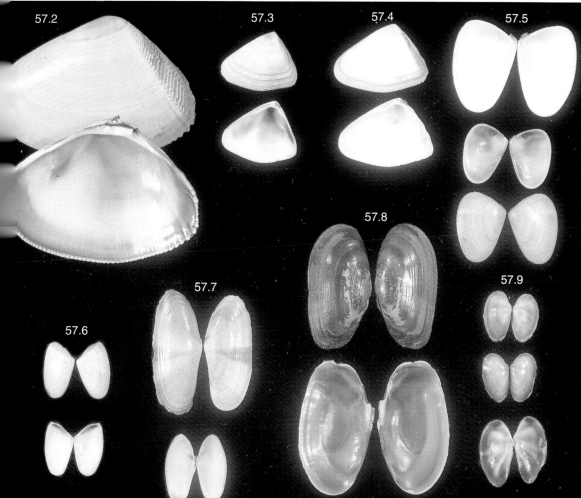

57.2

57.3

57.4

57.5

57.8

57.7

57.6

57.9

58.1 Ribbed venus *Gafrarium pectinatum alfredense*

IDENTIFICATION: Oval with well-developed radial ribs, which branch towards the margins of the shell. Colour pale, speckled or blotched with brown. SIZE: Up to 26 mm. BIOLOGY: Buries in coarse sand. Shells frequently washed up on Natal beaches.

58.2 Mottled venus *Sunetta contempta bruggeni*

IDENTIFICATION: Shells oval, with a deep depression spanning the joint just behind the hinge. Colour very variable, usually cream, mottled with pink or brown, interior tinged with pink. SIZE: Up to 38 mm. BIOLOGY: Common in coarse sand and gravel below 20 m depth, but shells often cast up on Natal beaches.

58.3 Warty venus *Venus verrucosa*

IDENTIFICATION: Shells oval and extremely heavy. Outer surface sculptured with strong concentric ridges which break up into rounded knobs near the margins. SIZE: Up to 60 mm long. BIOLOGY: Common in clean sand and gravel from low water downwards.

58.4 Heart clam *Dosinia lupinus orbignyi*

IDENTIFICATION: A heavy, almost perfectly circular white shell with a heart-shaped depression spanning the valves just in front of the hinge. Outer surface with fine concentric growth lines. SIZE: Up to 53 mm. BIOLOGY: Burrows in clean subtidal sands. The siphons can be up to three times the shell length, and are fused except at their tips. RELATED SPECIES:

58.5 *Dosinia hepatica* (Mossel Bay eastwards) is a smaller, entirely estuarine species with a smooth brown exterior and violet markings internally. Specimens seldom exceed 25 mm.

58.6 Streaked sand clam *Tivela compressa*

IDENTIFICATION: A large, fat triangular bivalve with a smooth shiny shell marked by concentric brown lines and radiating brown streaks. SIZE: Up to 61 mm. BIOLOGY: Burrows in clean sand below low-tide level. Washes up frequently in False Bay. RELATED SPECIES:

58.7 *Tivela polita* is smaller, pointed and triangular. It is common on wave-washed beaches from Transkei northwards. Interior usually edged with purple.

58.8 Beaked clam *Eumarcia paupercula*

IDENTIFICATION: A smooth, swollen clam with the posterior end produced into a 'beak'. Colour very variable, usually cream to brown with brown zigzags, flecks or rays. SIZE: Up to 42 mm. BIOLOGY: Burrows 2–3 cm below the surface of sheltered sand- or mud-banks. Abundant in Moçambique, where it is sold in markets.

58.9 Zigzag clam *Pitar abbreviatus*

IDENTIFICATION: A fat rounded clam with a large oval depression just in front of the hinge. White, usually with areas of brown zigzag markings. SIZE: Up to 38 mm. BIOLOGY: In clean sand, especially in Natal and Moçambique, where it is eaten.

58.10 Corrugated venus *Venerupis corrugatus*

IDENTIFICATION: Shell elongate oval, with hinge towards anterior end. Surface usually cream, with fine, wavy concentric growth ridges, sometimes attractively rayed in juveniles (58.10a), but becoming dull and corrugated in adults (58.10b). Interior purple towards margin. SIZE: Up to 77 mm. BIOLOGY: Commonly nestles amongst mussels in sandy areas or burrows shallowly in sheltered sandbanks. RELATED SPECIES:

58.11 *Meretrix meretrix* (Moçambique) resembles *Pitar abbreviatus*, but thick-shelled and with radial bands. Frequently sold in Maputo markets.

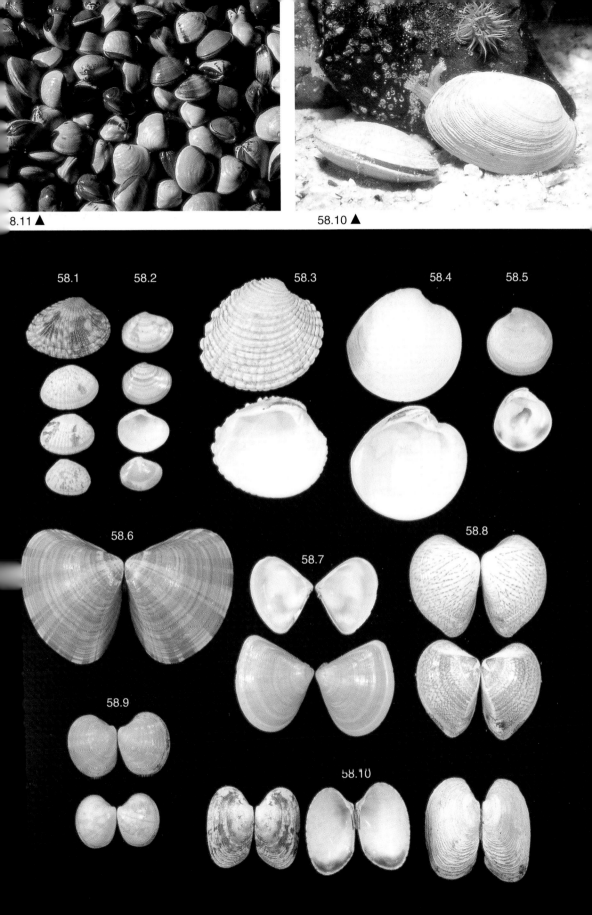

8.11 ▲

58.10 ▲

58.1

58.2

58.3

58.4

58.5

58.6

58.7

58.8

58.9

58.10

Polyplacophora : Chitons

Oval, flattened molluscs easily recognised by the eight overlapping shell plates or valves along the back. These are surrounded by a tough flexible girdle, often armed with protective scales, hairs or spines that can be useful aids to identification. The head is completely hidden beneath the girdle and lacks eyes or sensory tentacles. The remainder of the underside consists of a broad muscular foot rimmed with small gills. Chitons are sluggish creatures, usually found sheltering in crevices or beneath rocks. When active they creep slowly about, rasping encrusting plants or animals from the rock surface with a powerful file-like radula. There are some 26 southern African species, most of which are found intertidally.

59.1 Textile chiton   *Ischnochiton textilis*

IDENTIFICATION: Shell valves usually pale yellow or grey, central sections minutely pitted, lateral areas and end valves with fine radiating ridges. Girdle covered in small oval scales, each crossed by 12–24 fine ridges. SIZE: 20–40 mm. BIOLOGY: Found on the undersides of boulders in rock pools along the Cape coast. Detach and roll into a ball when the boulder is overturned.

59.2 Dwarf chiton   *Ischnochiton oniscus*

IDENTIFICATION: Colour often white but can occur in a wide variety of patterns and colours. Valves textured with fine pits, which are often arranged in rows, but lacking radiating ridges. Girdle narrow, covered in tiny scales that appear smooth to the naked eye, but reveal fine ribs under microscopic examination. SIZE: Usually about 10 mm. BIOLOGY: The smallest southern African chiton. Generally occurs in small groups on the undersides of rocks in sandy pools.

59.3 Ribbed-scale chiton   *Ischnochiton bergoti*

IDENTIFICATION: Colour usually off-white to brown, valves finely pitted, sometimes with vague radiating ridges or growth marks. Girdle scales with only 3–8 coarse radiating ribs. SIZE: Small, usually 10–20 mm. BIOLOGY: A rather drab species occasionally found under stones at low tide along the cold Atlantic coastline. Thought to brood its eggs beneath the girdle.

59.4 Tulip chiton   *Chiton tulipa*

IDENTIFICATION: An attractive chiton with smooth pink valves streaked and flecked with a wide range of brown patches and zigzags. Girdle striped, covered with large, smooth overlapping scales. SIZE: Typically 30–40 mm. BIOLOGY: A well-known and brightly-coloured chiton common under rocks near low tide, usually solitary.

59.5 Brooding chiton   *Chiton nigrovirescens*

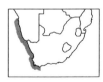

IDENTIFICATION: Valves brownish-black with vague radiating ridges, girdle dark with large smooth scales, sometimes with a coppery sheen. SIZE: 10–25 mm. BIOLOGY: A small species found in groups under stones from the mid-intertidal downwards. Unusual in that the eggs are retained under the girdle, where they develop into fully-formed baby chitons.

59.1 ▲

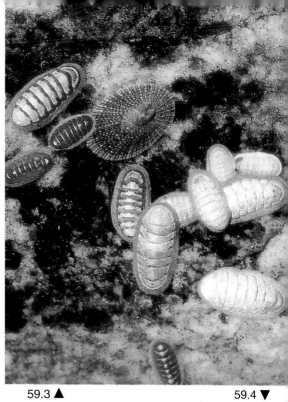

59.2 ▼

59.3 ▲

59.4 ▼

5 ▼

60.1 Broad chiton   *Callochiton castaneus*

IDENTIFICATION: Broad and flat with minutely-granular, dark-brown to orange valves. Girdle wide and densely covered with minute elongate, spine-like scales giving it a velvety texture. If the girdle is pulled away from the second to seventh valves, four characteristic slits are revealed in the margin of each shell. SIZE: 20–50 mm. BIOLOGY: Found singly under stones at low tide.

60.2 Black chiton   *Onithochiton literatus*

IDENTIFICATION: Shell valves usually brownish-black, sculptured with wavy radiating lines, often badly eroded in larger specimens. Girdle broad, brown to black, velvety with minute embedded spicules. SIZE: 25–50 mm. BIOLOGY: One of the few common chitons in Natal, found in the open on wave-exposed rocks or around the margins of rock pools.

60.3 Spiny chiton   *Acanthochiton garnoti*

IDENTIFICATION: Shell valves dull brown with oblique pale stripes, largely buried by the wide girdle. Girdle dotted with small spicules and with a series of nine pairs of characteristic tufts of long glassy spines. SIZE: 30–45 mm. BIOLOGY: Unusual in that it is abundant on exposed rock surfaces high in the intertidal. Should be handled with care, since the spines cause irritation if they become embedded in the skin.

60.4 Giant chiton or armadillo   *Dinoplax gigas*

IDENTIFICATION: Shell large with steeply arched grey or brown valves, usually badly eroded. Girdle brown and dotted with distinct tufts of small brown hairs. SIZE: 70–100 mm. BIOLOGY: Usually found partially or totally buried in sand on flat rocky reefs. Sometimes used as bait by fishermen and eaten by some people, although tough and leathery! RELATED SPECIES: Replaced in northern Transkei and Natal by another giant chiton, *Dinoplax validifossus*, in which the girdle hairs are uniformly distributed, not concentrated into tufts.

60.5 Hairy chiton   *Chaetopleura papilio*

IDENTIFICATION: Shell valves smooth and shiny, attractively marked with alternating light and dark brown stripes, sides often flecked with light blue. Girdle wide and brown with a sparse covering of long unbranched black bristles. SIZE: 40–70 mm. BIOLOGY: A distinctive West Coast species usually found singly under rocks in low-tide pools.

60.6 Orange hairy chiton   *Chaetopleura pertusa*

IDENTIFICATION: Shell valves sculptured with numerous beaded ribs, usually pink or orange. Girdle wide, often bright orange or pink and bearing both branched bristles and short simple hairs. SIZE: Up to 40–50 mm. BIOLOGY: A relatively uncommon but brightly-coloured species found under rocks at low tide or below.

60.1 ▲ 60.2 ▲ 60.3 ▲

▲ 60.5 ▼ 60.6 ▼

Gastropoda: I. Abalone

The shells of abalone form a very flattened spiral with an enormously enlarged aperture and a row of small holes along the left side. Water enters the gill cavity under the front shell margin and exits through the holes. As the shell grows, new holes are created and old ones filled in. The large muscular foot (which is much sought after as a seafood) grips tightly to the rock and allows for rapid locomotion. Abalone are herbivorous and shed their eggs and sperm into the water, where the larvae undergo a brief planktonic development.

61.1 Siffie or Venus ear   Haliotis spadicea

IDENTIFICATION: Shell ear-shaped with a concave outer lip (or growing edge). Interior mother-of-pearl, with a red stain beneath the spire; exterior fairly smooth, blotched reddish-brown. SIZE: Up to 80 mm long. BIOLOGY: A common but cryptic species found in rock crevices or amongst red-bait close to low water. Feeds on red algae. Collection is legally limited to 10 per person per day, and a minimum diameter of 32 mm.

61.2 Beautiful ear-shell   Haliotis speciosa

IDENTIFICATION: The outer lip of the shell is smoothly convex, and there is no red stain under the spire. The outer surface of the shell is mottled grey and red-brown with fine spiral ridges. SIZE: Up to 86 mm. BIOLOGY: An extremely rare species, about which little is known. Occurs in shallow waters but is seldom found in the intertidal zone.

61.3 Spiral-ridged siffie   Haliotis parva

IDENTIFICATION: Small, with a conspicuous spiral ridge running around the middle of the last turn of the shell. Holes on low humps, which arise from a second, less distinct marginal ridge. External colour brown or orange, sometimes mottled. SIZE: Up to 45 mm. BIOLOGY: Never abundant, lives under stones in rock pools or on shallow reefs. RELATED SPECIES:

61.4 Haliotis queketti (Port Alfred–Zululand) replaces *H. parva* to the east but is rare. The holes in the shell are raised on prominent turrets and there is usually one well-marked plus several weaker spiral ridges. Grows to 46 mm.

61.5 Perlemoen or abalone   Haliotis midae

IDENTIFICATION: Shell large and heavy with strong irregular corrugations running across the spire (parallel to the growing edge). Margins of foot with a dense fur of fleshy projections. SIZE: Up to 190 mm. BIOLOGY: Juveniles (61.5j) lack the corrugations and occur under intertidal boulders or beneath the spines of sea urchins. Adults occupy crevices or exposed positions on shallow reefs, reaching maximum densities in beds of kelp, *Ecklonia maxima*. The diet is made up of kelps and red algae, pieces of drift weed often being trapped by clamping down the foot. Sexual maturity is only reached after 8–10 years and minimum legal size after 13 years. Recreational divers must purchase a permit to collect perlemoen and are limited to four per day with a minimum diameter of 114 mm. A substantial commercial fishery for abalone is conducted in the SW Cape by divers using 'hookah' equipment from small boats. Most of the catch is frozen or canned and exported to the Far East, although the factories are compelled by legislation to sell 10% of their produce in South Africa. The annual commercial catch is approximately 640 tonnes whole wet weight. Recreational divers probably remove about half this amount.

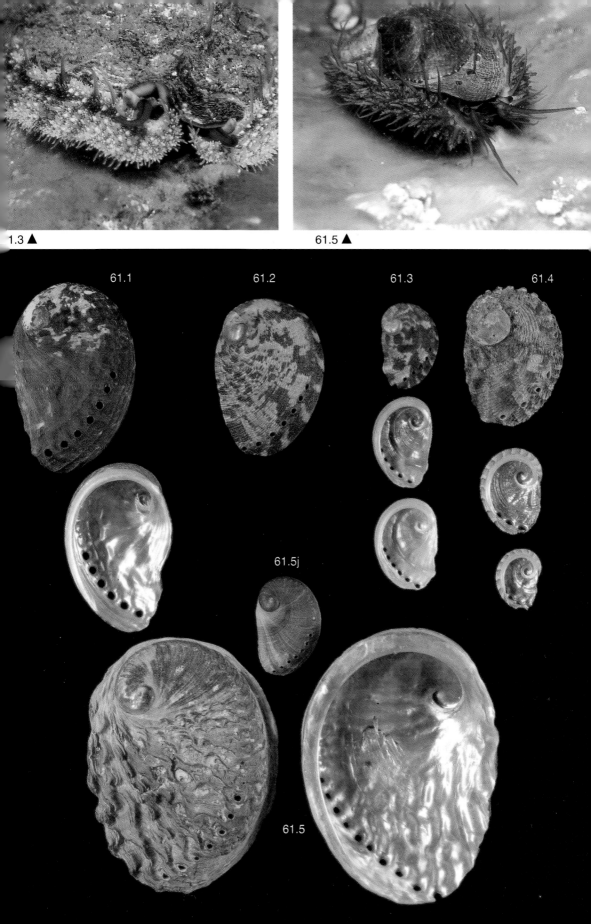

1.3 ▲

61.5 ▲

61.1

61.2

61.3

61.4

61.5j

61.5

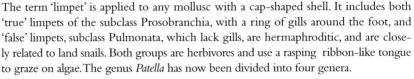

Gastropoda: 2. Limpets

The term 'limpet' is applied to any mollusc with a cap-shaped shell. It includes both 'true' limpets of the subclass Prosobranchia, with a ring of gills around the foot, and 'false' limpets, subclass Pulmonata, which lack gills, are hermaphroditic, and are closely related to land snails. Both groups are herbivores and use a rasping ribbon-like tongue to graze on algae. The genus *Patella* has now been divided into four genera.

62.1 Argenville's limpet   *Scutellastra argenvillei*

IDENTIFICATION: Shell large, tall and almost oval in outline, with fine radiating riblets and distinct teeth beneath the margin; inner surface porcellaneous white. SIZE: 90 mm. BIOLOGY: Occurs low on shores exposed to moderate wave action, achieving extraordinary densities and biomass on the West Coast (62.1a). Juveniles inhabit the shells of adults. Traps and feeds on kelp fronds, 'mushrooming' the shell upwards and then slamming it down to capture and break off fronds (62.1b). RELATED SPECIES:

62.2 *Scutellastra aphanes* (Transkei to Cape Vidal) resembles a miniature *S. argenvillei*. Its shell is oval, ribbed, pale with slightly darker rays; the edge never has teeth, although the ribs may crinkle the margin. Common on mussels; feeds on encrusting corallines. Reacts aggressively to others of its species, pushing them out of its territory.

62.3 Bearded limpet   *Scutellastra barbara*

IDENTIFICATION: Highly variable, but typically tall, with strong, spiky ribs. Sides of foot white, speckled grey. SIZE: 80 mm. BIOLOGY: Occurs on the low-shore and subtidally. On the East Coast it defends 'gardens' of filamentous red algae, thrusting away other grazers that intrude. On the more productive West Coast it lacks gardens and is not territorial. RELATED SPECIES:

62.4 *Scutellastra pica* (Zululand to Moçambique) has a slightly pear-shaped shell with radial ridges that throw the edge of the shell into irregular serrations, giving it a crumpled appearance. Often overgrown by algae. Sides of foot never flecked with grey.

62.5 Pear limpet   *Scutellastra cochlear*

IDENTIFICATION: Distinctively pear-shaped, often encrusted by coralline algae; inner surface white, tinged blue and often mottled with black; muscle scar U-shaped and black. SIZE: 70 mm. BIOLOGY: Lives in dense colonies low on exposed shores, forming a 'cochlear zone'. Very slow-growing, with a life span of 25 years. Associated with a paint-like coralline alga, *Spongites yendoi* (162.4), which it grazes. Narrow gardens of fast-growing fine red algae fringe and sustain larger individuals, and are territorially defended and fertilised by the limpets.

62.6 Duck's foot or long-spined limpet   *Scutellastra longicosta*

IDENTIFICATION: Shell star-shaped, with about 11 very long projecting ribs. Interior white or blue-white with a black edging. SIZE: 70 mm. BIOLOGY: Juveniles live on other shells, feeding on the encrusting alga *Ralfsia* (162.1), then move to the rock-face and eat encrusting coralline algae until they establish gardens of *Ralfsia,* which they defend against other grazers. The limpet cuts regular paths through the *Ralfsia,* increasing its growth rate and reducing the amounts of anti-herbivore chemicals produced by the alga. RELATED SPECIES:

62.7 *Scutellastra obtecta* (Transkei to Kosi Bay) is like a small *S. longicosta* but has less obvious ribs, lacks black edging, and is always white inside. It is easily confused with Natal specimens of *S. longicosta* which are stunted and have short spines.

62.8 Giant limpet   *Scutellastra tabularis*

IDENTIFICATION: The largest of African limpets. Shell with about 30 roughly equal-sized ribs that project slightly at the margin. Interior white, with an attractive pink margin. SIZE: 180 mm. BIOLOGY: Lives and feeds on large flat sheets of *Ralfsia* in the shallow subtidal zone; territorially defends its feeding area by violently thrusting against intruding herbivores.

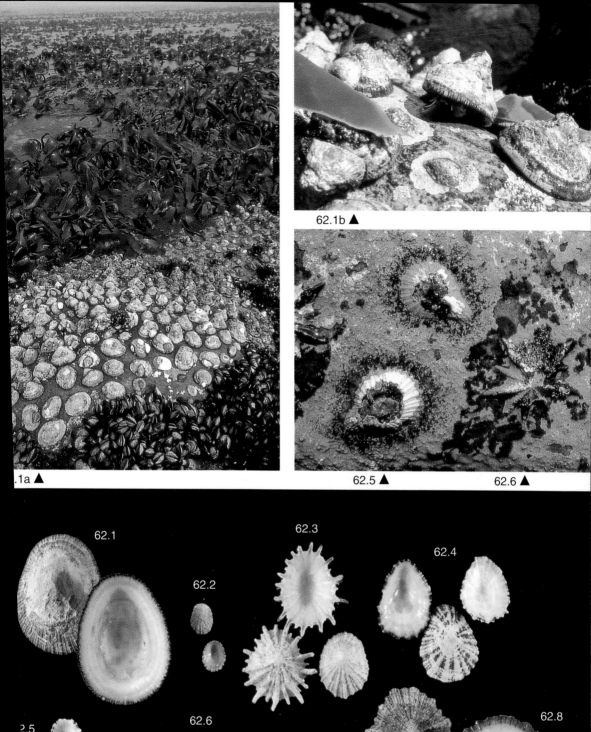

62.1b ▲

.1a ▲

62.5 ▲

62.6 ▲

62.1

62.2

62.3

62.4

2.5

62.6

62.7

62.8

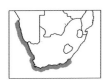

63.1a Pink-rayed limpet   *Cymbula miniata*

IDENTIFICATION: Shell thin, flat and nearly oval, with about 80 fine riblets that alternate in size and are slightly prickly. Pale, shot with numerous beautiful pink rays; interior pale in the centre and with a bluish sheen over the pink rays. SIZE: 80 mm. BIOLOGY: Lives low on the shore and in the shallows; feeds exclusively on encrusting coralline algae. VARIETIES: **63.1b *Cymbula sanguinans*** (Transkei to Natal) has a more robust, narrower shell that is almost smooth and has broad radiating purple-brown rays.

63.2 Goat's eye limpet   *Cymbula oculus*

IDENTIFICATION: Shell flat, dull brown above, with about 10 major ribs that project from the margin. Interior with a broad black margin and a pink-brown centre. Juveniles yellow, with iridescent green flecks. SIZE: 100 mm. BIOLOGY: Lives in the mid-shore and feeds on a range of algae. Aggressively attacks small predators, slamming its shell down to break their shells or cut off their feet. Changes sex, being first male, then female in the second or third year. Overexploited in Transkei, where large individuals (mainly females) are harvested for food. A flatworm, *Notoplana patellarum* (17.4), lives under its shell.

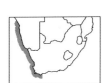

63.3 Granite limpet   *Cymbula granatina*

IDENTIFICATION: Similar in shape to *C. oculus,* but taller, with about 15 major ribs and a distinctly granite-like chevron patterning on the upper surface (although this is often eroded away). Interior pale, with a narrow edging that varies from granite-patterned to uniformly dark; central area shiny dark red-brown. Juveniles flecked green (63.3j). SIZE: 80 mm. BIOLOGY: Lives in remarkably high densities in Namaqualand, where its shells are tall and domed because the limpets are so closely packed. Dense colonies depend on drift seaweeds (mainly kelp) for food, and are concentrated in sheltered boulder-bays where drift weeds accumulate.

63.4 Granular limpet   *Scutellastra granularis*

IDENTIFICATION: Roughly oval, with approximately 50 fine ribs that are textured with white granules. Often eroded, smoothing the ribs and revealing a brown cap. Interior blue-white with a central brown patch and a dark border that is more obvious on West Coast (63.4a) than on South or East Coast specimens (63.4b). SIZE: 60 mm; largest sizes are reached on the productive West Coast. BIOLOGY: Abundant; occurs higher on the shore than other limpets. The calcium carbonate composition of the shell changes with temperature, aragonite predominating in warm seas and calcite in cold water. Archaeologists have used the ratio of the two in ancient shells to deduce prehistoric sea temperatures.

63.5 Kelp limpet   *Cymbula compressa*

IDENTIFICATION: Shell elongate, the sides compressed so that the opening is concave, designed to fit a cylindrical kelp stem; the shell 'rocks' on a flat surface. Outer surface brown, finely ribbed; interior blue-white to brown. SIZE: 90 mm. BIOLOGY: Found almost exclusively on the kelp *Ecklonia maxima*. Adults form scars on the 'stem' and territorially thrust away intruding limpets. Instantly drops off if its host kelp is detached and floats to the surface. One of the first South African shells ever described, being named by Carl von Linné in 1758.

63.6 Variable limpet   *Helcion concolor*

IDENTIFICATION: Shell flat, finely ribbed (but not granular); egg-shaped in outline. Colour extremely varied, but radiating streaks and dots are common. SIZE: 50 mm. BIOLOGY: Occurs in the mid-shore; very common in Natal. SIMILAR SPECIES: **63.7 *Cellana capensis*** (Port Alfred to Kenya) co-exists with it, and the two are easily confused. *C. capensis* has a more oval shell, its interior has a lustrous silky yellow or pearly sheen, shot with broad dark bands or spots, and the outer surface is finely granular when not eroded. Rolls its mantle over its shell to deter predators such as whelks (63.7b).

63.3j ▲

63.4 ▲

63.5 ▲

63.7b ▲

63.1a 63.1b 63.2 63.3

63.4

a b 63.5 63.6 63.7

64.1 **Prickly limpet** *Helcion pectunculus*

IDENTIFICATION: Shell tall, oval in outline, apex positioned very close to the anterior end; brown, with about 26 black ribs that are distinctively prickled. SIZE: 30 mm. BIOLOGY: Lives very high on the shore, concealed in crevices or beneath boulders.

64.2 **Rayed limpet** *Helcion pruinosus*

IDENTIFICATION: Shell low, fragile and oval; apex about one-third from front margin. Surface smooth, pale brown, shot with beautiful iridescent green spots and rays. SIZE: 25 mm. BIOLOGY: Hides beneath low-shore boulders; emerges by night. Large numbers are eaten by suckerfish (132.5). RELATED SPECIES:

64.3 *Helcion dunkeri* (Namibia to Natal) is similar but narrower, with fine radial ribs. West Coast specimens are white to yellow with broad black rays; South Coast examples are brown with darker rays and sometimes have green flecks like *H. pruinosus*.

64.4 **Dwarf limpet** *Patelloida profunda albonotata*

IDENTIFICATION: Small, but tall for its size, oval and with a central apex. Usually uniformly white although the interior has a greenish cast, and there can be marginal spots or blotches. SIZE: 10 mm. BIOLOGY: Very common but easily overlooked because of its small size. Occurs in the mid-to-high shore, living among barnacles or on the Natal rock oyster, *Saccostrea cucullata* (54.4).

64.5 **Cape false limpet** *Siphonaria capensis*

IDENTIFICATION: Apex of shell almost central; outline oval but slightly swollen on the right to accommodate a siphon that extends outwards from the lung cavity. About 40–50 low, flat ribs that are all roughly equal in size and scarcely project, so that the margin is smooth or, at the most, gently scalloped. Interior of shell a shiny brown, sometimes with white blotches, and with narrow dark-brown rays that arise near the centre and run to the margin. SIZE: 25 mm. BIOLOGY: Lives in pools and on exposed rock-faces in the mid-shore. Like other *Siphonaria* species, it is tolerant of sand-cover, and in areas where rocks are intermittently covered with sand, it replaces the less tolerant *Patella granularis* as the commonest mid-shore limpet. Each *S. capensis* has a fixed home-scar to which it returns after feeding. Feeds during low tide to avoid being washed away by waves. Its tissues are laden with a toxic milky mucus that repels predators. Eggs are laid in gelatinous rings that festoon pools. RELATED SPECIES:

64.6 *Siphonaria serrata* (Saldanha Bay to Zululand). Shell taller, with obvious siphonal lobe on the right; prickly ribs project from the serrated margin. Interior pale white to brown with pale rays radiating through the darker margin. Dark flecks on sides of foot. Previously called *S. aspersa*.

64.7 *Siphonaria concinna* (Cape Point to Zululand). Ribs well-spaced and never prickly, with 20–25 large ribs alternating with smaller ribs. Inside of shell has a large white central patch and a broad black margin shot with narrow white rays. Sides of foot with fine white spots. Juveniles have iridescent blue-green flecks on the outer surface of the shell. Previously called *Siphonaria deflexa*.

64.8 *Siphonaria oculus* (Cape Point to Zululand). Shell ribs numbering 46–58, raised and projecting from the margin, alternately small and large. Inside of shell dark brown with pale rays around the edge and (usually) a diffuse white bar across the apex; foot dotted with fine white spots.

64.9 *Siphonaria nigerrima* (Zululand to Moçambique). This is the smallest of the *Siphonaria* species, with an oval outline and a distinctively uniform dark-brown or black interior. Occurs high on the shore. Previously known as *S. carbo*.

64.10 *Siphonaria tenuicostulata* (Durban to Moçambique). Shell low, slightly pear-shaped, with 50–60 fine ribs; often shot with radiating iridescent blue-green flecks. Interior with a brown or brownish-orange blotch; margin dark brown with pale patches and white rays. Easily confused with *S. concinna*.

64.11 *Siphonaria anneae* (Durban northwards) is very similar but has 30–40 fine ribs, a more clearly-defined outer black edging and very short white rays.

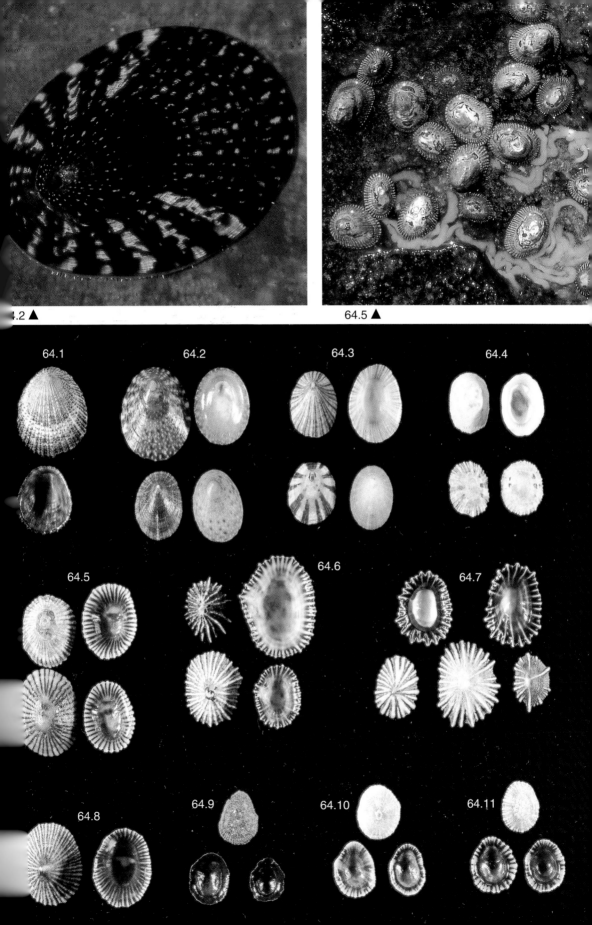

64.2 ▲

64.5 ▲

64.1

64.2

64.3

64.4

64.5

64.6

64.7

64.8

64.9

64.10

64.11

Gastropoda: 3. Keyhole and Slipper Limpets

Keyhole limpets have oval, flat shells with a central hole through which protrudes a siphon carrying the anus, so that faeces are voided well away from the gills, which are hidden under the shell. Slipper limpets are characterised by a partial shelf inside the shell, so that the empty shell resembles a slipper.

65.1 Saddle-shaped keyhole limpet *Dendrofissurella scutellum*

IDENTIFICATION: Shell much smaller than the body and saddle-shaped so that it rocks on a flat surface. 'Keyhole' smoothly oval. Front of foot with an elongate, frilly, branching protuberance. SIZE: 40 mm. BIOLOGY: The body cannot withdraw into the shell, and probably contains toxins so that predators avoid it. On contact with a predator, the mantle expands to cover the shell. Lives beneath boulders and in sheltered sandy areas. Herbivorous, and an important consumer of the commercial seaweed *Gracilaria*.

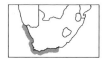

65.2 Conical keyhole limpet *Diodora parviforata*

IDENTIFICATION: Shell white, roughly oval in outline, with radiating riblets; animal completely covered by the shell. 'Keyhole' small and round, close to the front of the shell. SIZE: 25 mm. BIOLOGY: Seldom common; lives under boulders and in caverns low on the shore. Eats sponges and detritus.

65.3 Cape keyhole limpet *Fissurella mutabilis*

IDENTIFICATION: Shell flat, attractively marked with rays and scallops; smooth or with delicate radial ridges; almost large enough to cover the body. 'Keyhole' narrow and with undulating sides. SIZE: 20 mm. BIOLOGY: Common, hiding in groups under boulders or among mussels. Herbivorous. RELATED SPECIES:
65.4 *Fissurella natalensis* (Port Alfred–Moçambique) has heavier, dark-brown shells with paler rays and coarse radial ridges.

65.5 Mantled keyhole limpet *Pupillaea aperta*

IDENTIFICATION: Shell very small relative to the body, with a white rim and a large, bevelled oval 'keyhole'. The shell is almost completely covered by the mantle and the animal looks slug-like. The body is grey, striped or mottled black and the foot bright orange. SIZE: Shell 35 mm, body 80 mm. BIOLOGY: Solitary and never common; occurs subtidally on vertical rock-faces.

65.6 Slipper limpet *Crepidula porcellana*

IDENTIFICATION: Tear-drop shaped, with concave internal shelf. Surface smooth, usually brown. SIZE: 15 mm. BIOLOGY: Lives on other shells, frequently forming stacks, one on top of another. Changes sex as it matures: the larger (lower) individuals in stacks are female; the smaller, younger ones are male. Females brood yellow eggs beneath the shell. The gill is enlarged and filters plankton from the water. RELATED SPECIES:
65.7 *Crepidula dilatata* (Mossel Bay to Lambert's Bay) has a smooth oval shell with a convex shelf, and attaches to rocks. Previously called *C. capensis*.
65.8 *C. aculeata* (Namibia–Natal) is sculptured with curved rows of scales.

65.9 Chinese hat *Calyptraea chinensis*

IDENTIFICATION: Shell flat, smooth and round, with an internal spiral shelf from the apex to the margin. SIZE: 15 mm. BIOLOGY: Shell commonly washed ashore, but living animals seldom seen, hiding under rock ledges. Like the slipper limpet, it undergoes sex-change and is a filter-feeder. RELATED SPECIES:
65.10 *Calyptraea helicoidea* (Port Elizabeth to East London) has strong spiral ridges.

65.11 Horse's hoof *Hipponix conicus*

IDENTIFICATION: Shell conical but apex sloping backwards; usually with ribs that scallop the edge of the shell. SIZE: 15 mm. BIOLOGY: Attaches itself beneath other gastropods, eroding a depression in the shell. Gathers mucus and faeces from the host. Male much smaller than the female and attached on her shell. Commonest in Natal.

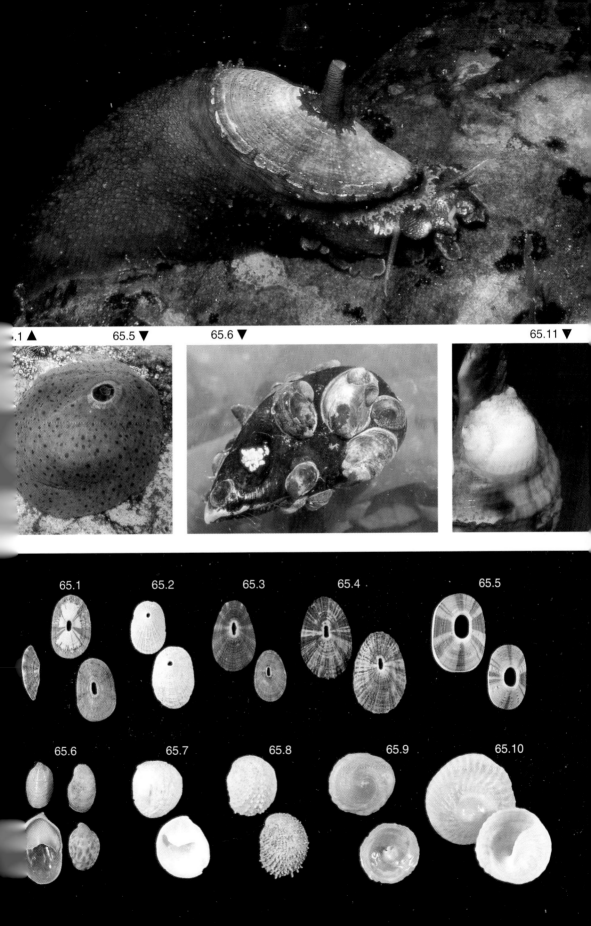

65.1 ▲ 65.5 ▼ 65.6 ▼ 65.11 ▼

65.1 65.2 65.3 65.4 65.5

65.6 65.7 65.8 65.9 65.10

Gastropoda: 4. Winkles

Winkles (or topshells) fall in the family Trochidae and have shells with a circular or semicircular mouth (aperture) that is blocked by a flexible, horny 'door' (operculum) when the animal withdraws into the shell. Almost all winkles have a spiral shell that is round and squat, and are herbivorous.

66.1 Ornate topshell *Calliostoma ornatum*

IDENTIFICATION: Top-shaped, the base sharply set off from the sides by an acute angle. About 2–3 strong granular ridges spiral around the lower whorls, alternating with weaker ridges. Aperture smoothly round, lacking teeth or nodules. Operculum round, thin and flexible. Shell colour brown to orange-red or violet; foot bright orange. SIZE: 20 mm. BIOLOGY: The commonest of several *Calliostoma* species. It occurs in the low intertidal and down to 50 m. Often found attached to gorgonians (sea fans) and possibly feeds on them.

66.2 *Calliostoma africanum* (Port Elizabeth–Transkei), is similar but not as broad at the base, and has about 6–8 thin, very finely granular spiral ridges, all of similar strength.

66.3 Black-chained topshell *Clanculus atricatena*

IDENTIFICATION: Shell conical, much wider than high, and with convex sides. Background colour grey-brown. Each whorl has about two well-developed, spiral, granular ridges with darker oblong spots. Between these ridges lie weaker ridges that lack this decoration. Operculum round, thin and flexible. The inner lip of the aperture has a strong tooth, and the outer lip a ridge-like tooth near its posterior edge. Viewed from below, the shell has a narrow tunnel (the umbilicus) that runs up its centre. SIZE: 20 mm. BIOLOGY: Live animals are seldom seen, hiding beneath boulders in sandy areas. All *Clanculus* species have a toothed inner lip and an umbilicus.

66.4 *Clanculus puniceus* (Natal South Coast to the Indo-Pacific tropics) is very similar but vivid red with darker spots and white dots on the spiral ridges.

66.5 *Clanculus miniatus* (Cape Point to Transkei) is recognised by being more obviously top-shaped, a sharp keel separating the base from the sides. Colour brown to reddish-pink, usually with spots around the keel.

66.6 Multicoloured topshell *Gibbula multicolor*

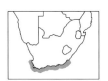

IDENTIFICATION: Shell small, wider than tall. Aperture rounded and lacks teeth. The shell whorls have 2–3 spiral ridges. The umbilicus is shallow or even closed completely. Colour extremely variable, but always bright, usually a background of red decorated with white bands and tiny green dots. SIZE: 10 mm. BIOLOGY: Lives beneath rocks, among seaweeds in rock pools and in the shallow subtidal. RELATED SPECIES: All *Gibbula* species have small shells (10 mm) that are wider than tall, an aperture that lacks teeth and a flexible, horny operculum. The umbilicus is developed to differing degrees.

66.7 *Gibbula capensis* (Saldanha to Agulhas) has a relatively flat shell; the sides form an acute angle with the base and have fine spiral threads. Colour pink to red with regularly-spaced white blotches or bars.

66.8 *Gibbula cicer* (Namibia to Transkei) is globular, only slightly taller than wide, the lowest whorl being smoothly rounded; 4–5 spiral ridges on each whorl. Colour usually pale cream with drab olive-brown to grey spots.

66.9 *Gibbula zonata* (Namibia to Agulhas) is easily recognised by its rounded shell, dull background colour and dark spiral lines.

66.10 *Gibbula beckeri* (Namaqualand to Cape Point) is dark grey, with irregular paler flames, and usually has 2–3 major spiral ridges as well as finer intermediary ridges.

66.11 *Cinysca granulosa* (Namibia to eastern Transkei): shell thick, rounded; uniformly pale on the West Coast, speckled on the South Coast. Umbilicus large and obvious. About eight strong equal-sized ridges spiral around each whorl and corrugate the outer lip. Operculum horny and flexible but has spiral rows of minute calcareous beads. Related to turban shells although it superficially resembles the winkles.

66.1 66.2 66.3 66.4 66.5

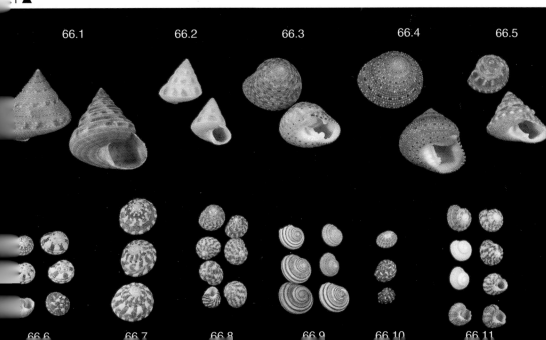

66.6 66.7 66.8 66.9 66.10 66.11

67.1 Black-spotted topshell *Trochus nigropunctatus*

IDENTIFICATION: Shell top-shaped, triangular when viewed from the side, wider than tall, and with spiral, granular ridges. The base has about five spiral ridges that run into the aperture; the umbilicus is shallow but has three spiral ridges encroaching into it. Colour is dull brown with darker spots which are most obvious on the base. SIZE: 25 mm. BIOLOGY: Hides in crevices or among seaweeds in tide-pools. Like many top-shells, it swivels its shell violently when touched by a predator.

67.2 Toothed topshell *Monodonta australis*

IDENTIFICATION: Easily recognised by its thick shell, rounded whorls, obvious spiral ridges, the single stubby tooth that projects inwards from the inner lip of the aperture, and the absence of an umbilicus. Colour usually pale brown with darker rectangular spots, but tints of green and pink often brighten the shell. Operculum round, flexible and horny. SIZE: 30 mm. BIOLOGY: A very common intertidal species, which is usually found in pools and grazes on microalgae.

67.3 Variegated topshell *Oxystele variegata*

IDENTIFICATION: Shell rounded, about as tall as wide, smooth or with very gentle spiral ridges. No trace of an umbilicus. Colour extremely variable but typically tabby-patterned with irregular dark bands that run obliquely across a pale-yellow or greenish background. The inside of the aperture is pearly, with a narrow margin that repeats the variegated tabby pattern. As is the case with all *Oxystele* species, the operculum is horny and flexible, yellow-brown, with spiral growth rings. SIZE: 25 mm. BIOLOGY: An abundant, intertidal rocky-shore mollusc. Uses its rasping file-like radula (67.3d) to graze on microalgae and encrusting algae. Juveniles live low on the shore, shifting upshore as they become more tolerant of the greater physical stresses experienced there. Upshore movement apparently shifts adults away from intense predation by starfish and whelks. If adults are experimentally transplanted downshore they navigate accurately and return to their original zone within the space of a day. RELATED SPECIES:
67.4 Oxystele impervia is almost identical to *O. variegata* in form, shape and distribution but is patterned with bead-like dots that run spirally around the whorls. Only recently, based on genetic studies, has *O. impervia* been recognised as being distinct from *O. variegata*. Their habits are also similar, although *O. impervia* tends to occur in more sheltered areas and slightly higher on the shore.
67.5 Oxystele tabularis (Port Alfred to southern Moçambique) is similar, but has clearly defined grey-brown to grey-green bands that run radially across each whorl. These bands alternate with thinner red and green lines.

67.6 Pink-lipped topshell *Oxystele sinensis*

IDENTIFICATION: Larger than the previous three species, *O. sinensis* has a round, blunt-spired shell with fine spiral sculpturing. Dark purple-black above; aperture with a distinctive pink inner lip. SIZE: 45 mm. BIOLOGY: Lives moderately low on the shore, extending down to about 5 m. Grazes on encrusting algae and microalgae. Together with *O. tigrina*, it is harvested as a source of food in Transkei. RELATED SPECIES:
67.7 Oxystele tigrina (Port Nolloth to Transkei) has a similar shell, with a dark (almost black) upper surface that sometimes has scattered white dots. The aperture is white with a narrow black border, lacking any traces of pink.

67.8 Variegated sundial shell *Heliacus variegatus*

IDENTIFICATION: Shell flat, height less than half the width, with four beaded spiral ridges on each whorl. Umbilicus well developed and lined with nodules. Colour grey-white, speckled with brown-to-black rectangular spots. Operculum calcareous and conical. SIZE: 15 mm. BIOLOGY: Slow-moving; it parasitises the zoanthid *Palythoa nelliae*, boring into its tissues with its proboscis, which is protected against the zoanthid's stinging cells by a rough cuticle. *Heliacus* is not a winkle, belonging to the family Architectonicidae.

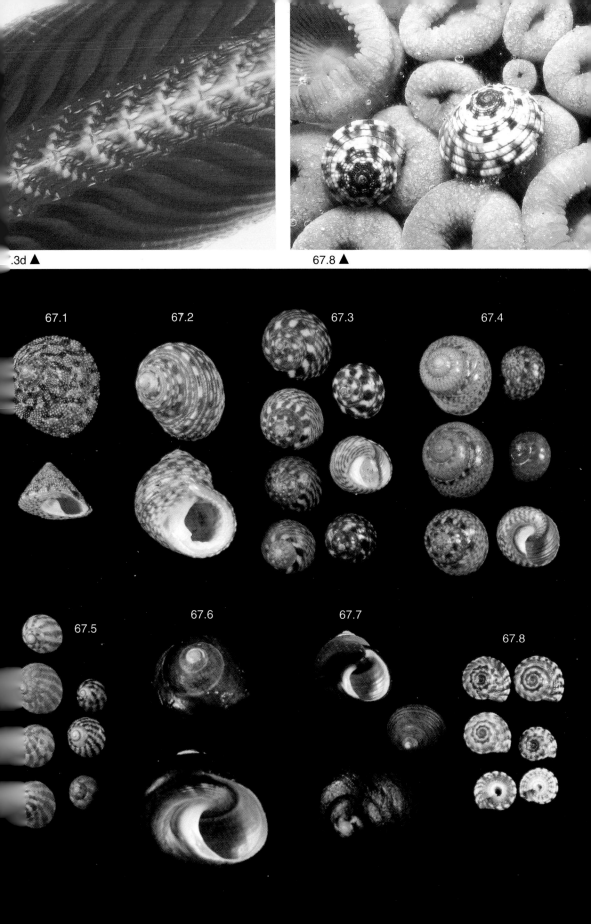

67.3d ▲

67.8 ▲

67.1

67.2

67.3

67.4

67.5

67.6

67.7

67.8

Gastropoda: 5. Turban Shells and Nerites

Turban shells (family Turbinidae) and nerites (family Neritidae) have a hard, calcareous operculum that blocks the opening of the shell when the animal withdraws. The aperture and operculum are rounded in the turban shells, but semicircular in the nerites. Both groups are herbivorous.

68.1 Alikreukel or giant periwinkle *Turbo sarmaticus*

IDENTIFICATION: Large and round, height less than width. About three rows of low nodules spiral around each whorl (though they may erode and disappear in later life). Aperture smoothly round, outer lip dark brown or black; inner lip white to bright orange. Operculum round, thick and calcified; outer surface with densely-packed coarse nodules. SIZE: 100 mm. BIOLOGY: Lives in pools and down to a depth of about 8 m. Although still common, it is becoming increasingly difficult to find large specimens in the intertidal zone except in marine reserves, despite regulations limiting its collection to 5 per day per person with a minimum size limit of 63.5 mm. Relatively slow-growing, it reaches this size at an age of about 3–4 years.

68.2 Crowned turban shell *Turbo coronatus*

IDENTIFICATION: Distinguished by its operculum, which has a greenish tinge and is smooth or has only a faint granulation. The shell has very strong, spirally-arranged nodules, the uppermost row on the body whorl being particularly well developed and forming distinct knobs. SIZE: 35 mm. BIOLOGY: Very common in the mid-shore where it is found in pools or under boulders. RELATED SPECIES:

68.3 *Turbo cidaris*. There are two distinctly different subspecies. The smooth turban shell, *T. cidaris cidaris* (Cape Peninsula to Port Elizabeth), has an attractive smooth, glossy shell, usually purple-brown to red-brown with paler radial flares (68.3a). Its operculum is sparsely granular and has a groove that spirals towards a central pit. *T. cidaris natalensis*, the Natal turban shell (Port Elizabeth to north of Durban), has 6–11 well-defined spiral ridges on the shell, but no nodules (68.3b), and its operculum is completely covered with coarse nodules. In certain areas it is gathered for food.

68.4 Blotched nerite *Nerita albicilla*

IDENTIFICATION: Shell semicircular when viewed from the side; spire sunken into the main body of the shell and not protruding. Upper surface smooth or with weak spiral grooves, most commonly black with white blotches or vice versa, but sometimes an attractive orange with pale radial stripes. The inner lip of the aperture forms a flat shelf that is dotted with pustules. Operculum calcareous, semicircular, pale and finely granular. SIZE: 30 mm. BIOLOGY: By far the most common nerite, forming dense aggregations in mid-shore pools and under damp boulders. It lays oval, white egg capsules about 2 mm in length, which prominently dot pools and the undersides of rocks. Several eggs are housed in each capsule and yield planktonic larvae. *N. albicilla* is active by night and grazes on lichens, encrusting algae and diatoms, even cutting into the rock-face with its powerful radula to remove embedded microalgae. Its shells are used as decorative beads by coastal people in Natal and the Transkei. RELATED SPECIES:

68.5 *Nerita polita* (East London to tropical Indo-Pacific) has a smooth, glossy upper surface, usually mottled brown but often with radial bands. Its operculum is smooth except for fine ridges that run across the outer margin.

68.6 Textile nerite *Nerita textilis*

IDENTIFICATION: Upper surface strongly ridged; off-white with conspicuous oblong black spots. Outer lip of aperture spotted black. Inner lip with three feeble teeth; the adjacent flat plate is granular and usually yellow. SIZE: 30 mm. BIOLOGY: Scarce. Lives high on the shore. Esteemed by Pondo people as an ornament. RELATED SPECIES:

68.7 *Nerita plicata* (western Transkei to the tropics) has a globular shell with about 14 radial ridges. The inner and outer lips are strongly toothed. Colour uniformly pale cream, sometimes with darker spots. Outer lip yellow.

1 ▼ 68.3a ▲ 68.4 ▼

68.1 68.2 68.3a 68.3b

68.4 68.5 68.6 68.7

Gastropoda: 6. Periwinkles and Small Species

Included here are the periwinkles (littorinids) – which dominate the uppermost tidal reaches of exposed rocky shores – and other small, round-apertured gastropods. All are herbivorous, feeding mainly on diatoms and microalgae.

69.1 African periwinkle   *Afrolittorina africana*

IDENTIFICATION: Aperture rounded and closed by a transparent, horny operculum. Shell smooth, with a short conical spire. Two species are recognised: *Afrolittorina africana* (69.1a) is blue-grey, with a brown aperture and is the dominant form in KZN, while *Afrolittorina knysnaensis* (69.1b) varies from flecked brown with a dark ring around each coil to almost pure black and ranges from Namibia to southern KZN. SIZE: Usually under 10 mm. BIOLOGY: Congregates in crevices, or hangs from the hot rock by a mucus thread in the heat of the day, emerging at night, or on moist days, to feed. Juveniles occur highest on the shore to avoid wave action. Adults have a stronger attachment and are found lower on the shore, where food is more plentiful. Previously classified as a *Littorina* species.

69.2 Striped periwinkle   *Littoraria glabrata*

IDENTIFICATION: Shell smooth with a fairly tall spire, flesh-coloured with oblique pink or brown lines or zigzags. SIZE: Up to 24 mm. BIOLOGY: An extremely desiccation-resistant species found at or above high spring-tide level on subtropical shores. Previously known as *Littorina kraussi*.

69.3 Estuarine periwinkles   *Littoraria scabra* group

IDENTIFICATION: Fine spiral ridges and irregular lines of dark dashes. There are three closely-related southern African species: *L. scabra* (white columella), *L. intermedia* (dark purple to pink columella, coarse ribs) and *L. subvittata* (narrow pink-brown columella, fine ribs). SIZE: 28 mm. BIOLOGY: Estuarine, often on mangroves and salt-marsh vegetation, well above high tide.

69.4 Nodular periwinkle   *Afrolittorina natalensis*

IDENTIFICATION: Shell blue-grey with fine spiral lines and about three rows of paler-coloured angular knobs running around each coil. Aperture dark brown. SIZE: Up to 12 mm. BIOLOGY: In crevices around the high-water mark on subtropical shores.

69.5 Tropical periwinkle   *Planaxis sulcatus*

IDENTIFICATION: Shell relatively thick and strong, and decorated with equal-sized spiral ridges. Colour grey-brown, with regular elongate dots on the spiral ridges. SIZE: Up to 20 mm. BIOLOGY: One of the dominant gastropods low on rocky tropical shores. Characteristic of sheltered shores.

69.6 Pheasant shell   *Tricolia capensis*

IDENTIFICATION: Attractive smooth glossy shells. Spire fairly tall; operculum white and calcareous. Colour variable; dull grey in Atlantic specimens (69.6a) to brilliant red or yellow with white markings. SIZE: 16 mm. BIOLOGY: Herbivorous; abundant amongst seaweeds and in beach drift. The larger, brighter East Coast varieties (69.6b) are popular with collectors. RELATED SPECIES:

69.7 *Tricolia neritina* (whole coast) has a small globular shell with a very short spire and smooth calcareous operculum. Colour pale pink with fine spiral pink or purple stripes. Herbivorous; found under rocks and amongst seaweeds.

69.8 Globular mud snail   *Assiminea globulus*

IDENTIFICATION: Tiny, globular. Spire short, operculum horny. Brown. SIZE: 4 mm. BIOLOGY: Countless thousands occur near the high-water mark on lagoonal mudflats (69.8a), feeding on diatoms and bacteria. RELATED SPECIES:

69.9 *Assiminea ovata* (Knysna–Moçambique) is very similar, but grows to 7 mm, usually has two pale bands and is the dominant form in East Coast estuaries.

8 ▲ 69.1a ▼ 69.6a ▼

69.1a 69.1b 69.2 69.3 69.4 69.5

69.6a 69.6b 69.7 69.8 69.9

Gastropoda: 7. Worm, Screw & Turret Shells

This rather miscellaneous group includes the worm shells, which have long irregularly twisted shells permanently attached to the substratum, and a variety of other elongate, tall-spired forms.

70.1 **Colonial worm shell** *Dendropoma corallinaceus*

IDENTIFICATION: A small gregarious species forming an intertwining mass of long, white, worm-like tubes, often sunken into encrusting coralline algae. Operculum present. SIZE: About 10 mm long. BIOLOGY: A filter-feeder which spins a mucus net to trap floating particles. RELATED SPECIES: Replaced in Transkei and KZN by the very similar *Dendropoma tholia*.

70.2 **Solitary worm shell** *Serpulorbis natalensis*

IDENTIFICATION: Shell is a solitary, irregularly-coiled, white tube cemented to the underside of rocks. Operculum absent. Animal pink or red. SIZE: 40 mm long. BIOLOGY: Spins and then eats a mucus net to capture planktonic food particles. Eggs are laid in capsules attached to the roof of the shell.

70.3 **Waxy screw shell** *Protomella capensis*

IDENTIFICATION: Very elongate, with numerous rounded whorls. A pair of fine spiral ridges frequently present on each whorl. Pink or pale brown, often with fine markings. SIZE: 33 mm. BIOLOGY: Buries shallowly in sheltered sand. Filters water drawn down a tube communicating with the surface. Extremely common in lagoons. Previously called *Turritella capensis*. RELATED SPECIES:

70.4 *Turritella carinifera* (Western Cape–S. Moçambique) has a long, pointed shell with a sharp spiral ridge running up the middle of each whorl. It is found on the open coast under rocks lying on sand, or embedded amongst sponges and ascidians.

70.5 *Turritella sanguinea* (Cape Point–KZN) has several fine spiral ribs on each whorl, these being coloured with red-brown spots.

70.6 **Truncated mangrove snail** *Cerithidea decollata*

IDENTIFICATION: Aperture notched and with a flanged outer lip. Spire elongate, but apex invariably broken off in adults. Each whorl has about 20 strong cross-ridges. SIZE: Up to 36 mm. BIOLOGY: Lives high on the trunks of mangroves out of reach of aquatic predators, descending to the mud to feed on detritus during low neap-tides.

70.7 **Knobbled horn shell** *Rhinoclavis sinensis*

IDENTIFICATION: Three or four granular ridges on each whorl, the first enlarged and with marked knobs. Anterior canal short, bent abruptly upwards. SIZE: 40 mm. BIOLOGY: Abundant in tropical sandy pools. Shells often occupied by hermit crabs. RELATED SPECIES: *Cerithium crassilabrum* (East London to Moçambique) is smaller, lacks the upturned anterior canal and has about four similar-sized, granular, spiral ridges.

70.8 **Mangrove whelk** *Terebralia palustris*

IDENTIFICATION: A large, heavy, dark-brown shell with three spiral grooves running up each whorl. Aperture notched, black inside. SIZE: Up to 120 mm. BIOLOGY: Crawls over wet mud, usually in association with mangrove swamps. Feeds on diatoms and mangrove leaves. The photograph shows a juvenile: adults have a thick lip.

70.9 **Ribbed turrid** *Clionella sinuata*

IDENTIFICATION: Shell dark brown with eroded apex. Whorls flattened and with about 16–18 coarse ribs across the last turn. Aperture with short anterior canal and a slight bend or notch in outer lip. SIZE: 65 mm. BIOLOGY: A scavenger found under loose rocks in sandy areas. Congregates to lay masses of purse-shaped egg cases under boulders, depicted in the photograph. RELATED SPECIES: **70.10** *Clionella rosaria* (Cape Point–Natal) also has oblique ribs, but is pinky-orange with brown and white dots along the suture.

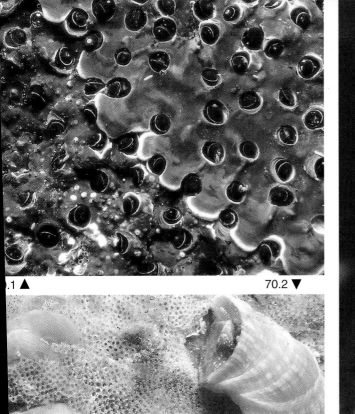

70.1 ▲

70.2 ▼

70.6 ▲

70.9 ▼

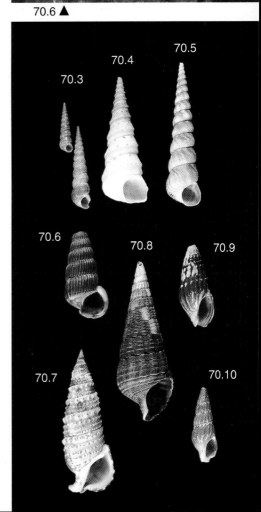

70.3

70.4

70.5

70.6

70.8

70.9

70.7

70.10

Gastropoda: 8. Cowries

Colourful, glossy shells with an unusual egg-like shape, the spire being enclosed within the last whorl. The aperture is a narrow slit running the full length of the shell, and both lips are usually ridged or toothed. Operculum absent. In life the mantle lobes, which may themselves be dramatically coloured and textured, extend over the shell, completely obscuring it and giving it the brilliant gloss so prized by shell-collectors.

71.1 Cape cowrie   *Cypraea capensis*

IDENTIFICATION: Easily distinguished by the numerous ridges running across the shell. Shell light brown with darker blotches dorsally. Mantle orange to black with small white projections and spots. SIZE: 18–37 mm. BIOLOGY: Browses on sponges on sandy reefs at depths of 8–110 m.

71.2 Toothless cowrie   *Cypraea edentula*

IDENTIFICATION: Pink to lilac with speckling of reddish-brown spots, base usually white. Both lips toothless and aperture relatively wide. Mantle smooth, usually orange, spattered with dark spots. SIZE: 12–30 mm. BIOLOGY: On reefs at 12–40 m depth. Probably feeds on sponges.

71.3 Dark-toothed cowrie   *Cypraea fuscodentata*

IDENTIFICATION: Shell blue or mauve marginally, densely marked with fused brown dots and dashes dorsally. A series of 15–19 prominent brown teeth form ridges which run right across the base. Mantle colour very variable, usually orange with black spots. SIZE: 22–44 mm. BIOLOGY: Usually found amongst thick black sponges on reefs 6–130 m deep.

71.4 Ring cowrie   *Cypraea annulus*

IDENTIFICATION: Shell pale, easily identified by the orange or yellow ring around the apex. Mantle greyish with branching projections. SIZE: 15–28 mm. BIOLOGY: Common in depressions on wave-washed rocks or under stones in pools, as well as on sheltered tropical seagrass beds. Often congregates in small groups.

71.5 Money cowrie   *Cypraea moneta*

IDENTIFICATION: Shell small but thick, with a rather irregular, knobbly outline; colour yellow, becoming purple when eroded. Mantle dark brown. SIZE: 15–28 mm. BIOLOGY: Abundant intertidally in the tropics, becoming very rare in Natal. Widely used in the curio trade, and formerly employed as currency in the slave and ivory trades.

71.6 Arabic cowrie   *Cypraea arabica*

IDENTIFICATION: Large, with a flattened base. Dorsal surface pale brown, densely patterned with an irregular network of darker brown markings, base pale with darker teeth; margins spotted. Mantle grey with small projections. SIZE: 51–102 mm. BIOLOGY: Occurs under rocks and in crevices, intertidally and on shallow reefs. An omnivorous grazer. Previously common but becoming scarce because of collectors.

71.7 Carnelian cowrie   *Cypraea carneola*

IDENTIFICATION: Shell elongate and deep; flesh-coloured and crossed by four darker bands. Edges of aperture violet. Mantle dark grey with paler conical projections. SIZE: 27–48 mm. BIOLOGY: Under rocks or in crevices on shallow reefs. An omnivorous grazer.

71.8 Tiger cowrie   *Cypraea tigris*

IDENTIFICATION: A large, well-known species with a grey shell densely covered with dark brown spots. Base white. Mantle mottled grey and black with long white-tipped processes. SIZE: 66–113 mm. BIOLOGY: A sponge-feeder common on sandy tropical reefs but rare in Natal and seldom seen in Transkei.

71.6 ▲

71.3 ▲

71.1 ▽   71.2 ▽   71.3 ▽   71.4 ▽   71.5 ▽

71.7 ▲

71.6 ▲

71.8 ▲

72.1 Snake's head cowrie   *Cypraea caputserpentis*

IDENTIFICATION: Thick, flattened shell with chocolate-brown margin; densely spotted with white dorsally; ends and underside pale. Mantle yellow with brown spots. SIZE: 20–37mm. BIOLOGY: Hidden on wave-washed shores by day, emerging to browse at night.

72.2 Honey cowrie   *Cypraea helvola*

IDENTIFICATION: Shell brown with numerous tiny white spots, underside orange-brown with lilac ends. Mantle red-brown with branched processes. SIZE: 12–37 mm. BIOLOGY: Under rocks on protected shores and subtidal reefs. Beach-worn shells are violet and are common on Natal beaches.

72.3 Orange cowrie   *Cypraea citrina*

IDENTIFICATION: A small orange-brown cowrie with round grey spots and a few brown ones. Underside orange. Mantle a dense bush of branching off-white processes. SIZE: 12–30 mm. BIOLOGY: Intertidal to 100 m depth; feeding habits unknown. Effectively camouflaged by the bush of processes on the mantle.

72.4 Eroded cowrie   *Cypraea erosa*

IDENTIFICATION: Olive-brown with small white spots, some of which show dark rings, a brown blotch mid-way along each side. Margin of shell prominent and ridged. Mantle pale brown with dense branched processes. SIZE: 19–52 mm. BIOLOGY: Under intertidal rocks.

72.5 Kitten cowrie   *Cypraea felina*

IDENTIFICATION: A small, elongate cowrie, blue-grey dorsally with dense brown speckles and darker cross-bands. Sides have large black spots. Underside yellowish. Mantle yellow with simple white projections. SIZE: 18–28 mm. BIOLOGY: Fairly common under boulders and in rock pools, particularly on the zoanthid *Palythoa nelliae* (5.4).

72.6 Stippled cowrie   *Cypraea staphylaea*

IDENTIFICATION: Shell light grey and usually textured with small raised nodules. Ends of shell orange-brown; teeth lined with orange. Mantle black with long projections. SIZE: 12–23 mm. BIOLOGY: Found in crevices and under rocks on shallow reefs. Rare in Natal and Transkei; common in the tropics.

72.7 Baby's toes   *Trivia aperta*

IDENTIFICATION: Fat pink shells with pale ridges running right around the shell, but halting at the mid-line. Mantle off-white to yellow or purple-grey, studded with pustules. SIZE: 14–27 mm. BIOLOGY: Feeds on compound ascidians, which the mantle matches closely in colour, and lays vase-shaped egg cases within cavities excavated in the ascidian colony. RELATED SPECIES:

72.8 *Trivia ovulata* (Cape Point–S. Transkei) has a smooth pink shell with a white underside; 13–20 mm. Mantle astoundingly variable: orange with brown spots, golden with black streaks, blue with white-ringed black spots, or white with dark squiggles.

72.9 *Trivia phalacra* (Port Elizabeth–East London) is pink and ridged below, but the ridges disappear half-way up the sides leaving the top smooth; 13–19 mm.

72.10 *Trivia millardi* (Cape West Coast) has a broad, smooth, white shell. Mantle highly variable: often yellow with pale streaks or dark dots; 14–23 mm.

72.11 Tear drops or riceys   *Trivia pellucidula*

IDENTIFICATION: Small and white, with 23–28 ridges running uninterrupted right around the shell. Mantle usually brown or green, spotted with short projections. SIZE: 4–8 mm. BIOLOGY: Feeds on compound ascidians on shallow tropical reefs. RELATED SPECIES: *Trivia oryza* (Port Alfred–Natal) is nearly identical but has a mid-dorsal longitudinal groove that interrupts the ridges.

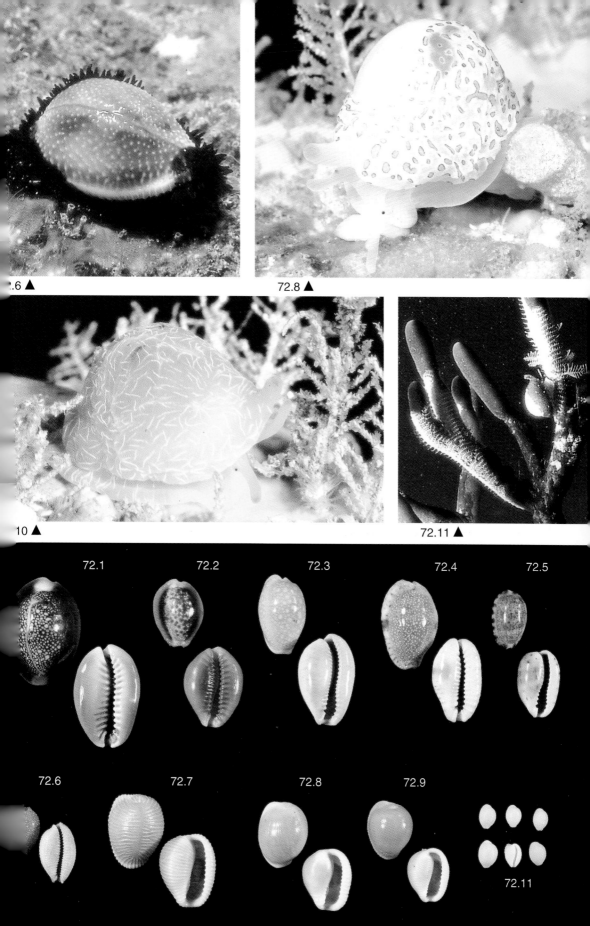

72.6 ▲

72.8 ▲

10 ▲

72.11 ▲

72.1

72.2

72.3

72.4

72.5

72.6

72.7

72.8

72.9

72.11

Gastropoda: 9. Necklace, Helmet & Violet Shells

These three families have broad, short-spired shells and predatory lifestyles. Necklace shells (so named because they lay collar-like egg masses) have a greatly enlarged foot, and plough through sand seeking out and drilling the shells of other molluscs. Helmet shells also occur in sandy areas but feed on sand-dwelling echinoderms. The fragile violet snails secrete rafts of bubbles and drift upside down on the ocean surface, where they consume bluebottles and their relatives.

73.1 Comma necklace shell   *Natica gualteriana*

IDENTIFICATION: Shell bluish- or brownish-grey, unmarked or with faint spiral patterning. Umbilicus present and comma-shaped. Operculum calcareous and semicircular. SIZE: 26 mm. BIOLOGY: Feeds on small bivalves and gastropods in East Coast estuaries or shallow marine sandbanks. Eggs laid in a coiled sandy collar, which lies unattached on the sand.

73.2 Mottled necklace shell   *Natica tecta*

IDENTIFICATION: Shell brown, densely marked with bands of darker spots or streaks. Umbilicus closed. Operculum smooth and calcareous. SIZE: 41 mm. BIOLOGY: Lives in clean sand on the open coast or in estuaries. Feeds on bivalves, including mussels, and drills a neat bevelled hole through their shells. Egg mass is a flattened, leaf-like strip attached to solid objects by a short stalk.

73.3 Moon shell   *Polinices didyma*

IDENTIFICATION: Shell smooth, globular; spire very short. Colour light brown, darker around callus and aperture. Operculum flexible, horny and brownish, with a spiral origin. SIZE: 50 mm. BIOLOGY: Burrows just below the surface of shallow sandbanks. Eggs laid in a coiled collar. RELATED SPECIES:

73.4 *Polinices mamilla* (Transkei eastwards) has a more elongate shell-shape and is pure white.

73.5 Helmet shell   *Phalium labiatum zeylanicum*

IDENTIFICATION: A shiny globular shell with a very short spire and a thickened outer lip. Last whorl usually with one or more rows of rounded knobs. Colour buff, with rows of paler spots and five pairs of dark bars on outer lip. Operculum smaller than aperture. Foot with a narrow yellow line around margin. SIZE: Up to 78 mm. BIOLOGY: A slow-moving predator living subtidally among rocks or on sand. Bores through the shells of sea urchins and pansy shells.

73.6 Bubble raft shell or violet snail   *Janthina janthina*

IDENTIFICATION: Thin, fragile, rounded shells grading from dark violet around the aperture to much paler at the apex. Aperture squarish, lacking an operculum. SIZE: Up to 34 mm. BIOLOGY: Hangs upside down from the sea surface, suspended by a raft of mucus-coated bubbles secreted by the foot. Often washed ashore in association with by-the-wind-sailors, *Velella* (16.3), and bluebottles, *Physalia* (16.1), upon which it feeds. The eggs are brooded and released as late stage larvae. RELATED SPECIES:

73.7 *Janthina prolongata* is uniform in colour and has an elongate, pear-shaped aperture that is produced to form an anterior 'spout'.

73.8 *Janthina pallida* is a very pale species with a broadly rounded aperture that lacks either a spout or notch.

73.9 *Janthina exigua* is a smaller form, reaching only 15 mm. The surface is covered in fine ribs angled to meet at mid-whorl, creating a V-shaped notch in the aperture.

73.10 *Janthina umbilicata* is another small form but is smooth, with a weak ridge at mid-whorl and no notch in the aperture.

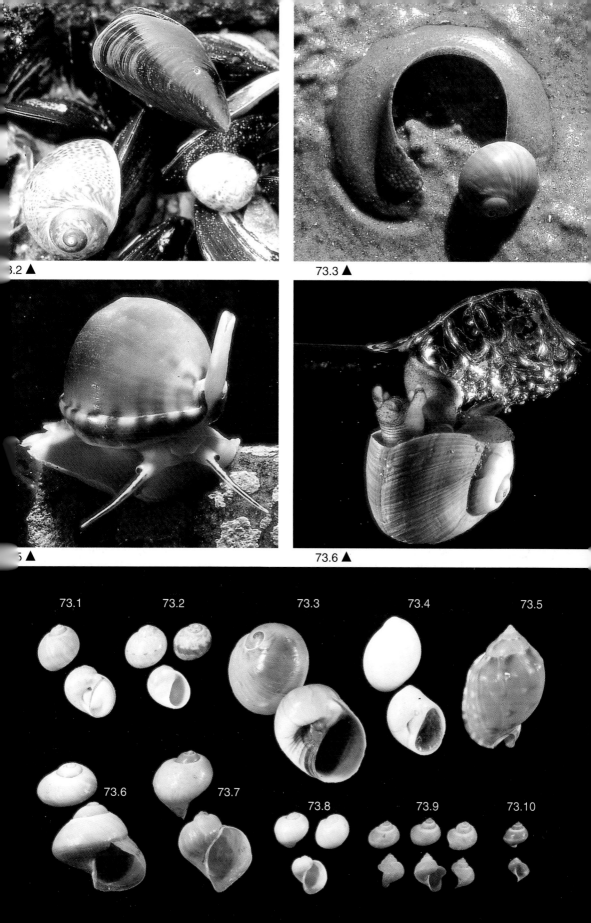

73.2 ▲

73.3 ▲

5 ▲

73.6 ▲

73.1   73.2   73.3   73.4   73.5

73.6   73.7   73.8   73.9   73.10

Gastropoda: 10. Whelks

Whelks are all predators or scavengers, capturing prey with a tubular proboscis. Their shells have an oval aperture which is notched anteriorly, or even extended into a tubular siphonal canal, to accommodate a cylindrical siphon.

74.1 Granular frog shell   *Bursa granularis*

IDENTIFICATION: Outer lip thickened, ridged, armed with small teeth. An anal canal notches the posterior end of the outer lip where it joins the body whorl. Two swollen longitudinal ridges run down the sides of the shell. About 5–7 spiral rows of granules traverse the body whorl. Brown with pale nodules. SIZE: 50 mm. BIOLOGY: Found under rocks in mid-shore pools; eats other gastropods.

74.2 Pink lady   *Charonia lampas pustulata*

IDENTIFICATION: Outer lip thickened and decorated with transverse brown bars; inner lip with transverse wrinkles. Fine spiral grooves, a spiral row of nodules and 2–3 longitudinal ridges decorate each whorl. Foot mottled brown-red; tentacles orange and usually barred. SIZE: Our largest whelk, reaching 20 cm. BIOLOGY: Most common subtidally, down to 40 m. Feeds on sea urchins, starfish and sea cucumbers.

74.3 Furry-ridged triton   *Cebestana cutacea africana*

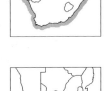

IDENTIFICATION: Shell usually short and squat, barrel-shaped; body whorl with about seven strong, grooved ridges. Outer lip corrugated or thickened with blunt teeth. SIZE: 50 mm. BIOLOGY: Often found among red-bait (*Pyura*), on which it may feed. Variable in shape, some forms being more slender than others. The shells of live animals are coated with a brown proteinaceous fur.

74.4 Pustular triton   *Argobuccinum pustulosum*

IDENTIFICATION: Thick-shelled with two longitudinal ridges; outer lip strengthened with about nine teeth. Inner lip with a strong nodule and 2–3 ridges. Pale brown with darker spiral ridges dotted with pale pustule-like nodules. SIZE: 60 mm. BIOLOGY: Common in shallow water; feeds on worms, particularly the Cape reef-worm (22.3). Secretes acidic 'saliva' that dissolves the worms' protective tubes. RELATED SPECIES: **74.5 *Ranella australasia gemmifera*** (Cape Point to Durban) has one row of large nodules, and its outer lip is toothed and white, with dark brown bars.

74.6 Branched murex   *Chicoreus ramosus*

IDENTIFICATION: A handsome, large shell. Mouth flanked by a row of spiky projections. Each whorl has three longitudinal rows of spikes, between which there are lower, nodular ridges. More delicate ridges spiral around the shell. The siphonal canal is as long as the aperture and forms an open gutter. SIZE: 18 cm. BIOLOGY: Lives in shallow water among rock rubble. Feeds on bivalves and other gastropods. Specimens from Durban have stunted spines.

74.7 Short-spined murex   *Murex brevispina*

IDENTIFICATION: Unmistakable with its extremely long siphonal canal and rows of short, blunt knobs. SIZE: 60 mm. BIOLOGY: Lives on protected intertidal sandbanks among eelgrass. Aggregates to mate, and produces communal balls of egg capsules. A mating pair is illustrated in the photograph.

74.8 Stag shell   *Pteropurpura graagae*

IDENTIFICATION: Small, white to pale brown. Body whorl with three rows of spines, including one row that edges the aperture. The shoulder spines are narrow, elongate and strongly bent, sometimes reaching the whorl above. SIZE: 25 mm. BIOLOGY: Hides beneath stones, and bores through the shells of other gastropods and barnacles to feed on them. RELATED SPECIES:
74.9 *Pteropurpura uncinaria* (Namibia to Port Alfred) is broader, the spines rarely bending around enough to reach the whorl above.

4.4 ▲

74.7 ▲

74.1

74.2

74.3

74.4

74.5

74.6

74.7

74.8

74.9

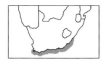

75.1 Fenestrate oyster-drill *Ocenebra fenestrata*

IDENTIFICATION: A spindle-shaped shell, readily recognised by having two spiral ridges that are pitted with oblong 'windows'. SIZE: 15mm. BIOLOGY: Lives subtidally beneath boulders, among red-bait (*Pyura*) or in the holdfasts of kelp. Diet unknown.

75.2 Mulberry shell *Morula granulata*

IDENTIFICATION: Shell grey-brown with spirally-arranged strong, black knobs aligned in nine longitudinal rows. Outer lip of aperture with four pronounced teeth. SIZE: 25 mm. BIOLOGY: Abundant in Natal and Transkei, extending up the shore as high as the oyster band. Drills holes in the shells of barnacles and molluscs; may influence the abundance of oysters by eating newly-settled spat.

75.3 Salmon-lipped whelk *Purpura panama*

IDENTIFICATION: Heavy and squat. Lip salmon-pink and armed with numerous low teeth. Body whorl has 3–4 low ridges with darker nodules. SIZE: 70 mm. BIOLOGY: Found under boulders in pools. Drills barnacles and oysters; also dislodges and consumes limpets. Females aggregate to spawn – the photograph illustrates a mating pair astride a mat of egg capsules. RELATED SPECIES:

75.4 *Thais bufo* (Transkei to Indo-Pacific) is very similar but has a shorter spire, and the body whorl often has a smooth callus where the outer lip joins it.

75.5 *Mancinella alouina* (Transkei to Indo-Pacific) is squat, with a low spire and four rows of strong nodules; inner margin of aperture yellow to pale orange.

75.6 Knobbly dogwhelk *Thais capensis*

IDENTIFICATION: Spire almost as long as the aperture. Grey, with 3–4 spiral rows of obvious, paler tubercles. SIZE: 40 mm. BIOLOGY: Hides under rocks in low-shore pools or shallow waters. Eats sea squirts (ascidians) by pushing its proboscis down their siphons, but will feed on gastropods. RELATED SPECIES:

75.7 *Thais wahlbergi* (Saldanha to False Bay) is off-white with fine spiral grooves.

75.8 *Thais savignyi* (Zululand to Indo-Pacific) has four spiral rows of angular tubercles and a mottling of dark flames; the mouth is edged with black.

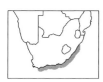

75.9 Girdled dogwhelk *Nucella cingulata*

IDENTIFICATION: White or blue-grey; usually girdled with 1–4 strong spiral ridges. SIZE: 30 mm. BIOLOGY: Common among mussels; drills neat cylindrical holes through their shells. The mussels can, however, retaliate by binding the whelks with their byssus threads. Lays pink egg capsules with a narrow stalk and two apical 'wings'.

75.10 Common dogwhelk *Nucella dubia*

IDENTIFICATION: Extraordinarily variable (hence *dubia*). Aperture usually large. Body whorl weakly ridged. Colour grey with black flames, red-brown with dotted bands, plain grey or brown. Aperture usually dark, often purple. SIZE: 20 mm. BIOLOGY: Extends almost to the high-tide mark on rocky shores. Eats limpets, barnacles and *Littorina*. Lays egg capsules from which crawling young eventually emerge. The absence of a planktonic larval stage reduces gene flow between populations, leading to enormous variations between populations.

75.11 Scaly dogwhelk *Nucella squamosa*

IDENTIFICATION: About 15 spiral ridges overlie longitudinal ridges, giving the shell a scaly (squamous) appearance. SIZE: 35 mm. BIOLOGY: Low-shore or subtidal. Lays egg capsules resembling flat clubs (75.11d). Shell usually coated by a commensal hydroid, *Hydractinia altispina* (14.1), making the shell prickly and orange. The hydroid's stinging cells repel several species of predators.

75.12 Elongate whelk *Afrocominella elongata*

IDENTIFICATION: Shell long and narrow; aperture about one-third the total length. Numerous fine spiral ridges. Pale brown, with delicate longitudinal darker streaks. SIZE: 50 mm. BIOLOGY: Subtidal, sometimes extending into the intertidal zone.

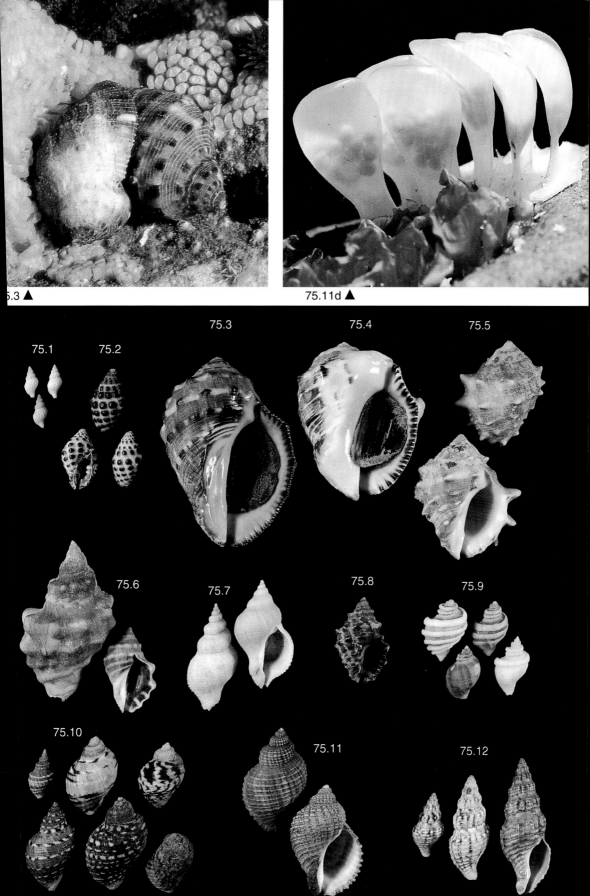

75.3 ▲

75.11d ▲

75.1 75.2 75.3 75.4 75.5

75.6 75.7 75.8 75.9

75.10 75.11 75.12

76.1 Ridged burnupena *Burnupena cincta*

IDENTIFICATION: Robust; outer surface with coarse spiral ridges, although West Coast specimens may lack these. Aperture about 1.5 times longer than the spire. Posterior end of outer lip strongly kinked inwards. Dull brown, often tinged green by algae. Aperture pale violet. SIZE: 40 mm. BIOLOGY: Scavenges on dead or injured animals low on the shore and subtidally. RELATED SPECIES:

76.2 *Burnupena lagenaria* (Saldanha to Zululand) is shorter (spire less than half aperture length), and often has wavy dark flames. The commonest species on the SE Coast, where it has obvious coarse spiral ridges (76.2a) and its aperture is violet-brown or yellow. On the W. Coast it has an even shorter spire, only a trace of spiral ridges, and a deep-purple aperture (76.2b). Lays scale-like egg capsules in domed clusters (76.2d).

76.3 Flame-patterned burnupena *Burnupena catarrhacta*

IDENTIFICATION: Shell elongate (length twice width), smooth but crossed by numerous fine spiral ridges. Juveniles often patterned with alternating dark and light flames (later obscured by erosion and algal overgrowth). Inside of aperture dark purple-brown. Outer lip pale, thin and only moderately kinked inwards at its posterior end. SIZE: 30 mm. BIOLOGY: A common scavenger. Rapidly congregates around dead or injured animals in rock pools. RELATED SPECIES:

76.4 *Burnupena* sp. An undescribed West Coast species. Smooth with fine spiral ridges like *B. catarrhacta*, but robust and squat. Its outer lip is thick and scarcely kinked inwards posteriorly. Shell dull brown; inside of aperture pale.

76.5 Papery burnupena *Burnupena papyracea*

IDENTIFICATION: Live specimens are coated by the purple or orange bryozoan *Alcyonidium nodosum,* which throws the surface into tiny bumps (see 48.5). Dead individuals (76.5a) lose this coating and are then dull brown with fine spiral ridges, and a papery outer layer that peels off. Outer lip thin, aperture white inside. SIZE: 50 mm. BIOLOGY: *Alcyonidium* is toxic and protects the whelk against predators. Abundant subtidally, *B. papyracea* can exceed densities of 200 per square metre, and may 'gang up' to consume rock lobsters (76.5b), excluding them from certain areas, with profound implications for species preyed upon by the rock lobsters. RELATED SPECIES:

76.6 *Burnupena pubescens* (NW Cape to Durban) is extraordinarily similar. In life it is similarly covered by *Alcyonidium*. Dead shells differ only in being smaller and having fine longitudinal ridges that cross the spiral ridges to create a checkered (cancellate) texture, most evident on the spire. Beach-worn shells are patterned with alternating white and brown streaks.

76.7 Long-siphoned whelk *Fusinus ocelliferus*

IDENTIFICATION: Shell elongate. Siphonal canal up to one-third the total length. Whorls decorated by delicate spiral ridges. Sometimes the shoulder ridge is strengthened and may carry nodules. Colour white to brown, sometimes flecked with darker spots. Foot bright orange-red. SIZE: 150 mm. BIOLOGY: Most common subtidally; feeds on polychaete worms, as evidenced by their bristles in its faeces. RELATED SPECIES:

76.8a *Fasciolaria lugubris lugubris* (Saldanha–False Bay) is foreshortened, with more delicate ribs and a shorter siphonal canal. The inner lip has two pleats near the siphonal canal, distinguishing it from *F. ocelliferus*. **76.8b *F. lugubris heynemanni,*** a subspecies, has a longer canal, pronounced shoulder knobs and no spiral ridges.

76.9 Forsskål's whelk *Peristernia forskalii*

IDENTIFICATION: A small but solid shell with a short siphonal canal, low ridges on the outer lip, and 11 bulging longitudinal ribs on the body whorl which are crossed by a much finer spiral texturing. Foot orange-red. SIZE: 25 mm. BIOLOGY: Very common on East Coast subtidal reefs and intertidal shores, often sheltering between oysters. Feeds on polychaetes. The normal darker subspecies *P. forskalii forskalii* (76.9a) is replaced south of Durban by the pure white *P. forskalii leucothea* (76.9b).

2d ▲   76.5b ▲

76.1   76.2   76.3   76.4   76.5a

a

b

76.6   76.7   76.8   76.9   a   b   a   b

Gastropoda: 11. Dogwhelks, Mitres & Strombs

Dogwhelks are scavengers, related to plough shells (Pl. 78), but tending to inhabit quieter sands or muds. The shells often bear strong axial (across the whorl) ribs and have a large shiny callus alongside the aperture. Mitres are sand-dwelling predators with spindle-shaped shells and 3–6 strong oblique pleats on the inner lip of the aperture. Wing shells, or strombids, are herbivores, which have a thickened, flared outer lip notched anteriorly to accommodate the stalked right eye.

77.1 Cape dogwhelk   Nassarius capensis

IDENTIFICATION: Shell narrow with a small, smooth, ventral callus and 12–14 strong axial ribs crossing the body whorl. Colour usually pale yellow-brown, with a darker spiral band on lower part of whorl and brown speckles. SIZE: Up to 17 mm. BIOLOGY: Found in sandy rock pools and to 30 m depth.

77.2 Shielded dogwhelk   Nassarius arcularius plicatus

IDENTIFICATION: Whorls swollen, with a distinct step between them, underside covered by a thick smooth callus shield; last whorl with about 15 strong axial ridges. Operculum with serrated margin. Colour grey-white with brown bands inside aperture. SIZE: Up to 27 mm. BIOLOGY: Scavenges on sheltered sandbanks. RELATED SPECIES: **77.3 Nassarius coronatus** (Durban northwards) has a much more restricted callus and fine ridges on the outer lip.

77.4 Lattice dogwhelk   Nassarius plicatellus

IDENTIFICATION: Shell cream to brown with broad axial and finer radial ribbing, creating a lattice-like pattern of paler ridges. Callus reduced. SIZE: Up to 25 mm. BIOLOGY: Found on sheltered sandbanks. RELATED SPECIES: **77.5 Nassarius albescens gemmuliferus** (Transkei northwards) has a shorter spire, wider callus and strong equal-sized spiral and longitudinal ribs.

77.6 Tick shell   Nassarius kraussianus

IDENTIFICATION: Shell almost globular with a short, smooth spire. Entire lower half of shell enveloped in a huge glossy-yellow callus. Upper surface sometimes matted by hydroids (*Hydractinia kaffraria*); otherwise purplish with yellow bands. SIZE: 7–10 mm. BIOLOGY: Abundant in estuarine or lagoonal mud-banks. Preys on small bivalves or scavenges.

77.7 Purple-lipped dogwhelk   Nassarius speciosus

IDENTIFICATION: Shell pale brown with purple anterior canal at the tip of the aperture. About 11 strong axial ribs on body whorl crossed by fine spiral threads to create knobbly ridges. Callus extends about half-way across the underside. SIZE: Up to 30 mm. BIOLOGY: Common in protected areas to 95 m depth. A scavenger.

77.8 Brown mitre   Mitra picta

IDENTIFICATION: A smooth spindle-shaped shell with a narrow aperture bearing 3–4 pleats on the inner lip. Off-white, streaked with brown flames, mauve when eroded. Operculum absent. SIZE: Up to 44 mm. BIOLOGY: A shallow-water predator of sipunculid worms. RELATED SPECIES: **77.9 Mitra litterata** (Moçambique–W. Transkei) has a broader yellowish shell with brown zigzag markings. It also feeds on sipunculid worms.

77.10 Variable stromb   Strombus mutabilis

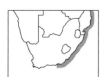

IDENTIFICATION: Shows the flared and notched outer lip that typifies the family Strombidae. Colour cream with spiral zones of yellow-brown flecks and spots. Aperture ridged and lined with orange. A row of nodules lies close to the suture line. SIZE: Up to 38 mm. BIOLOGY: In crevices or under stones and in tropical lagoons. Feeds on soft algae.

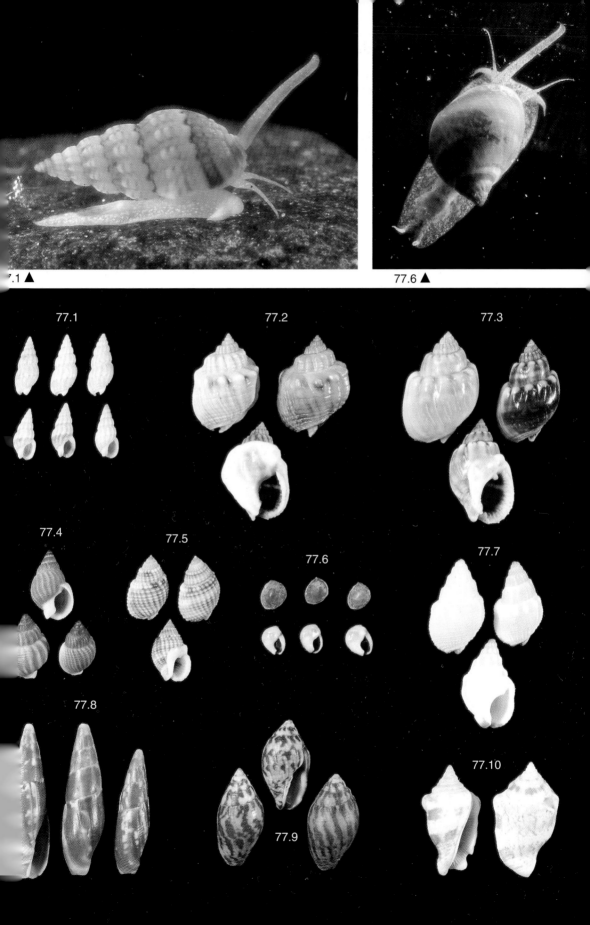

77.1 ▲

77.6 ▲

77.1

77.2

77.3

77.4

77.5

77.6

77.7

77.8

77.9

77.10

Gastropoda: 12. Plough Shells

Plough shells of the genus *Bullia* are a feature of wave-exposed sandy beaches all around southern Africa, and may be recognised by the extremely broad flat foot and reduced, often serrated operculum. All are blind and rely upon their keen sense of smell to detect carrion, around which they rapidly congregate. Some species remain permanently submerged, while others utilise the waves to 'surf' up the beach and crawl about on the wet sand with a characteristic 'rowing' movement in search of jellyfish, bluebottles and other animals cast ashore. The eggs are laid in oblong purses and are either buried deeply in the sand, or carried by the female and hatch directly into crawling young.

78.1 Annulated plough shell *Bullia annulata*

IDENTIFICATION: Pale brown with faint spiral ridges and a distinctive 'step' separating the whorls. Operculum oval and not serrated. SIZE: Up to 60 mm. BIOLOGY: Shells often washed ashore, but live animals are subtidal extending down to depths of about 100 m. RELATED SPECIES:

78.2 *Bullia callosa* (Mossel Bay–Natal North Coast) is also stepped between the whorls, but the step, aperture and heavy callus area around it are usually dark brown, while the spiral ridges on the shell vary from weak to absent.

78.3 Fat plough shell *Bullia laevissima*

IDENTIFICATION: A heavy, squat species with a short spire and very large aperture. The large shiny callus area covers most of the underside of the last whorl. Operculum tiny and oval, without serrated margins. SIZE: Up to 55 mm. BIOLOGY: Extremely common in deeper water below the surf zone, where it is exploited by a small commercial fishery. Extends into the intertidal in very sheltered bays and lagoons.

78.4 Finger plough shell *Bullia digitalis*

IDENTIFICATION: Narrow, with a long pointed spire and smooth cream shell, often tinged with violet. Operculum with serrated margins. SIZE: Up to 60 mm. BIOLOGY: The dominant plough shell on Atlantic shores. Lives buried low on the shore, emerging to 'surf' upshore (78.4a) in response to the smell of carrion. Often congregates in large numbers to feed on stranded jellyfish and bluebottles (78.4b).

78.5 Smooth plough shell *Bullia rhodostoma*

IDENTIFICATION: Similar to 78.4 but shell somewhat broader, aperture orange, and operculum having smooth margins. SIZE: Up to 55 mm. BIOLOGY: The dominant plough shell on the South and East coasts, congregating in enormous numbers around carrion cast up on exposed sandy beaches. Lighter than *B. digitalis*, and surfs higher up the shore. Slow-growing, reaching 10 mm after one year and 40 mm after 10 years.

78.6 Pure plough shell *Bullia pura*

IDENTIFICATION: Shell pale pinky-brown, sometimes with brown markings on lower whorls; sculptured with fine spiral ridges. Operculum has smooth margins. SIZE: Up to 34 mm. BIOLOGY: Usually subtidal, occasionally around low water of spring-tides. Broods its young under its foot, releasing them as fully formed juveniles.

78.7 Pleated plough shell *Bullia natalensis*

IDENTIFICATION: Shell smoothly tapering with flattened whorls, each whorl distinctly pleated just below the suture or join line. Colour light-brown, darker between the pleats. Operculum not serrated. SIZE: Up to 66 mm. BIOLOGY: The most abundant plough shell in the surf zone of sandy beaches in Natal. RELATED SPECIES:

78.8 *Bullia mozambicensis* (Moçambique–Natal South Coast) is ornamented by shallow spiral grooves, as well as pleats, which also extend further across each whorl than in *B. natalensis*.

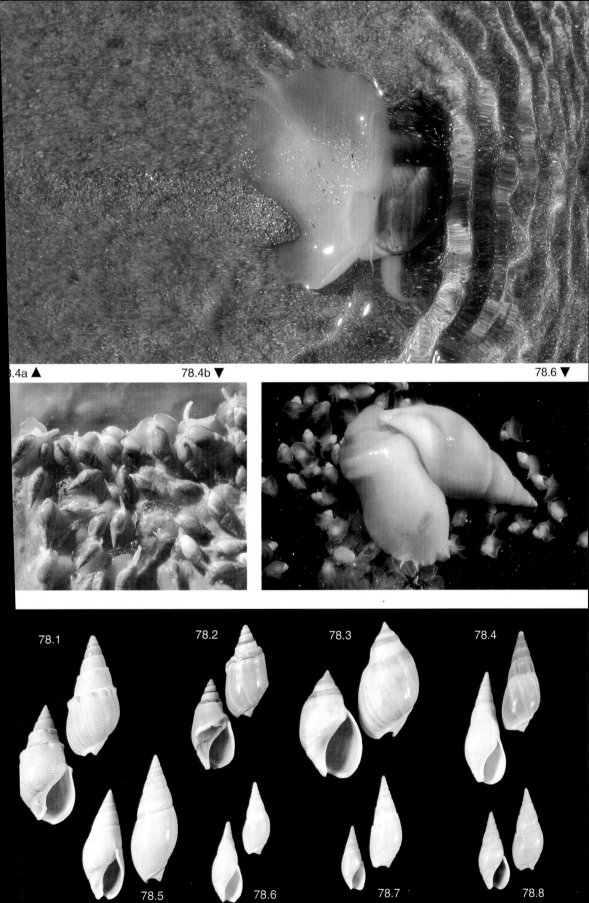

78.4a ▲   78.4b ▼   78.6 ▼

78.1

78.2

78.3

78.4

78.5

78.6

78.7

78.8

Gastropoda: 13. Olive Shells & Marginellas

Marginellas have smooth, shiny shells with an elongate, narrow aperture three-quarters or more of the total shell length. The spire is correspondingly short. The outer lip is strengthened, and the inner lip has 3–5 ridges or 'pleats'. Olive shells have a similar long thin aperture but lack these pleats.

79.1 Carolinian olive shell   *Oliva caroliniana*

IDENTIFICATION: Cylindrical, smooth and shiny, and attractively decorated with spiral dots and streaks. Spire sharply pointed but extremely short. Aperture violet; interrupted at its hind end (next to the spire) by a groove for the anus. SIZE: 45 mm. BIOLOGY: Occurs subtidally in sandy gulleys or on sheltered sandbanks. Scavenges on dead animal matter.

79.2 Pinch-lipped marginella   *Marginella rosea*

IDENTIFICATION: Ground colour white, exquisitely shot with pink or brown flares that run longitudinally or spirally. The most distinctive feature is a constriction of the outer lip near its posterior end: a 'pinching' that narrows the lip. Foot grey or cream, with vivid white streaks and red dots. SIZE: 30 mm. BIOLOGY: Lives low on the shore or in the shallow subtidal, usually concealed under boulders. Captures small gastropods with its proboscis, apparently paralysing them with a toxin. Often it will transport its prey attached to the posterior tip of its foot. RELATED SPECIES:

79.3 *Marginella piperata* (Cape Point to Natal North Coast) is smaller, and lacks the 'pinch' on the lip. Colour of shell extremely varied: usually white to cream with flares or bands of brown.

79.4 *Marginella ornata* (Port Elizabeth to Transkei) is more 'chubby', grey to red-brown, with three pale spiral bands. Shells of this species are quite commonly washed ashore, but no live animals have yet been recorded.

79.5 *Marginella musica* (Lüderitz to Cape Agulhas) is easily recognised, being grey to cream with about 10 narrow encircling black bands. The foot is cream-coloured and marked with scarlet lines.

79.6 Cloudy marginella   *Marginella nebulosa*

IDENTIFICATION: One of the larger marginellas, recognised by its size, the obviously angular shoulder on its shell, and the longitudinal, wavy bands of brown or grey. SIZE: 35–40 mm. BIOLOGY: Never abundant, but collected live by divers in depths of 5–75 m. It is usually found buried shallowly in sand around rocky outcrops. Eats other gastropods.

79.7 Cape marginella   *Volvarina capensis*

IDENTIFICATION: Shell small, oval, with a short, rounded apex; colour uniformly white to pale buff, occasionally golden. Foot grey with white spots and streaks. SIZE: 10 mm. BIOLOGY: Lives in sheltered sandy lagoons. Attracted to fish offal but also feeds on bivalves. Has round, domed egg capsules 2.5 mm wide, which it often attaches to other gastropods. RELATED SPECIES:

79.8 *Volvarina zonata* (Saldanha Bay to Port Elizabeth) forms small aggregations under stones low on the shore or in shallow waters, and has one broad red-brown spiral band or two narrow brown ones. Seldom achieves a size of more than 8 mm.

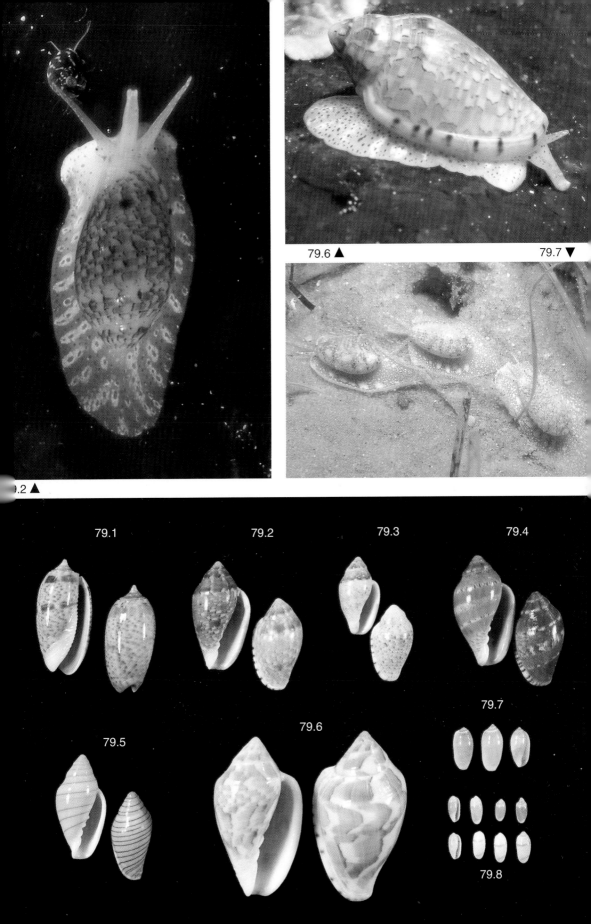

79.6 ▲   79.7 ▼

.2 ▲

79.1   79.2   79.3   79.4

79.7

79.5   79.6

79.8

Gastropoda: 14. Cone Shells

The family Conidae has conical shells, long narrow apertures, and short blunt spires. The outer lip is thin and smooth, the inner lip never toothed or pleated. Cones are predators and use hollow, harpoon-like teeth to inject a potent neurotoxin. *Conus textile* and *C. geographus* are potentially lethal to humans.

80.1 **Hebrew cone** *Conus ebraeus*

IDENTIFICATION: White to cream with 3–4 rows of dark, roughly rectangular, blotches. SIZE: 35 mm. BIOLOGY: Occurs commonly on rocky shores in low-shore pools, particularly those containing clean sand. Often buries itself in the sand. Feeds on polychaete worms.

80.2 **Natal textile cone** *Conus natalis*

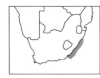

IDENTIFICATION: White to yellow with a fine brown network concentrated in spiral bands. Shoulder smoothly rounded. Spire largely brown. SIZE: 55 mm. BIOLOGY: Commonly washed up; rarely seen alive. Lives in muddy sand under boulders. Active by night, when it consumes gastropods. RELATED SPECIES:
80.3 *Conus textile*, tropical but occasionally found in Natal, has a similar 'tent' pattern, but the 'tents' tend to be flatter and the spire is more acute.

80.4 **Sponsal cone** *Conus sponsalis*

IDENTIFICATION: Shell with slightly rounded sides and a distinctly rounded shoulder. Colour white, with scattered orange-brown dots or blotches. SIZE: 25 mm. BIOLOGY: Common; aggregates in sandy pools where it buries itself, leaving only the siphon exposed. Feeds on polychaete worms. *Sponsalis* derives from the Latin for a betrothal – an allusion to its bridal-white colour.

80.5 **Livid cone** *Conus lividus*

IDENTIFICATION: Readily distinguished by the row of low knobs on the angular shoulder, and by its almost uniform yellow or brown colour, which is interrupted only by a pale, obscure band around the middle of the shell. SIZE: 50 mm. BIOLOGY: Moderately common in sandy or muddy pools. Eats worms.

80.6 **Algoa cone** *Conus algoensis*

IDENTIFICATION: Shoulder rounded and the spire 'stepped' between the whorls. Three subspecies exist. 80.6a: *Conus algoensis algoensis* (West Coast) is narrow, chocolate-brown with white blotches. Its spire is white, flamed with brown. 80.6b: *C. algoensis simplex* (Cape Point to Hermanus) is yellow with wavy, longitudinal brown bands that tend to unite on the shoulder. 80.6c: *C. algoensis scitulus* (Hermanus to Cape Agulhas) is smaller, white, yellow or pink with spiral bands of dark dots. SIZE: 50 mm. BIOLOGY: Lives in sand-covered intertidal pools, extending down to 50 m; eats polychaete worms.

80.7 **Elongate cone** *Conus mozambicus*

IDENTIFICATION: There are two subspecies. *C. mozambicus mozambicus* (80.7a) is slender and its spire acute. The shoulder has fine spiral ridges. Colour drab, plain brown or blotched with darker brown flecks concentrated on the spire and shoulder. *C. mozambicus lautus* (80.7b) has a shorter shell, less acute spire and is coloured red-brown, often with paler spiral streaks or spots. It can be confused with *C. tinianus*, but has a distinctly ridged shoulder slope. SIZE: 65 mm. BIOLOGY: *C. mozambicus mozambicus* (Lüderitz to Mossel Bay, but not Moçambique!) is the commonest cold-water cone and eats polychaete worms. *C. mozambicus lautus* replaces it east of Cape Agulhas.

80.8 **Variable cone** *Conus tinianus*

IDENTIFICATION: Sides of shell gently convex; shoulder strongly rounded; spire low and blunt. Colour extraordinarily variable – background colour pink, orange, yellow, olive or white, often blotched with brown or ringed with dots and streaks. SIZE: 50 mm. BIOLOGY: The most common and variable of southern African cones. Feeds on polychaete worms and possibly gastropods.

0.4 ▲

80.7 ▲

80.1

80.2

80.3

80.4

80.5

80.6a

80.6b

80.6c

80.7a

80.7b

80.8

Opisthobranchia : Sea Hares & Nudibranchs

This diverse group of gastropod molluscs includes the bubble shells, sea hares and nudibranchs (sea slugs), and is characterised by a reduction or total loss of the shell. With their soft body parts exposed, many opisthobranchs defend themselves by secreting toxic chemicals or re-utilising stinging cells derived from their prey. The flamboyant colour patterns of many nudibranchs warn potential predators of their unpleasant taste. Most opisthobranchs have a second pair of sensory tentacles, called rhinophores, on top of the head. The original gills are often lost and may be replaced by a plume of secondary gills around the anus, or by finger-like projections on the back called cerata. Unlike most molluscs, opisthobranchs are hermaphrodites, although incapable of self-fertilisation. Eggs are laid in jelly-covered strings or ribbons, and hatch into planktonic larvae. Most species are specialised predators, making them difficult to keep in aquariums. Nearly 300 species are found in southern Africa, many still unnamed.

81.1 Sand slug *Philine aperta*

IDENTIFICATION: A smooth, rather featureless creamy-white sea slug with a wedge-shaped body and thin translucent internal shell. SIZE: Up to 10cm. BIOLOGY: A widely distributed, active predator which glides just beneath the sand surface. Feeds mainly on small molluscs, which are swallowed whole and crushed by the muscular gizzard. Secretes sulphuric acid to deter predators.

81.2 Green bubble shell *Haminoea alfredensis*

IDENTIFICATION: Body flattened and yellowish-green with side flaps that fold over the sides of the relatively small, translucent external shell. SIZE: About 15 mm. BIOLOGY: Occurs in small colonies amongst algae and in seagrass beds in estuaries, sheltered pools and embayments. Feeds on filamentous green algae or diatoms scraped from the surfaces of rocks or seaweeds. RELATED SPECIES: Replaced by the closely related *H. natalensis* in Natal.

81.3 Striped bubble shell *Hydatina physis*

IDENTIFICATION: Shell external, delicate and oval, bearing numerous spiral brown lines. Foot large and pink with thin wavy margin edged in blue. SIZE: Averages 30 mm. BIOLOGY: A specialised predator found on protected sandflats or in rock pools. Feeds exclusively on burrowing polychaetes. RELATED SPECIES: *H. amplustre* has a white body and broad, black-edged pink-and-white bands across the shell.

81.4 Polka-dot bubble shell *Micromelo undata*

IDENTIFICATION: Shell external, white with thin wavy brown lines crossed by three thin spiral lines. Body an exquisite blue-green with white spots and a yellow margin. SIZE: 15 mm. BIOLOGY: Common in intertidal pools in KwaZulu-Natal, where it feeds on polychaete worms.

81.5 Spotted sea hare *Aplysia oculifera*

IDENTIFICATION: A typical smooth-skinned sea hare with prominent tentacles and ear-like rhinophores. Green-brown, speckled with small, white-centred black spots. SIZE: 150 mm. BIOLOGY: Common in shallow bays and estuaries. Hides by day; emerges at night to graze on seaweeds. RELATED SPECIES:
81.6 *Aplysia parvula* is smaller (60 mm) and plain brown or with tiny white dots. *A. juliana* has black splashes and a sucker at the posterior end of the foot. *A. dactylomela* has larger black circular markings and grows to 400 mm.

81.7 Wedge sea hare *Dolabella auricularia*

IDENTIFICATION: A large dull-brown sea hare with a shaggy surface. Posterior end is a flat oblique disc, tail absent. SIZE: 150–400 mm. BIOLOGY: A slow-moving nocturnal grazer found in sheltered pools or weed-beds. By day it lies partially buried in sand or mud. Emits a purple dye if disturbed.

81.1 ▲

81.2 ▼

81.3 ▲

81.4 ▼

.5 ▼

81.7 ▼

6 ▼

82.1 Shaggy sea hare *Bursatella leachi*

IDENTIFICATION: Body covered in shaggy tassels, which are short and dense in Cape animals, but much longer in Natal specimens, which also have blue circles on the body. SIZE: About 100 mm. BIOLOGY: Common in estuaries and tidal pools. Forms dense breeding aggregations, laying eggs in long green stringy tangles. Herbivorous. Emits a purple dye if disturbed.

82.2 Lemon pleurobranch *Berthellina citrina*

IDENTIFICATION: Body smooth, yellow to orange, often with white spots; shell fragile and internal; single gill under the mantle edge on the right side. SIZE: About 30 mm. BIOLOGY: Commonly found in pairs under boulders intertidally and on shallow reefs. Believed to be a scavenger, although also takes larger prey such as shrimps.

82.3 Warty pleurobranch *Pleurobranchaea bubala*

IDENTIFICATION: Rhinophores widely separated on corners of head; body grey with elevated opaque white markings; shell internal; single gill present under the mantle edge on the right side. SIZE: 60 mm. BIOLOGY: A voracious predator on other opisthobranchs on shallow reefs. RELATED SPECIES: Closely resembled by the smaller (25 mm) *P. tarda*, which has a smooth dorsal surface. *P. xhosa* is a deep red colour.

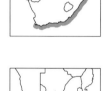

82.4 Sea swallow *Glaucus atlanticus*

IDENTIFICATION: Readily recognised by its unusual habits and lateral tufts of silverblue processes. SIZE: 30 mm. BIOLOGY: Floats upside down on the surface of the ocean, gulping air bubbles to help keep afloat. Feeds on bluebottles and their relatives. Normally occurs in the open ocean, but often cast up on the shore following onshore winds. Can utilise the stinging cells of its prey for its own defence.

82.5 Airbreathing sea slug *Onchidella capensis*

IDENTIFICATION: Body is a flattened, mottled brown hemisphere with a thick warty skin, resembling a shell-less limpet. The head has two short tentacles, and a lung opens posteriorly behind the anus. SIZE: Under 10 mm. BIOLOGY: Airbreathing slugs common in the upper intertidal. Large numbers share a communal shelter, emerging to feed on diatoms when the air is moist and the tide low. They then return faithfully to the same shelter. RELATED SPECIES: The much larger *Peronia peronii* (40 mm) occurs in northern Natal.

82.6 Cowled nudibranch *Melibe rosea*

IDENTIFICATION: The unusual hooded or cowled form of the head is unique. Colour ranges from white to orange-red, often with opaque white patches. SIZE: Typically 30–40 mm. BIOLOGY: The most common intertidal nudibranch in the Cape. The cowl surrounds the mouth and is used like a basket to trap crustacean prey. Lays a tall, collarlike egg mass on the underside of boulders in rock pools, often high in the intertidal.

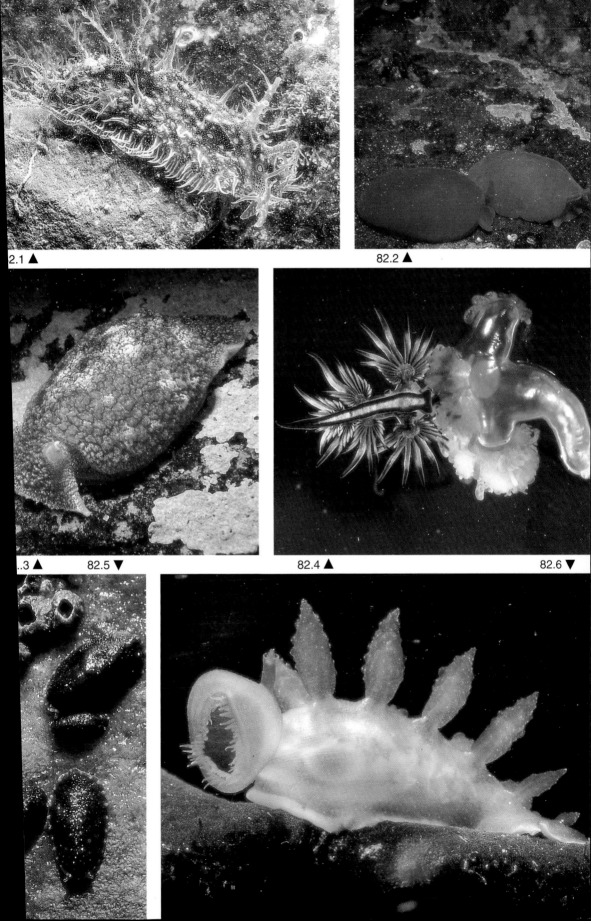

82.1 ▲

82.2 ▲

..3 ▲   82.5 ▼

82.4 ▲

82.6 ▼

83.1 Gas flame nudibranch *Bonisa nakaza*

IDENTIFICATION: A large and abundant species with a dense covering of yellow or pink cerata, which may or may not have blue tips. SIZE: Typically 50–80 mm. BIOLOGY: Conspicuous on shallow reefs in the Western Cape, where it is a favourite subject of underwater photographers. Feeds on bryozoans. Lays a highly convoluted mass of white eggs.

83.2 Silver nudibranch *Janolus capensis*

IDENTIFICATION: Back surrounded by silver-topped, grey, club-shaped projections or cerata. SIZE: 15–25 mm. BIOLOGY: Common intertidally and on shallow reefs; feeds on bryozoans, particularly upright, bushy species such as *Menipea triseriata* (49.5). The egg mass is a globular, convoluted string of beautiful white egg capsules, each containing 30–40 eggs.

83.3 Four-colour nudibranch *Godiva quadricolor*

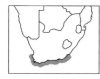

IDENTIFICATION: The paired bunches of painted cerata with their striking bands of brown, orange, blue and yellow, readily characterise this beautiful nudibranch. SIZE: 20–30 mm. BIOLOGY: A voracious predator that feeds on anemones and on other nudibranchs. Occurs intertidally and on shallow rocky reefs, extending down to a depth of about 15 m.

83.4 Coral nudibranch *Phyllodesmium serratum*

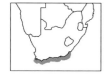

IDENTIFICATION: Recognised by the paired bunches of long, bright-orange or red cerata with their opaque off-white or luminous blue surface sheen. A white stripe runs along the length of the back and head. SIZE: Typically 30–40 mm. BIOLOGY: Locally common, from the extreme low-shore down to about 30 m. Feeds on the soft tissues of sea fans.

83.5 White-tipped nudibranch *Cratena capensis*

IDENTIFICATION: Cerata arranged in rows across the back and are brown or red with white tips. Body white, with characteristic red or orange patches on the side of the head. SIZE: 10–20 mm. BIOLOGY: A small but common nudibranch that feeds on a range of hydroid species. The colour of its cerata varies, depending on the prey species that has been consumed.

83.6 Indian nudibranch *Aeolidiella indica*

IDENTIFICATION: A pair of orange lines extends backwards from the head, separating on the back to surround white patches. Tentacles short, cerata club-shaped, grey or pink with pale tips. SIZE: 10–40 mm. BIOLOGY: A common intertidal and shallow-reef nudibranch, which preys on the anemone *Anthothoe stimpsoni* (3.2).

1 ▲

83.2 ▼

83.4 ▼

83.6 ▼

84.1 **Crowned nudibranch** *Polycera capensis*

IDENTIFICATION: Head with a fan of six gold-tipped finger-like processes. Further gold processes arise on either side of dorsal tuft of gills. Body white with a variable pattern of longitudinal dark and yellow stripes. SIZE: 25–50 mm. BIOLOGY: Feeds on bryozoans of the genus *Bugula* intertidally and on shallow reefs.

84.2 **Orange-clubbed nudibranch** *Limacia clavigera*

IDENTIFICATION: White with orange spots on the back, and orange tips to the gills and club-like marginal projections. SIZE: 10–25 mm. BIOLOGY: Found on large algae in shallow waters, where it feeds on encrusting bryozoans of the genus *Membranipora* (48.6 & 48.7). Lays white egg masses in the form of flat, spiralling ribbons.

84.3 **Spanish dancer** *Hexabranchus sanguineus*

IDENTIFICATION: A well-known tropical nudibranch recognised by its large size and flat orange-to-red body. Noted for its ability to swim using undulating movements of the brightly-coloured body margin, which is unrolled when the animal is disturbed. One of many species of the suborder Doridacea – recognised by the daisy-like tuft of gills that surrounds the anus near to the hind end of the body. SIZE: 70–100 mm or more. BIOLOGY: An unselective predator found on coral reefs.

84.4 **Cape dorid** *Hypselodoris capensis*

IDENTIFICATION: Body pale with white lines along the back and a broken blue margin. Rhinophores and tips of gills orange. SIZE: About 40 mm. BIOLOGY: Feeds on a light-blue sponge intertidally and on shallow reefs. RELATED SPECIES: Replaced in Natal by the very similar *H. carnea*. The purple band around the front of the head is continuous in *H. carnea* but usually broken in *H. capensis*. Microscopic details of the radular teeth allow more certain identification, *H. carnea* having tiny denticles on most of its teeth.

84.5 **Mottled dorid** *Hypselodoris infucata*

IDENTIFICATION: Extremely variable in colour. Some individuals have a blue-black background colour with a mottling of bright yellow spots and streaks. Others are much paler (as illustrated here). The rhinophores and edges of the gills are always red. SIZE: 35 mm. BIOLOGY: Found intertidally and in shallow water on tropical reefs.

84.6 **Polka-dot chromodorid** *Chromodoris annulata*

IDENTIFICATION: Easily distinguished by yellow spots and purple rings around the gills and rhinophores. SIZE: 20–30 mm. BIOLOGY: Found in intertidal rock pools. Habits unknown. RELATED SPECIES: One of the more than 100 members of this genus of usually brightly-coloured, tropical nudibranchs, of which at least 16 species occur in southern Africa.

84.7 **Gaudy chromodorid** *Chromodoris vicina*

IDENTIFICATION: Background colour cream to golden brown, attractively spotted with purple and yellow and lined with a blue-to-purple margin. Gills and rhinophores brown to purple-blue. SIZE: 20–30 mm. BIOLOGY: Little is known about this species, which has been recorded only from Natal and Tanzania. It occurs subtidally in shallow water, usually among sponges.

84.8 **Warty dorid** *Doris verrucosa*

IDENTIFICATION: Body is a low, warty yellow hump with 8 gills arranged in a circle around the anus. SIZE: Typically 20–30 mm. BIOLOGY: Feeds on the sponge *Hymeniacedon perlevis* (1.4), which it resembles closely in colour. RELATED SPECIES: One of numerous rather featureless humpy dorids distinguished by surface texture and colour pattern.

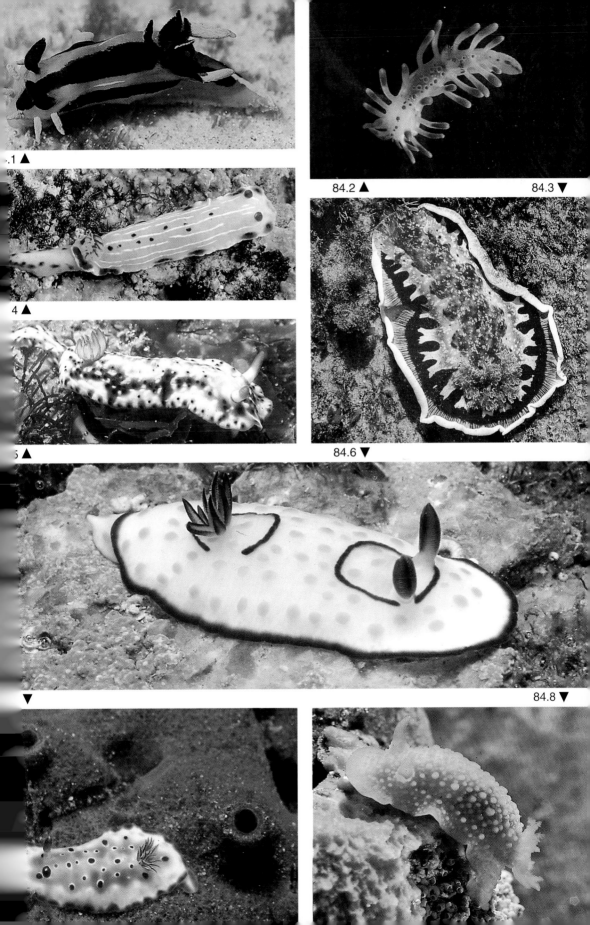

.1 ▲

84.2 ▲

84.3 ▼

4 ▲

84.6 ▼

5 ▲

▼

84.8 ▼

85.1 **Dotted nudibranch** *Jorunna zania*

IDENTIFICATION: A large white dorid with prickly black spots and a prominent plume of black-veined gills. SIZE: About 60 mm. BIOLOGY: A conspicuous resident of rock pools on the Natal coast, where it feeds on sponges. The egg mass is a tall, white, undulating collar.

85.2 **Black nudibranch** *Tambja capensis*

IDENTIFICATION: A tall, elongate nudibranch easily recognised by its blue-black body colour and green outline. SIZE: Typically 30–40 mm. BIOLOGY: One of the more common nudibranchs seen by divers in the Cape. Feeds on bryozoans, especially the bushy *Bugula dentata* (49.1). Lays a rose-like spiral ribbon of bright yellow eggs.

85.3 **Iridescent nudibranch** *Notobryon wardi*

IDENTIFICATION: Brown with iridescent green spots. The two sets of fine branching gills and flanking pairs of flattened body lobes are unique. SIZE: 20–40 mm. BIOLOGY: A common intertidal species which feeds on hydroids. Swims readily by thrashing its body from side to side.

85.4 **Ridged nudibranch** *Phyllidia* sp.

IDENTIFICATION: Black with a white, knobbly ridge along the mid-line and others radiating down the sides of the body. Tips of knobs and rhinophores yellow. Body firm to the touch. SIZE: 40–60 mm. BIOLOGY: This apparently undescribed species is one of many gaudily-coloured species in this genus which are known to produce toxic chemicals from glands in the skin. The bright colours probably warn predators that they are poisonous to eat. Thought to feed on sponges. RELATED SPECIES: *Phyllidia coelestris* (previously called *P. varicosa*) has two longitudinal blue ridges on the back. Two other undescribed pink-ridged species also occur in Sodwana. Some *Phyllidia* species are so toxic that their mere presence in an aquarium can kill the fish confined with them.

85.5 **Plant-sucking nudibranch** *Elysia viridis*

IDENTIFICATION: A small green nudibranch covered in tiny metallic spots. The sides of the body are extended into two leaf-like flaps. SIZE: Up to 10 mm. BIOLOGY: Pierces the cell walls of the alga *Codium* and sucks out all contents, including the chloroplasts. These continue to photosynthesise inside the mollusc's tissues, providing it with nutrition for several weeks and colouring it green. RELATED SPECIES: At least fifteen *Elysia* species with different colour patterns occur in southern Africa, many of which remain unnamed.

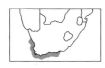

85.6 **Umbrella pleurobranch** *Umbraculum umbraculum*

IDENTIFICATION: Unmistakable large orange foot covered in pale knobs and topped by an unusual flat external shell which does little more than protect the gills. SIZE: 40–60 mm. BIOLOGY: Usually found clinging to rocks in low-shore pools and gullies but often buries in sand or mud during the day. Feeds on sponges.

85.7 **Wing-footed opisthobranchs** *Cavolinia* spp.

IDENTIFICATION: Tiny bubble-like transparent shells with a curved mouth and three stubby horns. SIZE: 3 mm. BIOLOGY: Planktonic, swimming by undulation of two flat expansions of the foot. Shells litter the drift-line of sandy shores. RELATED SPECIES: *Creseis acicula,* another member of the order Pteropoda, has a slender transparent tusk-like shell.

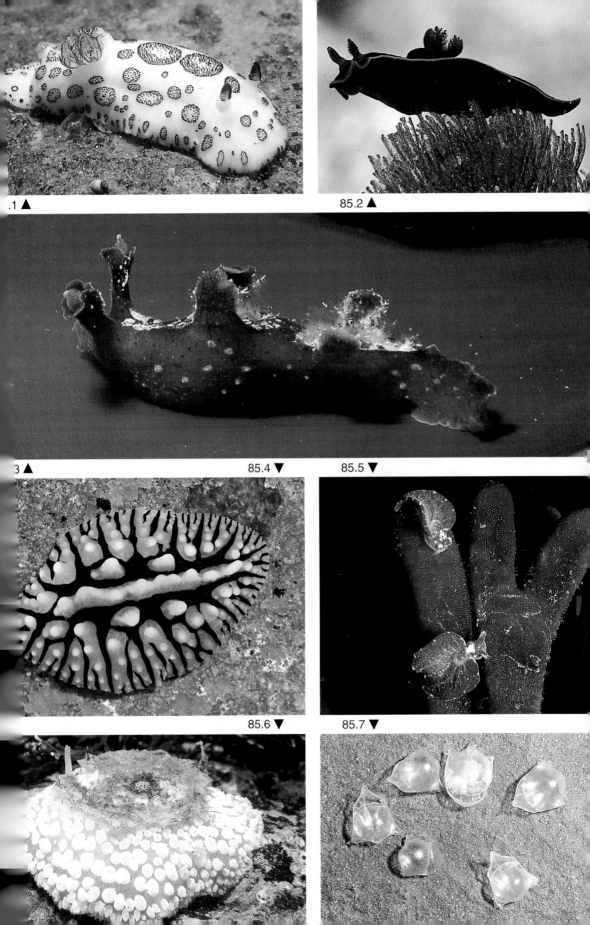

85.1 ▲

85.2 ▲

85.3 ▲

85.4 ▼

85.5 ▼

85.6 ▼

85.7 ▼

Cephalopoda : Octopus, Squid & Cuttlefish

The cephalopods are the most advanced molluscs and rank amongst the most sophisticated of all invertebrates. Many have reduced their shells to an internal flotation device or a thin internal skeleton. Octopuses have virtually lost their shell. The head and foot are merged, and the foot is divided into 8 or 10 arms with suckers. The mouth lies between the arms and has a strong parrot-like beak that tears into the prey and may inject a toxin. No southern African species are dangerous to humans, but one Australian species is lethal. The mantle that covers the gut and gonads forms a sheath into which water is drawn and then forcefully expelled, jetting the animal backwards. In prehistoric times the ancestors of squid and cuttlefish dominated the seas. Even today the abundance of squid has made them a target for commercial fisheries. They include the largest invertebrates known: the giant squids, which reach a length of 20 m.

86.1 Paper nautilus *Argonauta argo*

IDENTIFICATION: Body with eight arms, two of which are flattened and responsible for secreting and holding the shell. The shell is flat, with two keels around the edge which are studded with low, conical knobs. The sides are rippled with radiating ridges. SIZE: 10 cm. BIOLOGY: Only the female produces a shell. The evolutionary history of the paper nautilus indicates that its ancestors lacked a shell. The ability to manufacture a shell has been re-evolved, but the 'new' shell is not at all comparable with that of other molluscs. The female floats near the surface of the sea, and deposits thousands of eggs in the shell, which serves primarily as a brood chamber. The male is minute, shell-less, and planktonic. RELATED SPECIES: Two more rare species occur in our region: *Argonauta nodosa* has lumpy radial ridges. *A. boettgeri* is smaller, with coarser radial ridges and large, rounded nodules on the keels, which are usually smoky-black.

86.2 Brush-tipped octopus *Aphrodoctopus schultzi*

IDENTIFICATION: Shell-less, with eight arms that have only a single row of suckers. There are also tiny brush-like structures at the tips of the arms, but they are not obvious. SIZE: 20 cm. BIOLOGY: Little is known about this species. It occurs in shallow water and feeds on small crustaceans. Previously known as *Eledone thysanophora*.

86.3 Common octopus *Octopus vulgaris*

IDENTIFICATION: All *Octopus* species have eight arms, each with two rows of suckers, and lack a shell. The species are distinguished by the tip of the male's right third arm (the hectocotylus arm), which is suckerless and transmits sperm to the female. In *O. vulgaris* this structure is tiny (less than 2% of the arm length) and spoon-shaped. SIZE: 60 cm. BIOLOGY: The most common shallow-water octopus, extending down to 200 m. Exploited as a source of bait. Preys on crabs, rock lobsters and shellfish. Hides in cavities, which it defends territorially against other octopuses. The entrance to its hole is often littered with the discarded shells of its prey. The female lays clusters of eggs which she defends and aerates by blowing water over them. There is no larval stage: miniature octopuses emerge directly from the eggs. They grow rapidly, become sexually mature in a few months and reach a maximum mass of about 6 kg within a year. Few live longer. RELATED SPECIES: Octopus species are difficult to tell apart.

86.4 *Octopus magnificus*, the giant octopus, is one of the largest in the world, with arms of 3 m. It is a regular by-catch for South Coast trawlers. Its hectocotylus arm ends in a long groove that exceeds 10% of the arm length. Another characteristic is that the outer half of its gills has 13 filaments (there are 9–11 in *O. vulgaris*).

86.5 Ram's horn shell *Spirula spirula*

IDENTIFICATION: Shell cylindrical, forming a loose, flat spire, and made up of a series of chambers (86.5d). Animal rarely seen: squid-like, with 8 short arms and two long tentacles. SIZE: 25 mm. BIOLOGY: The live animal hangs head-down and varies the gas in its shell to regulate its buoyancy. The gas-filled shells wash ashore frequently.

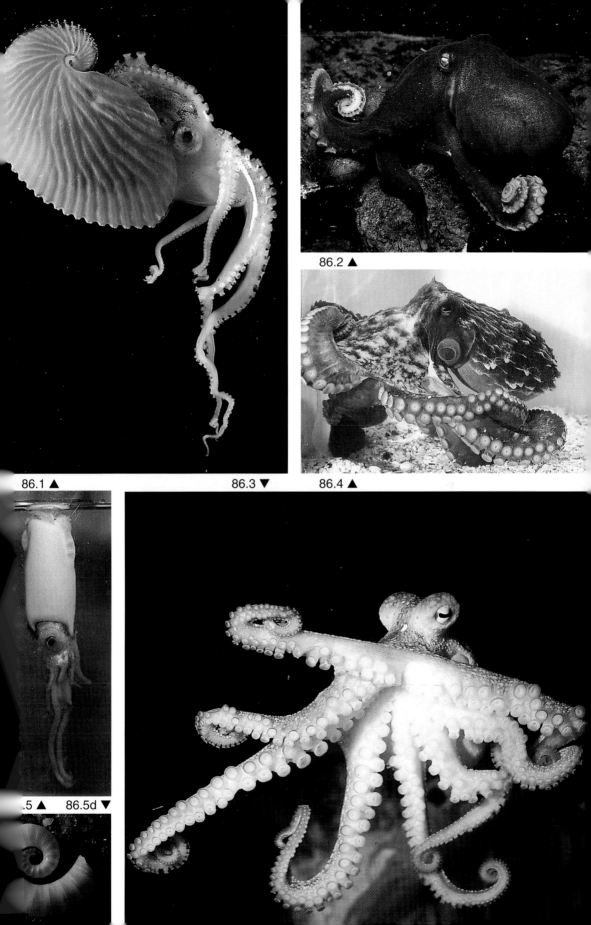

86.1 ▲

86.2 ▲

86.3 ▼   86.4 ▲

.5 ▲   86.5d ▼

87.1 **Common cuttlefish** *Sepia vermiculata*

IDENTIFICATION: Body elongate, with a fin running down each side. Pale below; upper surface smooth but decorated with constantly changing rippling bars of colour. Head with eight short arms, and two very long tentacles which can be retracted and concealed in 'pockets' below the eyes. The tentacles end in a club, which has about eight oblique rows of suckers including one row of enlarged suckers. All the suckers on the short arms are similar in size. Shell reduced to a chalky cuttlebone housed inside the body. It is flat and smoothly oval with a sharp posterior spine. SIZE: Body 15 cm. BIOLOGY: The largest of the southern African cuttlefish, and particularly common in sheltered lagoons and estuaries. Captures fish by shooting out its tentacles with speed. Lays small bunches of pea-sized black eggs. The cuttlebone is used to regulate buoyancy by modifying its gas and liquid content. RELATED SPECIES: *Sepia papillata* has three extremely large suckers on the club of the tentacles, equal in width to the club itself. Its shells are often washed ashore, and resemble those of *S. vermiculata* except that the spine is reduced to a rounded knob. *Sepia simoniana* has a similar shell but very long tentacular clubs with numerous, minute, equal-sized suckers.

87.2 **Tuberculate cuttlefish** *Sepia tuberculata*

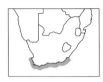

IDENTIFICATION: Upper surface of body roughened by soft papillae and the lower surface by two oval glandular patches. Some of the suckers on the club of the tentacles are enlarged, but only to about half the width of the club. The cuttlebone is oval and has no posterior spine. SIZE: 80 mm. BIOLOGY: Lives mainly in shallow water, often in intertidal pools or gulleys; sticks to rocks with its glandular patches. RELATED SPECIES: The southern cuttlefish, *Sepia australis* (Namibia to Port Alfred), is abundant in deeper waters (50–100 m). It is recognised by its smooth body, purple pigmentation on the lower surface, a narrow red-brown band down the base of the fins, and a light-emitting organ in the mantle cavity. The central suckers on the short arms are much larger than those in the rows on either side. The shell tapers very obviously to a sharp spine at the rear end. *Sepia typica* (Saldanha to False Bay) is easily recognised by its small size at maturity (30 mm) and two rows of large pores that run down the sides of the 'belly'.

87.3 **Chokka** *Loligo vulgaris reynaudii*

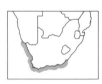

IDENTIFICATION: All squid are more slender and delicate than cuttlefish, and have reduced their shells to a transparent 'pen' that lies just under the skin of the dorsal surface. Eight short arms and two very long retractable tentacles ring the mouth: all have suckers. The chokka is distinguished by its relatively long diamond-shaped fins, which cover more than half the length of the mantle, and by the fact that the largest of the suckers on the tentacle are smoothly ringed, not finely toothed. Unlike other squid, its eyes are covered with a membrane that is continuous with the skin covering the head. SIZE: 20–30 cm. BIOLOGY: The most abundant squid in southern African waters. Becomes concentrated in bays between Cape Point and Port Elizabeth in summer when it breeds, and sustains an important squid fishery between Plettenberg Bay and Port Elizabeth. It is caught by 'jigging' with lures, but a significant by-catch is also made by trawlers. Squid are active predators and feed mainly on small fish. Eggs are laid in sausage-like strings on the sea floor. The young are transported westwards by the Agulhas Current and mature on the Agulhas Bank; adults actively migrate eastwards to the spawning grounds the following summer. The chokka is a significant predator on small fish, and is itself immensely important as a food-source for several sharks, fish, seabirds and marine mammals. RELATED SPECIES: The Indian Ocean squid, *Loligo duvaucelii* (Port Alfred to Moçambique), has shorter fins, less than half the mantle length, and two light-emitting organs on the ink sac in the mantle cavity. Prawn-trawlers catch it off Durban. The diamond squid, *Thysanoteuthis rhombus,* is one of the largest squid in our waters, with a body length of 1 m and a mass of 20 kg. Its fins cover the full length of the mantle and have a diamond-shaped outline.

1 ▲

▲   87.3 ▼

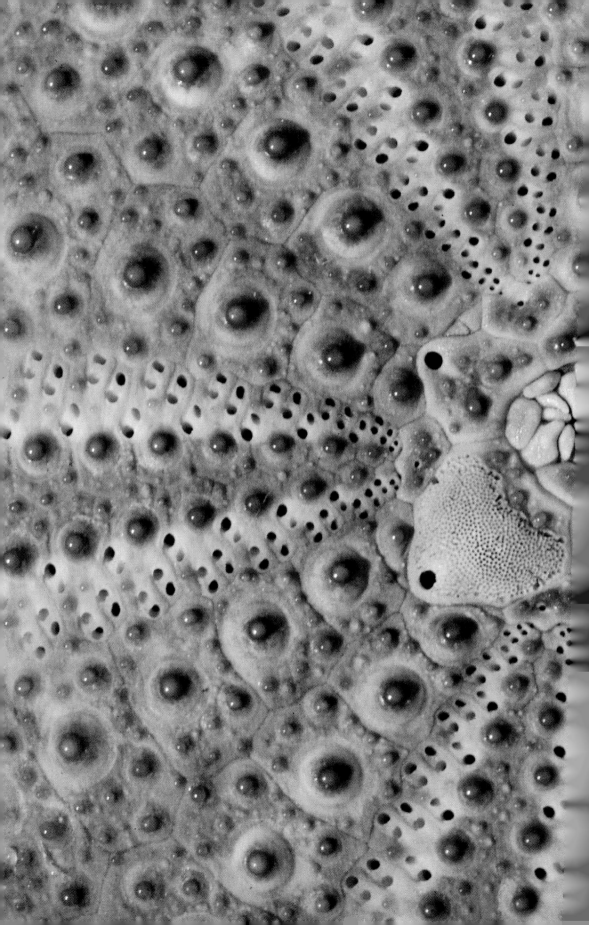

Echinoderms

The phylum Echinodermata includes the starfish, sea urchins, brittlestars, sea cucumbers and feather stars. As their name implies, they have spicules or spines in their skins (*ekhinos*, a hedgehog, and *derma*, skin), although this feature is developed to varying degrees in the different groups. Echinoderms have a radial symmetry, ie. they have no font or rear end but radiate from a central point. Curiously, this radial symmetry is acquired only during their adult life. The larvae are biliaterally symmetrical like most other animals, having similar left and right sides. Echinoderms reproduce sexually. Most release eggs and sperm into the water where fertilisation occurs. Such external fertilisation is a wasteful affair, and huge quantities of gametes have to be shed to compensate for this. The larvae are attractive, planktonic creatures, transparent and with long slender arms.

The starfish, class Asteroidea, are the most familiar echinoderms and have a flattened central body with five or more projecting arms. Under each arm lie delicate cylindrical tubefeet with suckered tips, which provide the main mode of locomotion. Some starfish feed on detritus or microalgae, but many are predators. Among the predators, some control mussels and prevent them from monopolising the shore, and the crown of thorns, *Acanthaster planci*, is infamous for the devastating effects it can have on corals.

Feather stars, class Crinoidea, are a striking feature of the subtidal world explored by divers. Their tiny, round bodies are attached to rocks by short hook-like limbs, and support a handsome crown of feather-like appendages that sift the water for food particles.

Brittlestars, class Ophiuroidea, are aptly named because of the five or more slender, brittle arms that arise from their centrally placed round bodies. These arms are flexible and propel the animal with a snake-like undulation. Most brittlestars are detritus-feeders.

Sea urchins, class Echinoidea, are also well known. Their bodies are encased in a hard, calcium carbonate shell, the test, formed by fusion of the spines in the skin. Through tiny pores in this shell, tubefeet protrude and serve to move urchins around. Long protective spines project from the test. These are mounted on a ball-and-socket, and can be swivelled to point directly at any threat. In most cases the spines are harmless to humans, if irritating. A few species have long, lance-like spines with backward-pointing serrations; they can penetrate deeply and then break off, and may even require surgical removal. Some long-spined species discharge toxins through the tips of their broken spines. Like starfish, urchins are dotted with tiny pedicellaria that resemble grappling irons. These three-jawed structures are used in defence against predators, and are usually armed with poison glands. They are harmless to humans in almost all cases, but one species, *Toxopneustes pileolus*, is potentially lethal. Urchins are grazers, and the more mobile and abundant species control the growth of seaweeds. In the tropics, where predatory fish abound, most urchins shelter in crevices and catch drift-weed. Some are able to burrow into rock, creating holes in which they shelter.

The sea cucumbers, class Holothuroidea, have become sausage-shaped and lie on their sides. They retain a reminder of their ancestral radial symmetry in having five rows of tubefeet with which they attach themselves to rocks. They have also become flexible because the spines in their skin are reduced to spicules, which are easily sea cucumbers requires microscopic examination of thse spiclues, which are easily revealed by macerating a fragment of the body wall in caustic soda (potassium hydroxide). At the front of the body, sea cucumber have a series of branching feeding tentacles that gather organic particles.

Asteroidea : Starfish

Familiar echinoderms recognised by their flattened bodies, which merge imperceptibly into five (or sometimes more) relatively stout, tapering arms. The mouth is situated centrally on the under surface, while the anus is on the upper side. Each arm contains its own set of respiratory, digestive and reproductive organs, while a groove running from the mouth along the underside of each protects rows of tiny hydraulically-operated tubefeet. These are responsible for the slow creeping movement of starfish. Most species are mobile scavengers or predators, though a few are herbivorous. Some 90 species are found around southern Africa.

88.1 Sand starfish   *Astropecten irregularis pontoporaeus*

IDENTIFICATION: Body off-white to apricot, with fairly long, triangular arms becoming mauve at the tips. Each arm is edged by a double tier of large, tile-like plates. Those in the upper row each bear a single spine, while on the lower tier each have 3–4 longer spines. The tubefeet are pointed and lack suckers. SIZE: 90 mm. BIOLOGY: Lives in sand, into which it digs in search of small molluscs or crustaceans. The commonest of eight *Astropecten* species distinguished by spination patterns. RELATED SPECIES: *A. granulatus* (Durban to Maputo) lacks spines on the upper row of the tile-like side-plates.

88.2 Beaded starfish   *Pentaceraster mammillatus*

IDENTIFICATION: Colours very variable, including yellow, green, brown, red and grey. Centre of body elevated, tapering down to triangular arms. Rows of conical bead-like knobs, often of contrasting colour, run down each arm and around the body margin. SIZE: 20 cm across. BIOLOGY: A tropical species typical of seagrass beds in sheltered sandy lagoons. RELATED SPECIES: *Protoreaster lincki* (Moçambique) has similar habits and body form but more slender arms with one or two very large spines that project from the sides of each arm near its tip.

88.3 Spiny starfish   *Marthasterias glacialis*

IDENTIFICATION: Body orange or blue-grey, and covered in conspicuous spines each surrounded by a halo of minute pincer-like organs called pedicellaria. SIZE: To 20 cm or more across. BIOLOGY: A voracious predator, particularly of mussels. Hunches over its victim and extrudes its stomach through the mouth to digest the prey externally. Common on rocky shores in the Cape, where it can form large feeding aggregations.

88.4 Crown of thorns starfish   *Acanthaster planci*

IDENTIFICATION: Instantly recognisable by its great size, multiple arms and formidably spined surface. SIZE: 40 cm. BIOLOGY: A specialist predator of corals. Very dense populations have been noted on certain reefs, destroying the coral over huge areas. Originally it was believed that these outbreaks were caused by pollution, or overexploitation of predators, but examination of marine sediments for spine fragments has now shown that periodic population explosions have occurred for thousands of years, long before human interference became significant. These starfish should be handled with care since the spines are coated with an irritant poison.

88.5 Brooding cushion star   *Pteraster capensis*

IDENTIFICATION: A large, fat, orange-brown starfish with short, blunt arms. Upper surface covered by a unique soft outer skin supported by short, turret-like spines. The cavity created under the skin is ventilated through a central hole on the back and acts as a brood chamber. SIZE: About 10 cm. BIOLOGY: Found subtidally on both rock and sand. The eggs are brooded in the space under the dorsal skin and emerge as tiny starfish. Secretes copious amounts of mucus when disturbed in order to deter predators. Can be observed 'breathing' in and out to inflate the brood chamber.

88.1 ▲

88.2 ▲

88.3 ▲　　　　　　88.4 ▼　　　　　　88.5 ▼

89.1 Dwarf cushion star   *Patiriella exigua*

IDENTIFICATION: Tiny, flattened, pentagonal starfish; arms poorly developed. Dorsal surface made up of tile-like plates, each with a cluster of tiny knob-like spines. Uniformly khaki-green on the West Coast; variegated patterns of orange, brown, green and white on the SE coasts. SIZE: 20 mm. BIOLOGY: Occurs in the intertidal zone; abundant but well camouflaged. Feeds by extruding the stomach through the mouth and plastering it onto the rock to digest microscopic algae. Eggs are laid under rocks and hatch directly into tiny starfish, without an intervening planktonic larval stage.

89.2 Subtidal cushion star   *Patiriella dyscrita*

IDENTIFICATION: Very similar to *Patiriella exigua* (89.1) in shape and colour, but larger and distinguished by having an evenly granular surface texture rather than clusters of spines. SIZE: 40 mm. BIOLOGY: Because of their similar appearance, *P. dyscrita* and *P. exigua* have long been confused. However, *P. dyscrita* occurs only subtidally, in contrast to *P. exigua,* and also differs in its life cycle, for its eggs develop into normal planktonic larvae.

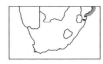

89.3 Pincushion starfish   *Culcita schmideliana*

IDENTIFICATION: A large, plump, blotchy-orange cushion star with a pentagonal body and virtually no arms. SIZE: Up to 14 cm across. BIOLOGY: A tropical form which feeds on both corals and detrital material, and is fairly common on reef flats.

89.4 Red starfish   *Patiria granifera*

IDENTIFICATION: Arms clearly tapering and flattened on their lower surface, giving them a semicircular cross-section. Colour deep orange to red, with a surface texture resembling small overlapping tiles. SIZE: Up to 80 mm across. BIOLOGY: A detritus feeder. Often found with *Henricia* (89.6), but distinguished by its tapering arms.

89.5 Blocked starfish   *Fromia elegans*

IDENTIFICATION: Colour orange, the surface granules arranged into blocks, defined by black lines (reminiscent of giraffe skin). Arms fairly long, slightly swollen close to base, then tapering evenly. SIZE: About 10 cm across. BIOLOGY: Found on shallow tropical reefs.

89.6 Reticulated starfish   *Henricia ornata*

IDENTIFICATION: Arms long, very gradually tapering and cylindrical. Colour orange to maroon with an irregularly-honeycombed surface texture. SIZE: Up to 90 mm across. BIOLOGY: Thought to feed on sponges by everting the stomach and digesting the sponge tissue *in situ*, although may also consume detritus.

89.7 Granular starfish   *Austrofromia schultzei*

IDENTIFICATION: Surface coated in granules and regularly raised into small humps. Ground colour orange, humps and tips of arms yellow. Arms broaden slightly just beyond base, before tapering evenly to blunt tip. SIZE: About 10 cm across. BIOLOGY: A resident of shallow reefs in the Southern Cape.

89.2 ▲

89.3 ▲

89.4 ▲

.1 ▲

89.6 ▲

89.7 ▼

Crinoidea : Feather Stars

Graceful echinoderms with small, soft bodies, surrounded by 10 or more elongate, upraised arms, each consisting of a central axis with numerous side branches or pinnules. Food particles trapped by the arms are passed along ciliated grooves back to the mouth, which is situated on the upper surface. A simple U-shaped gut fills most of the body cavity and ends in an anal cone close to the mouth. The gonads are in the pinnules of the arms closest to the body, and these become greatly expanded during the breeding season, eventually rupturing to release the eggs or sperm into the water. Crinoids are essentially sedentary, gripping the substratum with a ring of claw-like segmented limbs (the cirri) beneath the body, but they can crawl or even swim using their arms. Attached stalked crinoids, called sea lilies, are abundant in the fossil record, but few survive today. There are 17 southern African species, but only three of these are commonly encountered.

90.1 Common feather star   *Comanthus wahlbergi*

IDENTIFICATION: Colour very variable, from white to pink, blue or orange, often variegated. Arms rather ragged, with few side branches, varying from 10 to 22 in number and usually curled above the body. The anus is central and the mouth off-centre. Cirri number from 12 to 25. SIZE: Arm length up to about 80 mm. BIOLOGY: The most abundant southern African feather star, found under rocks or in crevices at low tide and often forming dense clusters coating shallow reefs. Tiny pentacrinoid larvae can sometimes be seen clinging to the adults. SIMILAR SPECIES: *Annametra occidentalis* (Saldanha to Knysna) has 35–40 cirri and a central mouth with an off-set anus.

90.2 Elegant feather star   *Tropiometra carinata*

IDENTIFICATION: A striking species with 10 long arms (90.2a) evenly lined with tapering rows of straight pinnules (90.2d). Arms usually purple near base, becoming yellow at tips, although the colour does vary considerably. Cirri number 20–30. SIZE: Arm length up to 15 cm. BIOLOGY: Usually found singly on shallow reefs, although becomes more abundant on deeper reefs, crowding crevices and overhangs. If displaced, it may swim with a slow beat of the arms, or fold the arms overhead, dropping elegantly to the bottom (90.2b).

90.1 ▲   90.2a ▼   90.2b ▼

90.2d ▼

Ophiuroidea : Brittlestars

Brittlestars have a flat, circular body (called the disc), with five or more long thin arms. The arms are jointed and flexible but break off easily, hence the name 'brittlestars'. The segments of the arms are each covered by 1–3 tiny plates, and their shape and number help distinguish species. The sides of the arms often have spines. Brittlestars move by snake-like undulation of their legs. The mouth lies on the lower surface of the body and is surrounded by five toothed jaws. The texture of the disc allows division of brittlestars into three groups: those with (1) granules, (2) short spines, or (3) a leathery or scaly texture. Just above the origin of each arm there are usually two enlarged scales called radial shields. Most brittlestars have minute planktonic larvae, but a few brood their young in their bodies and give birth to miniature replicas of themselves. There are 120 species in southern Africa, but only about a dozen are common.

91.1 Basket star *Astrocladus euryale*

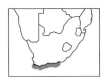

IDENTIFICATION: Immediately recognisable because the arms branch into ever-finer, delicately striped tendrils. The disc is decorated with coarse knobs that are very obvious because their colour contrasts with the disc and they are often ringed with black. SIZE: 30–50 cm. BIOLOGY: Holds its arms outstretched like a basket to catch passing animals. Often attaches itself to sea fans.

91.2 Banded brittlestar *Ophiarachnella capensis*

IDENTIFICATION: Disc granular, arms strikingly patterned with alternating dark grey and pink (or red) bands. Very short spines flank the arms. The lower surface of the disc has a single long genital slit on either side of each arm. SIZE: 45 mm. BIOLOGY: Lives under boulders near the low-tide mark. Feeds on detritus.

91.3 Serpent-skinned brittlestar *Ophioderma wahlbergi*

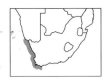

IDENTIFICATION: Disc granular; arm spines short and closely applied to the arms. Upper surface uniformly black-brown; arms occasionally banded. SIZE: 90 mm. BIOLOGY: Lives gregariously on subtidal gravelly or sandy substrates. Very common in certain areas, thousands massing together. On either side of the base of each arm there are two genital slits that lead into chambers in which juveniles are brooded until they emerge as fully-formed miniature brittlestars.

91.4 Spiny snake-like brittlestar *Ophiocoma valenciae*

IDENTIFICATION: Disc granular, the granules being taller than broad near the edge of the disc. Disc mottled, arms usually indistinctly banded. Arm spines about as long as the arm width and projecting very obviously. SIZE: 100 mm. BIOLOGY: Very common on rocky shores in Natal. Hides in crevices and extends its tentacles to feed, picking up organic particles and dead animal matter or preying on tiny animals. RELATED SPECIES: *Ophiocoma erinaceus* is black with tiny black or orange tentacles on the arms. Its arms are about 4–5 times the disc diameter. *Ophiocoma scolopendrina* has mottled or banded arms that are more than 6 times the disc diameter. Both occur from Durban northwards, and both have disc granules that are shorter than broad.

91.5 Hairy brittlestar *Ophiothrix fragilis*

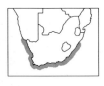

IDENTIFICATION: Upper surface of disc covered with spines; even the radial shields have spines. Arm length less than 6 times the disc diameter. Arms are fringed with long spines that project at right-angles and taper towards their tips. SIZE: 50 mm. BIOLOGY: Very common on rocky shores beneath boulders, and extends down to depths of about 100 m. Often lives in dense aggregations. Holds its arms up into the water to catch detrital particles. RELATED SPECIES: *Ophiothrix foveolata* (Zululand northwards) lacks spines on its radial shields, but is otherwise very similar. *Macrophiothrix hirsuta cheneyi* (Durban northwards) has longer arms (7–10 times the disc diameter) and its arm spines are characteristically expanded and flattened towards their tips. A thin, pale line runs down the upper surface of each of its arms, often flanked with blue.

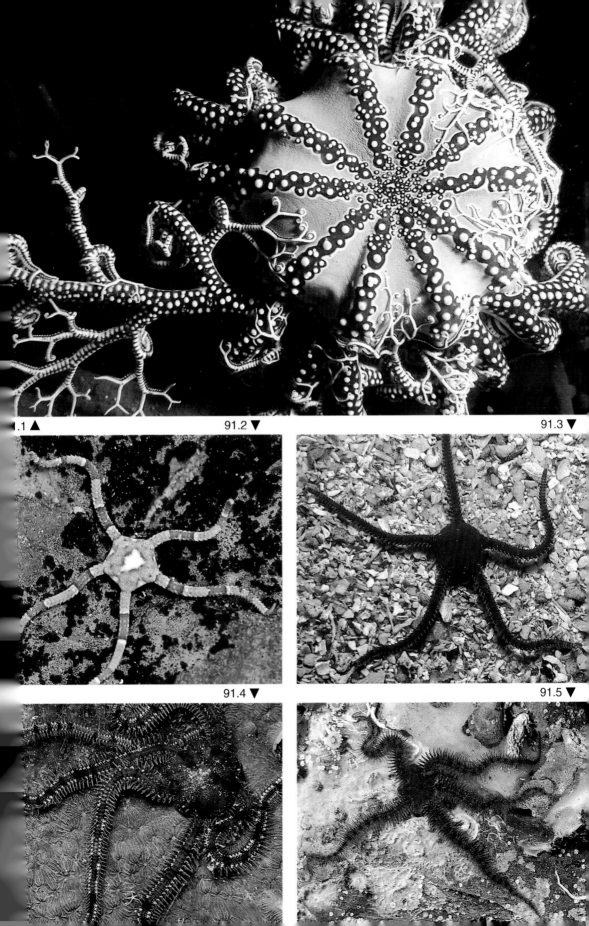

91.1 ▲

91.2 ▼

91.3 ▼

91.4 ▼

91.5 ▼

92.1 **Striped brittlestar** *Ophionereis dubia*

IDENTIFICATION: Disc appears smooth to the naked eye (although it does have scales, they are microscopic) and often has a net-like pattern including irregular 'Y' markings near the base of each arm. Arms very long and narrow, each 'joint' with a central plate and two flanking accessory plates (only visible when magnified). Arms pale, often white, with a narrow stripe across them about every fifth joint. SIZE: 70 mm. BIOLOGY: Seldom abundant; conceals its body in rock crevices and waves its arms over the rock-face. RELATED SPECIES:

92.2d *Ophionereis porrecta* (Cape Agulhas to Zululand) also has accessory plates on its arms, which are long and thin, but its disc is clearly scaly and the bands on its arms broader.

92.3 **Equal-tailed brittlestar** *Amphiura capensis*

IDENTIFICATION: Upper surface of disc lacks spines but covered with small scales. Radial shields touch one another only at their outermost tips: nearer to the centre of the disc, they diverge from one another, resembling a fat 'V' (92.3d). The individual joints of the arms are covered by a single scale. Disc usually grey to black; arms yellow to orange, sometimes with broken bands. Colour is, however, not a reliable means of identification, some individuals being brown, others red or purple (see 92.3a,b,c for an indication of the variability in colour). SIZE: 25 mm. BIOLOGY: Common under stones embedded in gravel or coarse sand; often aggregates in dense colonies. RELATED SPECIES:

92.4d *Amphioplus integer* (Port Nolloth to Maputo) shares most of the features described for *Amphiura capensis,* but its radial shields are in contact with one another for their full length.

92.5 **Scaly-armed brittlestar** *Amphipholis squamata*

IDENTIFICATION: A very small species, uniformly white or pale grey, distinguished by the arrangement of the scales on the upper surface of the arms. In most brittlestars the scales of adjacent segments are clearly in contact with one another for most of their width. In this species, the scales are small, triangular with rounded corners, and either separated from one another or only just touching. Disc scaly. Radial shields arranged in pairs that abut against one another for their full lengths. SIZE: 10 mm. BIOLOGY: Extremely common and ranging from the intertidal zone down to 175 m. Most often found in areas with gravel, sand or mud. Feeds on minute detrital particles.

92.6 **Snake-star** *Ophiactis carnea*

IDENTIFICATION: Upper surface of disc scaly, but towards the edge of the disc there are scattered short spines. The radial shields are triangular and scarcely touch one another. Often red to white in colour, sometimes with irregular bands on the arms. SIZE: 25 mm. BIOLOGY: Usually found in gravel or shell deposits. RELATED SPECIES: *Ophiactis savignyi* (Durban northwards) is closely related, but usually has six arms instead of five, is bright green, and has conspicuous dark patches on its radial shields and a white patch at the outer end of each radial shield.

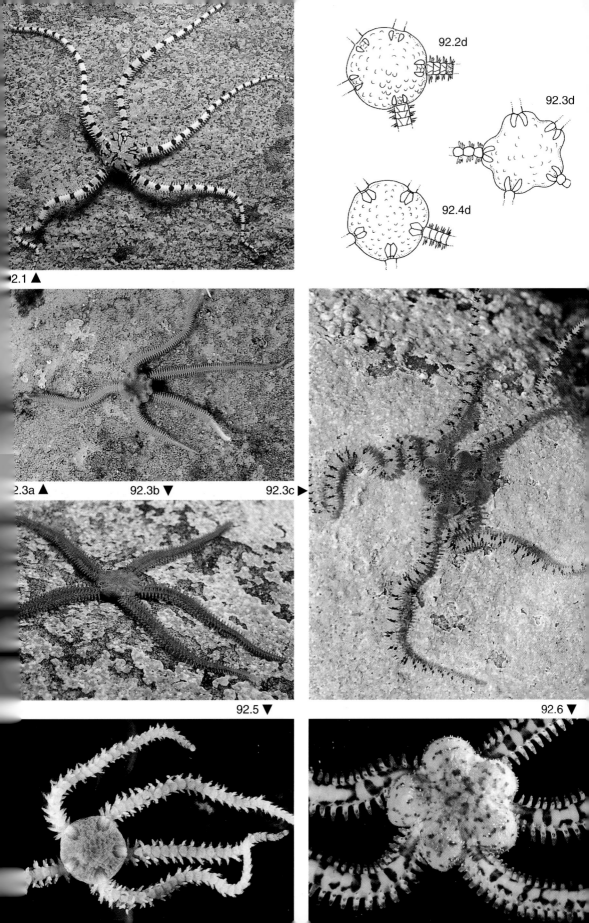

92.2d

92.3d

92.4d

92.1 ▲

2.3a ▲   92.3b ▼   92.3c ▶

92.5 ▼   92.6 ▼

Echinoidea : Sea Urchins

Most sea urchins have globular bodies encased in a calcium carbonate shell, or test, covered with spines. The mouth is situated centrally on the underside, and the anus is usually on the upper surface. Five double rows of tubefeet run from the apex down the sides of the shell. Minute defensive pedicellaria, resembling stalked pincers, are dispersed over the body. Most rocky-shore urchins are grazers, but the more flattened sand-dwelling forms feed on detritus. Some 58 species occur in southern Africa.

93.1 Needle urchin   *Diadema setosum*

IDENTIFICATION: Black urchins with extremely long, needle-sharp spines, which they wave menacingly at any intruder. Iridescent dotted blue lines run down the body between the spines; and a red ring surrounds the anus. The spines of juveniles are banded. SIZE: Test diameter up to 70 mm. BIOLOGY: Algal grazers found amongst rocks or coral. Dangerous to handle: the spines are hollow, penetrate the flesh easily and contain an irritant toxin. RELATED SPECIES: *D. savignyi* (Transkei northwards) is almost identical, more common, but lacks a red ring around the anus.

93.2 Oval urchin   *Echinometra mathaei*

IDENTIFICATION: Test oval when viewed from above, purple to black. Spines fairly long and stout, tapering evenly to a sharp tip. Spine colour purple, brown or green, sometimes with white tips. SIZE: Diameter 70 mm. BIOLOGY: The most common urchin in East Coast rock pools. A mobile grazer which occupies hollows in exposed reefs, catching drifting algae or emerging to graze at night.

93.3 Pot-hole urchin   *Stomopneustes variolaris*

IDENTIFICATION: Test circular in outline; spines about half test diameter, strong, cylindrical and obviously tapering. Colour shiny black or dark purple. SIZE: Width 100 mm. BIOLOGY: Shelters on wave-swept shores in crevices or hollows which it may excavate over many generations. Feeds on drift algae. Strong wave action erodes its spines, reducing feeding efficiency and body size.

93.4 Bicoloured urchin   *Salmacis bicolor*

IDENTIFICATION: Test circular in outline, with a dense covering of short thin spines; those around the mouth becoming markedly flattened. Spines red or purple at their bases, becoming attractively banded with yellow or white towards the tips. SIZE: Test diameter up to 70 mm. BIOLOGY: Found amongst algae in sheltered lagoons and reefs, but seldom common.

93.5 Short-spined urchin   *Tripneustes gratilla*

IDENTIFICATION: A large round urchin evenly coated with short white spines, all of roughly the same length. The tubefeet are long and numerous, often projecting well beyond the spines. SIZE: Test diameter up to 145 mm. BIOLOGY: Found in weed-beds, often concealed by pieces of algae held over the body by the tubefeet.

93.6 Cape urchin   *Parechinus angulosus*

IDENTIFICATION: Test round, densely covered in shortish, pointed spines which vary in length but never exceed about one-fifth test diameter. Colour variable, most often purple, but also green, red or pale. SIZE: 60 mm. BIOLOGY: Abundant on rocky shores in the Cape. It is an important grazer that controls the survival of newly settled kelp plants. Uses dead shells as a 'sunshade'.

93.7 Flower urchin   *Toxopneustes pileolus*

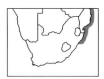

IDENTIFICATION: Distinctive, easily recognised by its beautiful flower-like pedicellaria, or pincers. Spines very short; tubefeet elongate, reaching well beyond spines. SIZE: Up to 150 mm. BIOLOGY: Its enormous pedicellaria are equipped with potentially lethal poison glands; they should not be touched with bare hands. *Toxopneustes* is a tropical reef species which often carries seaweeds or shells to shade it from the sun.

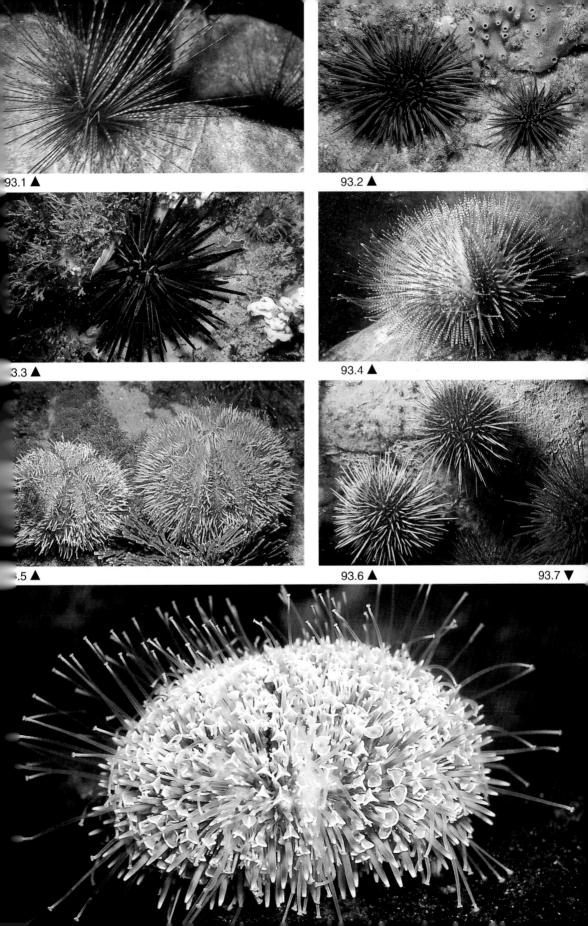

93.1 ▲

93.2 ▲

3.3 ▲

93.4 ▲

.5 ▲

93.6 ▲

93.7 ▼

94.1 Rough pencil urchin *Prionocidaris pistillaris*

IDENTIFICATION: Spines broad and flattened, the shafts covered with rows of pointed knobs. Colour initially purple, but often encrusted with algae so that colour shows only at the base of the spines. SIZE: Test up to 60 mm across. BIOLOGY: A tropical grazer. The massive spines probably provide protection from predators. RELATED SPECIES: The slate pencil urchin, *Heterocentrotus mammillatus,* has massive, smooth, attractively-banded spines often sold as ornaments.

94.2 Tuft urchin *Echinostrephus molaris*

IDENTIFICATION: A small, easily-overlooked urchin with a circular, but peculiarly top-heavy test. The lateral spines are short, while those on the top are elongate and needle-sharp, making an erect tuft. SIZE: 28 mm. BIOLOGY: Excavates a cylindrical burrow from which the long dorsal spines are extended to intercept drift food. Excess food may be stored at the bottom of the burrow.

94.3 Banded urchin *Echinothrix calamaris*

IDENTIFICATION: Test circular, armed with two distinct types of black-and-white-banded spines: longer, thicker, primary spines as long as the test diameter, and short-er, needle-like, secondary spines, with downward-pointing barbs. SIZE: Test diameter up to 130 mm. BIOLOGY: A nocturnal grazer with an irritant toxin in the hollow spines. The eye-like anal cone is often inflated like a balloon and dotted with white spots.

94.4 Heart urchin *Echinocardium cordatum*

IDENTIFICATION: Fragile test an irregular oval with a well-developed frontal depres-sion leading back to the mouth, which is positioned near the front and has a scoop-like posterior lip. Spines white and arranged around a series of depressed 'petals' con-taining the tubefeet. SIZE: 40 mm long. BIOLOGY: A detritus-feeder which burrows in fine sand, using specialised paddle-shaped spines on the underside. RELATED SPECIES: **94.5 *Spatogobrissus mirabilis*** (Lüderitz–Agulhas) is a larger, smooth-shelled heart urchin (typically 90 mm long), without a frontal depression. Shells are often found by divers in areas of coarse sand.

94.6 Pansy shell *Echinodiscus bisperforatus*

IDENTIFICATION: Flat, biscuit-like urchins with two closed slits in the back half of the test. The short fur-like spines drop off after the animal dies, revealing a characteristic petal-like pattern of holes, through which the tubefeet of the live animal extended. SIZE: 90 mm. BIOLOGY: Lies buried just below the sand surface in sheltered waters, feeding on fine organic particles sorted from the sediment. RELATED SPECIES: In the trop-ical *E. auritus* (Maputo northwards) the slots in the test extend to the edge of the shell, like deep notches.

94.7 Lamp urchin *Echinolampas crassa*

IDENTIFICATION: A large urchin with a strong, heavy, flat-bottomed shell densely coat-ed with uniformly short, yellowish spines. Mouth central. SIZE: Up to 120 mm across. BIOLOGY: Ploughs along just below the surface of coarse sands, feeding on organic particles, which it ingests along with large amounts of sand. Regularly collected by divers, especially in False Bay, where dense populations can be found.

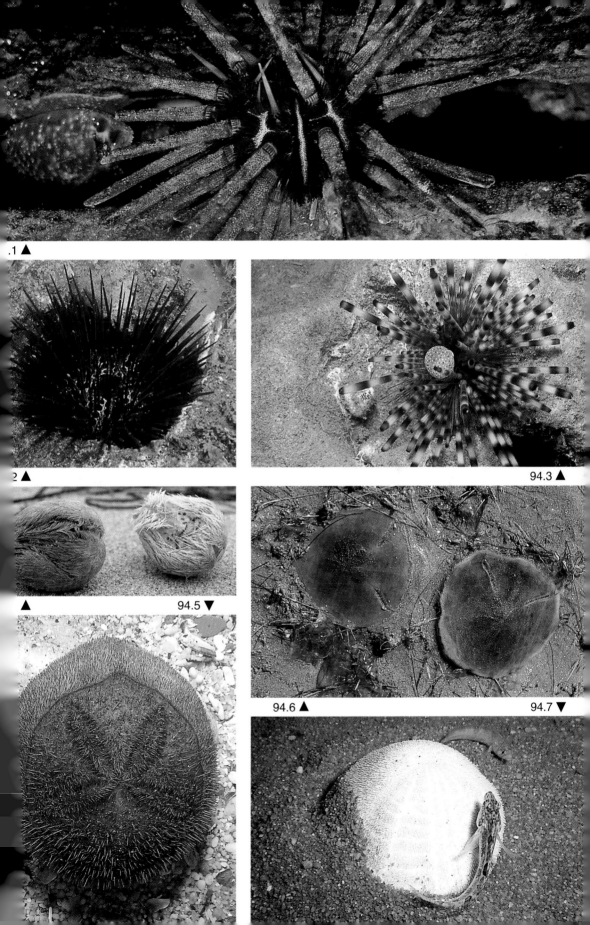

.1 ▲

2 ▲

94.3 ▲

▲   94.5 ▼

94.6 ▲   94.7 ▼

Holothuroidea : Sea Cucumbers

Sea cucumbers have lost the star-shaped symmetry and skeletal structures typical of other echinoderms. Instead they have evolved elongate, sausage-shaped bodies with soft leathery skins, and have taken to lying on their sides. The mouth is situated at one end, and is surrounded by 10–20 retractable feeding tentacles, while the anus lies at the other. Up to five bands of tubefeet run the length of the body. Sea cucumbers use their sticky tentacles to gather detritus from the seabed or to catch plankton floating overhead. When disturbed, some species eject long sticky threads from the anus, while others disgorge part or all of the gut, which they subsequently regenerate.

95.1 Snake sea cucumber   *Synapta maculata*

IDENTIFICATION: Body extremely elongate and rope-like, with a mottled brown coloration. Tubefeet absent. The twelve feather-like tentacles each consist of a central axis with regular side branches. SIZE: May exceed 1 m in length. BIOLOGY: Lies amongst seagrass or coral debris on tropical shores, picking detritus from the sand using the sticky tentacles, which are then licked clean. Skin packed with anchor-like spicules that adhere to everything they touch – almost impossible to detach from a diver's wet-suit.

95.2 Golden sea cucumber   *Thyone aurea*

IDENTIFICATION: A floppy, soft-skinned, golden-orange cucumber with ten irregularly branched tentacles. The tubefeet are weak and scattered randomly over the entire body surface. SIZE: Up to 13 cm long. BIOLOGY: Lives buried in sand, with only the tentacles visible (95.2a), or between mussels, red-bait or other holothurians. Regularly washed ashore because of their poor powers of adhesion: their sausage-shaped pinky-orange bodies can litter the shore after storms (95.2b). Disgorges its gut at the slightest provocation. RELATED SPECIES: The commonest of seven *Thyone* species known from southern Africa.

95.3 Mauve sea cucumber   *Pentacta doliolum*

IDENTIFICATION: Body tough, upper surface mauve-black, becoming pale grey on the flattened underside. Tubefeet strongly suctorial and restricted to five regular bands, each consisting of a double row. Ten irregularly-branched tentacles. SIZE: Typically 70 mm. BIOLOGY: A filter-feeder which forms dense beds on shallow reefs along the West Coast. *Thyone* (95.2) often occurs in mixed beds with this species, relying on it for its stronger powers of adhesion to avoid being washed away. RELATED SPECIES: *Pentacucumis spyridophora* (Saldanha to Durban) is similar but uniformly pale grey.

95.4 Horseshoe sea cucumber   *Roweia frauenfeldii*

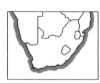

IDENTIFICATION: Body arched into a U-shape, dark above and grey below. Tubefeet grouped into five bands, each made up of 4–6 rows (2–3 rows near the ends of the body). Ten irregularly-branched tentacles. SIZE: Typically 70–80 mm. BIOLOGY: A crevice-dwelling species. Cape specimens are usually found under sand with only the tentacles and anus projecting, the remainder of the body being firmly attached to a rock buried beneath the sediment surface.

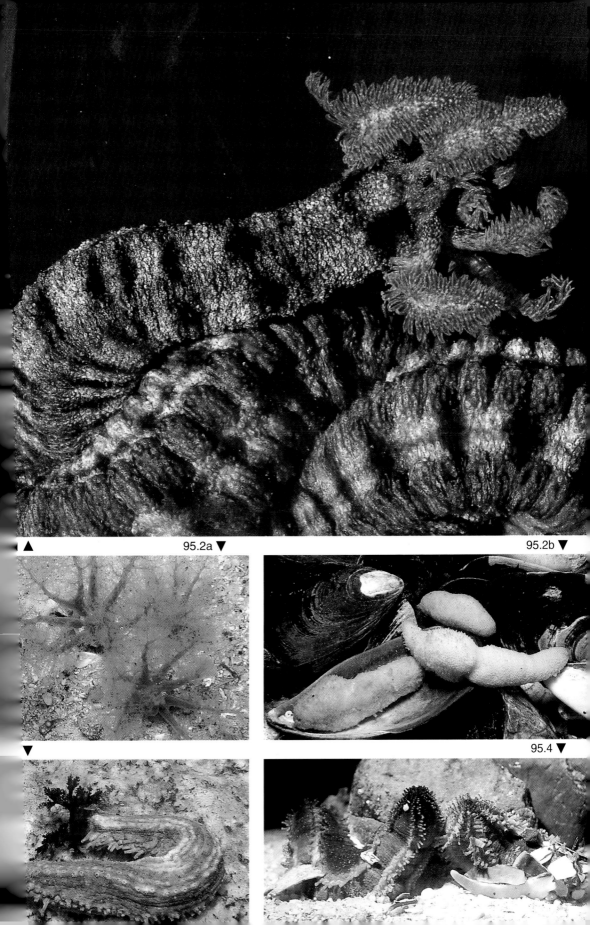

95.2a ▼

95.2b ▼

95.4 ▼

96.1 Stephenson's sea cucumber *Roweia stephensoni*

IDENTIFICATION: Uniformly black. Upper surface with irregularly-scattered tubefeet and soft papillae. Tubefeet on lower surface arranged in distinct bands of six rows each. Ten irregularly-branched tentacles. SIZE: Typically 60–70 mm. BIOLOGY: Aggregates around the sides of rocks or in sandy crevices. Common in intertidal gullies. Previously known as *Cucumaria stephensoni*.

96.2 Red-chested sea cucumber *Pseudocnella insolens*

IDENTIFICATION: Colour diagnostic. Intertidal specimens may be dark above with a red 'chest', but subtidal ones are uniformly bright red. Tubefeet scattered dorsally, but in three bands of two or three rows each on the underside. Ten irregularly-branched tentacles. SIZE: 40 mm. BIOLOGY: A small species found in dense colonies covering shallow reefs. The young are brooded in pockets in the skin and can often be seen clinging to the surface of the parent. Previously known as *Cucumaria insolens* and as *Trachythyone insolens*.

96.3 Black sea cucumber *Pseudocnella sykion*

IDENTIFICATION: Uniformly black, like *Roweia stephensoni*. Distinct bands of tubefeet are visible on the dorsal surface, but odd tubefeet can be found scattered between these bands. The bands on the ventral 'sole' have no more than four rows of tubefeet. Tentacles 10 in number, branching in a tree-like fashion. SIZE: 60 mm. BIOLOGY: Common to abundant on rocky shores. Insinuates itself into rocky crevices near the low-tide mark. Previously known as *Cucumaria sykion*.

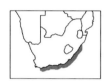

96.4 Warty sea cucumber *Neostichopus grammatus*

IDENTIFICATION: Moderately large. The dorsal surface lacks tubefeet but is covered with obvious pointed wart-like papillae. Colour uniformly red or cream to brown. At least 18 tentacles (although the tentacles are closely crowded and difficult to count). Each tentacle arises from a short stalk that divides into short, horizontal branches. SIZE: Average 15 cm. BIOLOGY: Relatively uncommon; found in rock pools and shallow waters down to about 5 m.

96.5 Tufted sea cucumber *Holothuria cinerascens*

IDENTIFICATION: A large but cryptic species with a rough sand-covered, brown-to-pink dorsal surface (96.5a) and scattered tubefeet on the rosy underside. The 18 tentacles each have a distinct stem ending in a tuft of short, button-like branches (96.5b). SIZE: Average 15 cm. BIOLOGY: Common in low-shore rock pools and gullies in Natal but not easily seen because of its sand coating. RELATED SPECIES: One of several large *Holothuria* species with similarly tufted tentacles from Natal and Moçambique. *H. scabra* (Durban to central Moçambique) is almost black above and mottled grey and black below. It lives on sheltered sandbanks and has been heavily exploited in parts of Moçambique, being exported to the Far East for human consumption. Its exploitation has indirectly led to profound changes in the ecology of nearby terrestrial systems because large quantities of wood are burnt to dry the cucumbers. *Holothuria atra* is very large (220 mm) and uniformly black, and confined to Moçambique.

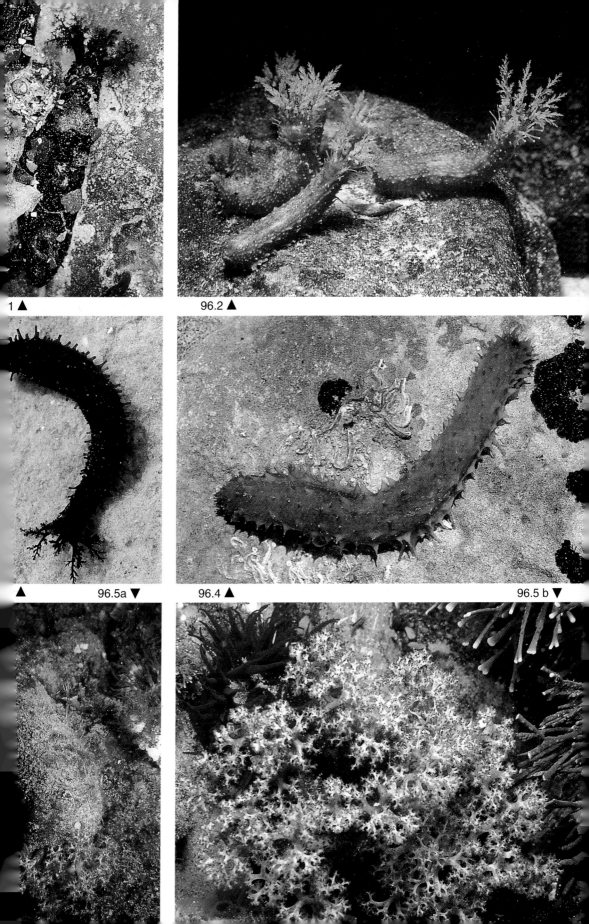

1 ▲

96.2 ▲

▲   96.5a ▼

96.4 ▲

96.5 b ▼

Ascidiacea : Sea Squirts

Although sea squirts, or ascidians, might appear somewhat featureless, they are in reality highly advanced animals, closely related to the vertebrates. These affinities are most evident in the larvae, which are tadpole-like creatures with a primitive backbone, nerve cord and tail. These are lost by the adults, which become sessile, sac-like animals topped by a pair of tubular siphons and enclosed in a cellulose casing or test. Water sucked in through the inhalant siphon is filtered through the enlarged sieve-like pharynx and then expelled through the exhalant siphon. Most larger ascidians are solitary animals, but many of the smaller species are colonial, either branching from a common base or consisting of numerous tiny individuals (zooids) embedded in a shared mass or sheet of jelly-like tissue. Such individuals may maintain separate siphons or share common exhalant siphons. Sea squirts are extremely common on most rocky shores and reefs, but have not been well studied: thus, names are available for only a few of the more common forms.

97.1 **Red-bait** *Pyura stolonifera*

IDENTIFICATION: A very large and well-known solitary ascidian with a thick, opaque-white test covered in a wrinkled, dark-brown skin. The flesh, which can be pink to bright red, is commonly used by anglers as bait, hence the common name. SIZE: Typically 15 cm height, but can reach the size of a rugby ball. BIOLOGY: Very common on open coasts, where it forms extensive sheets coating the rocks from low tide to about 10 m depth. Also forms loose boulder-like masses on the seabed or even develops root-like extensions and buries in sheltered sands. Commensal copepods, amphipods and pea crabs (45.5) can be found living in the pharynx.

97.2 **Angular ascidian** *Styela costata*

IDENTIFICATION: A tall solitary ascidian with a tough, leathery, six-sided test which arises from a thin tapering stalk. Both siphons are terminal. SIZE: 5–10 cm tall. BIOLOGY: Occurs singly on rock surfaces where the water is clean and the current fairly strong. Often partially concealed by seaweeds or other marine growths on the test.

97.3 **Transparent ascidian** *Ciona intestinalis*

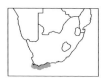

IDENTIFICATION: A tall, cylindrical, solitary ascidian with a soft, transparent, off-white to green tinged test, through which the pharynx and digestive organs can clearly be seen. The larger inhalant siphon is terminal, while the smaller exhalant one projects off the shoulder. SIZE: 5–10 cm tall. BIOLOGY: Common in sheltered waters, especially harbours and quiet bays, often forming dense colonies on jetties and piers.

97.4 **Choirboys** *Podoclavella* sp.

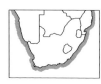

IDENTIFICATION: A colonial form in which clusters of independent zooids are attached together by a common base. A darker patch between the two siphons is conspicuous against the otherwise almost completely transparent, pale test. SIZE: Colony height 5–10 cm. BIOLOGY: A popular subject of underwater photographs, this attractive ascidian is most common along vertical rock-faces and under overhangs on Cape reefs. Colour variable, usually blue (97.4) or yellow (97.5) but more rarely light blue-green.

97.6 **Fan ascidians** *Sycozoa* spp.

IDENTIFICATION: Colonial forms with flattened fan-shaped heads arising from a short, often much-branched stalk. Large specimens can be very convoluted. The individual zooids are arranged in double or multiple rows along the margin of the fan. SIZE: Colony up to 10 cm across and 5 cm tall. BIOLOGY: Found on vertical rock-faces where there is a strong water flow. Several species have been observed, including a spectacular fluorescent-blue fan-shaped species (**97.7**) found on Eastern Cape reefs.

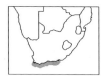

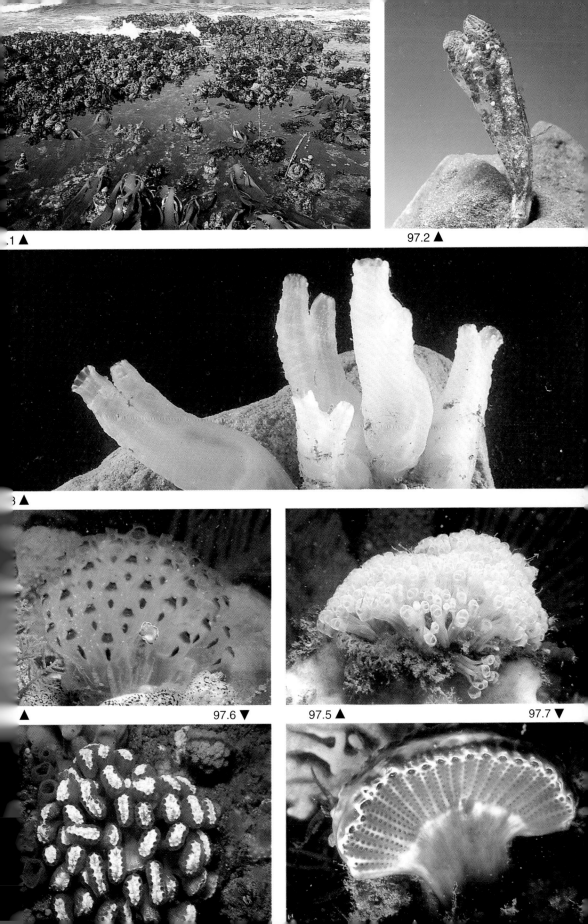

.1 ▲

97.2 ▲

3 ▲

▲   97.6 ▼

97.5 ▲   97.7 ▼

98.1 Green urn ascidian *Didemnum molle*

IDENTIFICATION: Forms urn-shaped colonies on tropical reefs. The star-shaped inhalant openings of the individual zooids dot the outside wall of the colony, while the exhalant openings all open into a large central cavity. SIZE: Colony height about 5 cm. BIOLOGY: Easily misidentified as a sponge because of the single exhalant siphon. The green interior coloration is caused by symbiotic blue-green algae which grow around the inhalant siphons of each zooid. RELATED SPECIES: A bright pink species with similar growth form has been reported from the Eastern Cape.

98.2 Lattice ascidians *Didemnum* spp.

IDENTIFICATION: Colonial ascidians forming thin, firm incrustations over rocks or other surfaces. The star-shaped inhalant openings of the minute zooids dot the surface, giving a beautiful lattice-like appearance. The larger turret-like exhalant siphons are shared by numerous zooids. SIZE: Colonies only 1–2 mm thick, but can be 50 cm across. BIOLOGY: Zooids of this group are the smallest of all ascidians. Several different colours and colony forms have been recorded, which may represent separate species. Abundant but often overlooked because of the sponge-like appearance of the colonies.

98.3 Star ascidian *Botryllus magnicoecus*

IDENTIFICATION: Individual zooids form distinct star-like groups, surrounding a larger, common exhalant opening or atrium. Zooids buried in a soft gelatinous sheet that encrusts rocks, seaweeds or other animals. SIZE: Zooids 1–2 mm, colony 5–10 cm. BIOLOGY: Common, but has a bewildering variety of colours and forms. Eggs are brooded in the atrium; the tadpole larvae emerge for only about two hours before settling to establish a new colony. RELATED SPECIES:
98.4 *Aplidium* sp., a greeny-blue species, is abundant around Durban.

98.5 Ladder ascidian *Botrylloides leachi*

IDENTIFICATION: Colonial forms with zooids embedded in a thin gelatinous sheet. Zooids typically in two parallel rows, with a long, slit-like exhalant cavity, or atrium, between them. These 'ladder' systems are frequently obscured by crowding. SIZE: Zooids 1–3 mm, colonies 10 cm or more across. BIOLOGY: Occurs world-wide on rocky coasts. *Botrylloides* species are often difficult to distinguish from *Botryllus* (98.3) as both are very variable in colour and form.

98.6 Seaweed ascidian *Botryllus elegans*

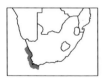

IDENTIFICATION: A colonial form consisting of separate opaque-white, golden or pink zooids regularly spaced within a transparent jelly-like matrix spread over fronds of flat-bladed algae, usually *Epymenia obtusa* (154.1). Opaque dots between the zooids form part of a common blood-circulation system that permeates the matrix. SIZE: Zooids 2 mm across, colony 10 cm. BIOLOGY: As the name implies, an anomalous species in that all other *Botryllus* species share common exhalant openings (98.3). Common in gullies and kelp forests. Previously knows as *B. anomalus*.

98.7 Blue lollipop ascidian *Eudistoma coeruleum*

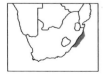

IDENTIFICATION: A colonial ascidian consisting of a dark-blue club-like lobe studded with minute, independent zooids and supported by a short, stout stalk. SIZE: Colony height about 30–40 mm. BIOLOGY: Abundant in rock pools along the East Coast, and often grows in pools that are periodically inundated by sand. Easily recognised by its dramatic colour.

98.8 Encrusting ascidian *Diplosoma* sp.

IDENTIFICATION: Forms small, flat encrusting pads. Pale blue to blue-green. Surface dotted with a large number of small inhalant openings, amongst which are a relatively small number of exhalant openings about twice the size. SIZE: 30 mm. BIOLOGY: Often grows on other animals, such as the solitary ascidian *Polycarpa* shown here (recognisable by the dark, blue-spotted interior to its inhalant siphon).

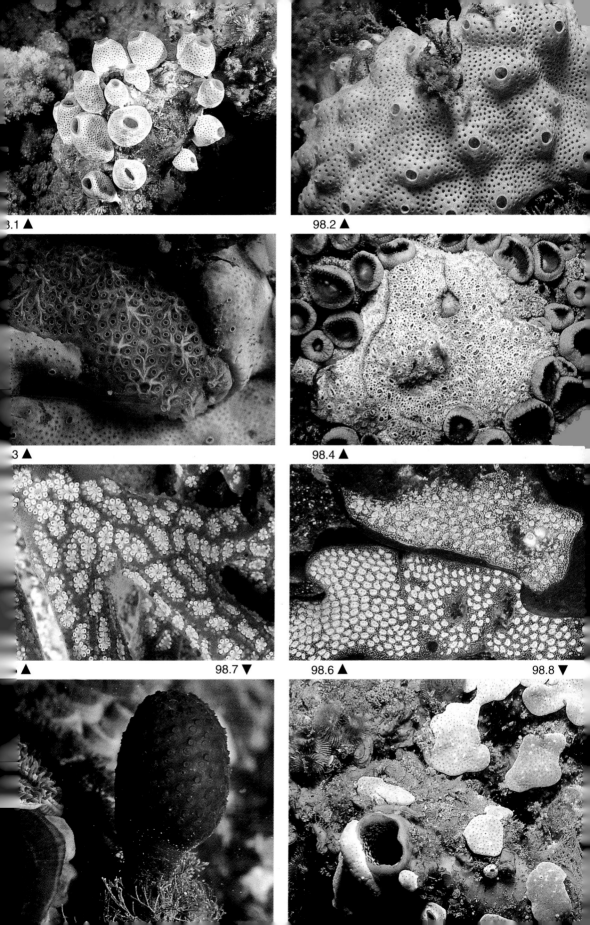

98.1 ▲

98.2 ▲

98.3 ▲

98.4 ▲

98.5 ▲

98.7 ▼

98.6 ▲

98.8 ▼

Over two thousand species of fishes occur in the seas around our coast, ranging in size from tiny, cryptic cling fishes to huge, but harmless, whale sharks. The following pages describe and illustrate the common coastal species. Over half of the fishes found along our coast are tropical and subtropical Indo-Pacific species. The Atlantic Ocean representatives constitute only 4%. Endemic fishes, restricted to southern Africa, make up 13%.

Fishes can be broadly divided into elasmobranchs, such as sharks, with skeletons of cartilage, and teleosts, such as hake, sardines or red steenbras, with bony skeletons. Elasmobranchs are usually cylindrical (sharks) or flattened (rays) and have 5–7 gill slits. Though lacking scales, the skin has the texture of sandpaper because of tiny tooth-like structures embedded in it. The teeth can have sharp, cutting edges or can be flattened for grinding food. Adult males may be recognised by a pair of claspers on the ventral surface. Elasmobranchs have an exceptionally well-developed sense of smell.

Teleosts also vary greatly in shape. On each side of the head they have a single gill opening, protected by a gill cover. The skin is usually covered by overlapping scales. Unless there are different colour patterns in males and females, it is generally difficult to tell the sexes apart without dissecting dead specimens. Teleosts have sensory receptors in the lateral line along the flanks of the body which are responsible for sound detection.

In the text that follows, features for the identification have been confined, as far as possible, to characters which are visible to the naked eye. The numbers of spines and soft rays in the dorsal and anal fins are usually characteristic of each species. Spines are hard and spiky; rays are soft and flexible, usually divide near their tips, and are often crossed by vein-like stripes. The numbers and arrangement of spines and rays are given in formulae in which Roman numerals indicate spines and Arabic numerals indicate soft rays. For example, the fin count for the flathead mullet is DIV + I 8, AIII 8. Decoded, this means that it has two dorsal fins (D), the first one with four spines (IV) and the second one with a single spine (I) and eight soft rays (8). The anal fin (A) has three spines (III) and eight soft rays (8).

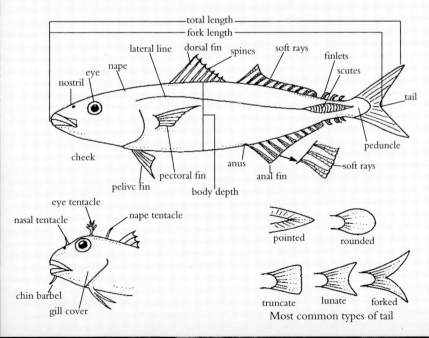

Most common types of tail

Sharks

About 100 species of sharks occur off southern Africa, although many occupy deep-water habitats and are rarely seen. Most are harmless, but a few attack bathers. Shark nets do protect bathers but have detrimental effects on other sea life.

99.1 Great white shark   *Carcharodon carcharias*

IDENTIFICATION: A large powerful grey-blue shark with a conical snout. Teeth very large, triangular, serrated on both sides. Pectoral fins long, with a black patch behind the fin base. Tail fin sickle-shaped. Anal and second dorsal fins small. SIZE: 6.4 m and 1208 kg. BIOLOGY: Occurs world-wide in temperate offshore and coastal areas, often in proximity to islands with seals. Ranges widely between the Cape and KZN. Dangerous: attacks swimmers, divers, surfers and boats. Sexual maturity achieved at 3.5 m. Legally protected in South Africa.

99.2 Zambezi or bull shark   *Carcharhinus leucas*

IDENTIFICATION: Grey, robust, with a rounded snout and broad triangular saw-edged upper teeth. High first dorsal fin; no markings on fins. No mid-back ridge. SIZE: 3.2 m and 214 kg. BIOLOGY: Extremely dangerous; several recorded attacks on people. Circumglobal, frequenting estuaries and coastal waters, this species has even been recorded far upstream in the Zambezi River. Feeds on small sharks, fishes and dolphins. Sexually mature at 1.8 m; bears up to eight pups in summer.

99.3 Dusky shark   *Carcharhinus obscurus*

IDENTIFICATION: Dusky grey; snout broad and rounded. Upper teeth triangular. Characteristic mid-back ridge; first dorsal fin small, second dorsal low. SIZE: 4.2 m; 327 kg. BIOLOGY: Occurs world-wide. Eats fish and squid. Frequently caught in fishing tournaments. Proliferation of its juveniles in KZN possibly due to removal of larger predatory sharks by nets. Bears up to 10 young in summer off KZN. RELATED SPECIES: The copper shark, *Carcharhinus brachyurus*, has no dorsal ridge; upper teeth with narrow pointed cusps. Common in Eastern Cape.

99.4 Tiger shark   *Galeocerdo cuvieri*

IDENTIFICATION: Vertical dark bars; ridge along mid-back. Upper tail-lobe long. Low keels on tail peduncle. Snout blunt. Large saw-edged cockscomb teeth. SIZE: 5.5 m and 559 kg. BIOLOGY: Coastal and offshore, world-wide. A voracious predator on fishes, birds, mammals and turtles; implicated in attacks on humans. Sexually mature at around 3 m; bears litters of 23–46 young.

99.5 Ragged-tooth shark   *Carcharias taurus*

IDENTIFICATION: Robust, plump-bodied, brown; frequently has darker spots along the sides. Snout pointed; numerous rows of fang-like teeth. Dorsal and anal fins equal-sized. SIZE: 3 m and 294 kg. BIOLOGY: Widespread; common on coastal reefs. Mates off Zululand in summer; two young are born nine months later. Feeds on fishes, small sharks and crustaceans. Only dangerous if provoked.

99.6 Scalloped hammerhead shark   *Sphyrna lewini*

IDENTIFICATION: Head peculiarly flattened, protruding on each side; front edge scalloped (99.6d). Smooth, slanting triangular teeth. Rear edge of pelvic fin straight. SIZE: 3 m. BIOLOGY: Pelagic; occurs world-wide in tropical and temperate seas. Mature at 1.6 m. Has litters of about 30 young. Juveniles form large schools near the surface. RELATED SPECIES: The shape of the snout differs in **Sphyrna zygaena** (99.7d) and **Sphyrna mokarran** (99.8d); the latter has serrate teeth, and the hind edge of the pelvic fin is concave.

99.1 ▲

99.2 ▲

.3 ▲

99.4 ▲

▲

99.6 ▼

99.6d

99.7d

99.8d

100.1 **Whitetip reef shark** *Triaenodon obesus*

IDENTIFICATION: A slender, greyish-brown shark with conspicuous white tips on the first dorsal fin and upper lobe of the tail fin. Snout extremely broad. No ridge between dorsal fins. Teeth sharp, cusps oblique. SIZE: 2 m. BIOLOGY: Indo-Pacific; inhabits coral reefs and often seen resting in caves. Feeds on fish, octopus and crustaceans. Sexually mature at 85 cm; females bear 1–5 young. Usually harmless; shy of divers.

100.2 **Leopard catshark** *Poroderma pantherinum*

IDENTIFICATION: Elongate with long nasal barbels and small pointed tricuspid teeth. Black spots which can vary in size and shape occur along the body and also on the fins. Some animals have uniform round spots, others a mix of dots and stripes. SIZE: 70 cm. BIOLOGY: Endemic. Leopard catsharks are similar to pyjama catsharks in habit.

100.3 **Pyjama catshark** *Poroderma africanum*

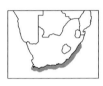

IDENTIFICATION: An elongate shark with long nasal barbels and small pointed tricuspid teeth. Seven longitudinal black stripes extend along body. SIZE: 1 m. BIOLOGY: These endemic catsharks are usually found in shallow rocky areas. They are nocturnal and spend much time lying on the seabed. Feed on small reef fish and octopus. Females produce two brown egg cases, which are attached to seaweed.

100.4 **Puffadder shyshark** *Haploblepharus edwardsii*

IDENTIFICATION: A yellowish-brown elongate shark with brown saddles that have dark margins. Numerous pale spots. Eyes narrow; nostrils connected to mouth by grooves. Fins small, marked with brown blotches and pale spots. SIZE: 60 cm. BIOLOGY: An endemic coastal shark. Females lay two eggs which are protected by tough egg cases. SIMILAR SPECIES: *Halaelurus* species have nostrils that are entirely separate from the mouth. The two commonest species, *H. lineatus* (Natal and Moçambique) and *H. natalensis* (Saldanha to Port Elizabeth), both have brown saddles like the puffadder shyshark, but differ in having a pointed, upturned snout. Apart from differences in distribution, *H. lineatus* is distinguished from *H. natalensis* because its saddles are narrower and more numerous.

100.5 **Dark shyshark or skaamoog** *Haploblepharus pictus*

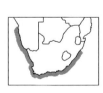

IDENTIFICATION: Small brown shark with about seven dark saddles extending over the head and body. Deep grooves connect the nostrils to the mouth. SIZE: 60 cm. BIOLOGY: Shysharks curl themselves into a circle when disturbed, covering the eyes with the tail (hence 'skaamoog'). They are endemic and live in shallow inshore areas. Feed on bottom-dwelling invertebrates. Females produce two egg cases, often called 'mermaid's purses' (100.5d). RELATED SPECIES: *Haploblepharus fuscus* (Agulhas to southern Natal) is pale brown with faint dorsal saddles.

100.6 **St Joseph shark** *Callorhinchus capensis*

IDENTIFICATION: The unusual shape, silvery and scaleless skin, long tail and trunk-like appendage on the snout characterise this curious cartilaginous fish. Males have a spiny knob on the head. SIZE: 1 m. BIOLOGY: An endemic, bottom-dwelling species found in shallow water, particularly in the south-western Cape, where it is fished commercially by set nets in St Helena Bay. Feeds on bottom-dwelling invertebrates and small fish. Females deposit brown, spindle-shaped, hairy egg cases on the seabed in summer.

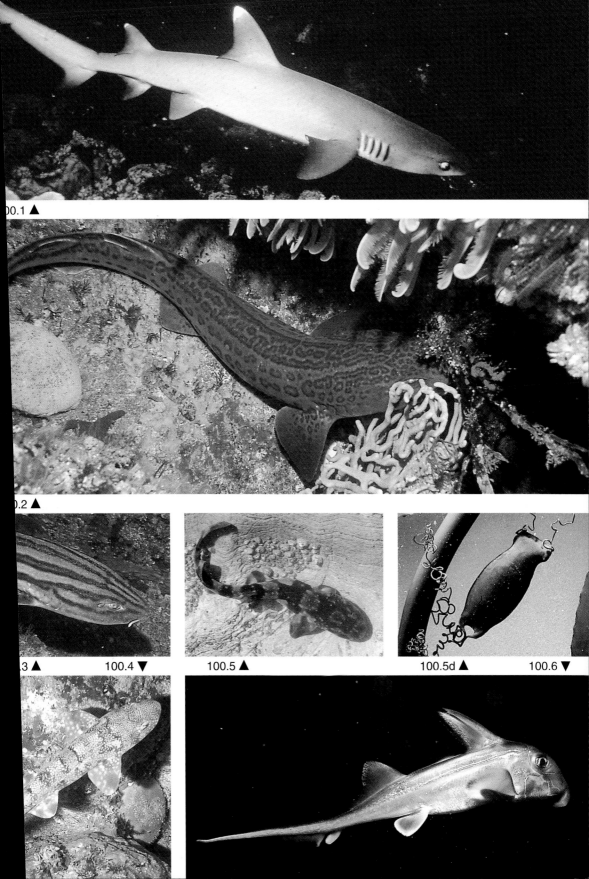

100.1 ▲

100.2 ▲

100.3 ▲   100.4 ▼   100.5 ▲   100.5d ▲   100.6 ▼

Rays

There are about 60 species of rays in southern African waters and most live on the sea-bed. Their pectoral fins are greatly expanded into wing-like structures and the gill openings are on the underside of the head.

101.1 Blackspotted electric ray   *Torpedo fuscomaculata*

IDENTIFICATION: A rounded, disc-like ray. Numerous black spots on the dorsal surface. Two dorsal fins; tail fin rounded. Five pairs of gill openings beneath the body. SIZE: 64 cm. BIOLOGY: Found in western Indian Ocean estuaries and shelf areas. Kidney-shaped electric organs at the base of the pectoral fins generate powerful shocks, which are used to stun prey or for defence. RELATED SPECIES: *Torpedo sinuspersici* is brownish-red with a yellow reticulated pattern on the dorsal surface. It occurs in KwaZulu-Natal and Moçambique.

101.2 Lesser sandshark   *Rhinobatos annulatus*

IDENTIFICATION: Guitar-shaped, with a flattened, pointed head and blunt pavement-like teeth. Colour usually brown with dark spots on the dorsal surface. SIZE: 1.2 m and 28 kg. BIOLOGY: Common in shallow sandy areas, particularly the surf zone of beaches. Preys on bottom-dwelling molluscs and crabs which it crushes with its pavement teeth. Females bear 2–10 young in summer. Endemic to southern Africa.

101.3 Giant sandshark   *Rhynchobatus djiddensis*

IDENTIFICATION: Flattened, pointed head with elongate snout. Teeth blunt and pavement-like. Grey with rows of white spots. A pair of darker eyespots occurs on the bases of the pectoral fins. SIZE: 3 m and 227 kg. BIOLOGY: Indian Ocean. Inhabits shallow sandy areas, from the surf zone down to 30 m. Abundant in summer in KZN. Feeds on crabs, molluscs and small fish. Matures at 1.5 m; bears four young in summer.

101.4 Blue stingray   *Dasyatis chrysonota*

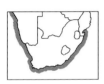

IDENTIFICATION: Brown, with a mottled dorsal surface. Snout pointed. One serrated spine on the whip-like tail. SIZE: Width 75 cm. BIOLOGY: Coastal; found in shallow, sandy areas. Often caught by beach anglers. Buries under a thin layer of sand and feeds on small fish and crustaceans. One to five young are born in summer after 9 months gestation. Previously called *Dasyatis marmoratus*.

101.5 Spotted eagleray   *Aetobatus narinari*

IDENTIFICATION: A diamond-shaped ray with a distinct duckbill snout. Tail whip-like, 1–3 spines at the base. Dorsal surface dark with numerous white spots; underside white. SIZE: Width of 2.3 m and 98 kg. BIOLOGY: Free-swimming circumtropical; inhabits shallow coastal waters. Feeds on bottom-dwelling invertebrates which it crushes with its chevron-like pavement teeth. Bears up to four young. RELATED SPECIES: *Myliobatis aquila* (Namibia to KZN) is plain brown and has two serrated tail-spines.

101.6 Honeycomb stingray   *Himantura uarnak*

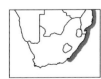

IDENTIFICATION: Brown with a pale honeycomb pattern on the upper surface; white below. Snout pointed; one spine on whip-like tail, which can be up to three times longer than the body. SIZE: Width of 2 m and 120 kg. BIOLOGY: An Indo-West Pacific species. Inhabits estuaries and coastal areas down to 50 m. Feeds on bottom-dwelling invertebrates and bears 3–5 young in summer.

101.7 Largetooth sawfish   *Pristis pectinata*

IDENTIFICATION: A conspicuous saw-like snout with 21–28 pairs of teeth. Body flattened with angular pectoral fins. SIZE: 6 m and 276 kg. BIOLOGY: Lives in shallow coastal areas in the tropical Atlantic and Indo-Pacific. Uses its saw to dig crustaceans and molluscs from the seabed or to catch fish. Females bear 15–20 young, and used to be found in St Lucia and Richard's Bay in summer. Juveniles occur in estuaries and bays.

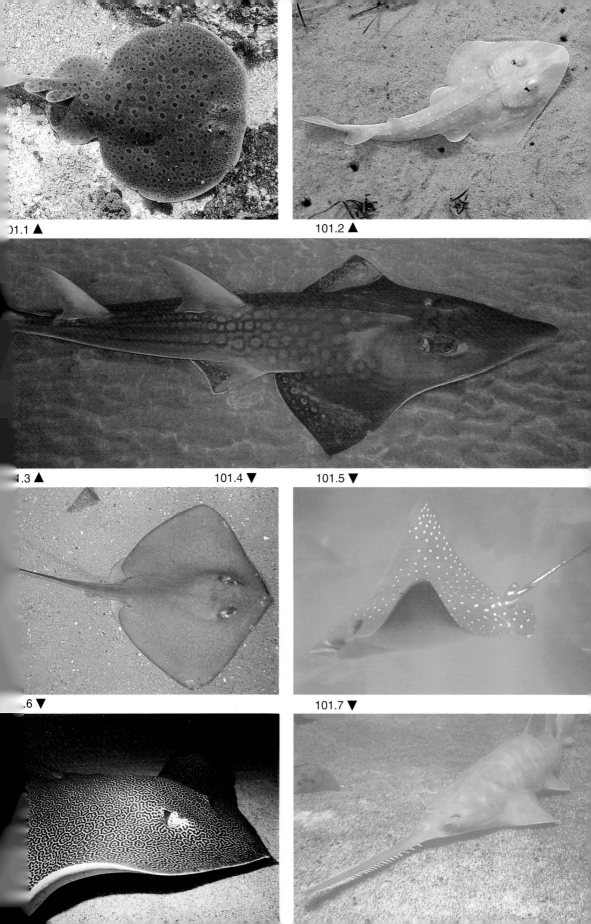

01.1 ▲

101.2 ▲

.3 ▲

101.4 ▼

101.5 ▼

.6 ▼

101.7 ▼

Eels & Hagfish

Eels have long, snake-like bodies that lack scales and pelvic fins. Thirteen different families of eels occur in southern Africa. Although the large moray eels are probably the best known, there are many deep-water species that are rarely seen. Eels have peculiar, flattened transparent larvae called leptocephalus larvae, which are widely dispersed by ocean currents. Eels belonging to the genus *Anguilla* live in freshwater but migrate to the ocean to spawn. Their larvae metamorphose into 'glass eels' (a flat, almost transparent stage) and enter estuaries before migrating upstream into rivers. Hagfish are among the more primitive fish, lacking jaws or paired fins.

102.1 Blackedged conger   *Conger cinereus cinereus*

IDENTIFICATION: A long, thin, grey eel with white underside. Pectoral fins well developed; other fins have distinct black edges. A black stripe extends along the lower edge of the eye. SIZE: 1 m. BIOLOGY: An Indo-Pacific species inhabiting crevices in coral reefs. Common in the tropics, extending from the shore down to 30 m.

102.2 Ocellated snake-eel   *Myrichthys maculosus*

IDENTIFICATION: A scaleless, elongate, cream eel with a series of black, oval spots along the body. The tail is hard and pointed. SIZE: 50 cm. BIOLOGY: Widespread in Indo-Pacific, where it is usually found buried in sandy areas near reefs. Sometimes mistaken for a sea snake, but easily distinguished by absence of scales and lack of a flattened, paddle-like tail.

102.3 Honeycomb moray eel   *Gymnothorax favagineus*

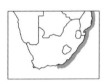

IDENTIFICATION: One of 20 species of moray eels in South African waters, this dark-brown eel has a characteristically yellow honeycomb pattern. The dorsal and anal fins are connected and lack spines, and there are no pectoral fins. Prominent teeth in jaws. SIZE: 1.5 m. BIOLOGY: An Indo-Pacific species which lives in caves and holes in coral and rocky reefs. These cryptic predators favour octopus in their diet. They can inflict a serious bite if threatened.

102.4 Floral moray   *Echidna nebulosa*

IDENTIFICATION: Creamish colour, with longitudinal double rows of dark blotches along body, each of which usually has a yellow centre. The snout is typically white. The jaws lack sharp fangs. SIZE: 70 cm. BIOLOGY: Inhabits shallow coral and rocky reefs in the Indo-Pacific region. Juveniles often found in tidal pools in Natal.

102.5 Zebra moray   *Gymnomuraena zebra*

IDENTIFICATION: Black with 40–80 narrow white rings encircling the head and body. Jaws bear granular teeth but do not have any sharp fangs. The anus is positioned well behind the mid-point of the body. SIZE: 1.5 m. BIOLOGY: Inhabits shallow coral and rocky reefs in the Indo-Pacific region. Usually conceals itself in crevices, only its head protruding.

102.6 Six-gill hagfish   *Eptatretus hexatrema*

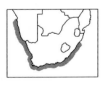

IDENTIFICATION: Eel-shaped, grey, with a cartilaginous skeleton. No paired fins. There are six barbels on the head; 5–8 pairs of gill openings. Two rows of slime pores along underside of body. SIZE: 80 cm. BIOLOGY: Closely related to lampreys. Endemic. Occurs down to 400 m. Frequently caught in lobster-traps on the West Coast. Secretes copious amounts of sticky slime when irritated.

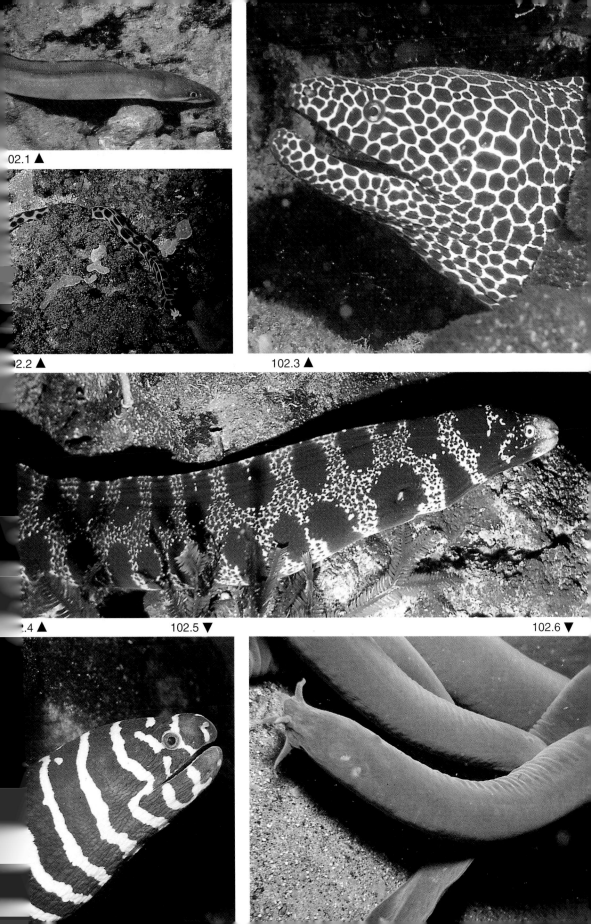

02.1 ▲

2.2 ▲

102.3 ▲

.4 ▲   102.5 ▼   102.6 ▼

Sardines, Anchovies & Herrings

The clupeiform fishes are small, primitive, pelagic fish that often form huge shoals. About 17 species of anchovies and sardines occur in South Africa, but microscopic examination is often necessary to distinguish between them.

103.1 Estuarine roundherring   *Gilchristella aestuaria*

IDENTIFICATION: D 14–15, A 20. Tiny, translucent, with a bright silver lateral stripe. The single dorsal fin lacks spines. SIZE: 7 cm. BIOLOGY: An endemic fish that forms large shoals in estuaries and coastal lakes on the South and East coasts. Tolerates fresh to extremely saline conditions. Spawns in estuaries. Filter-feeds on plankton and is important in the diet of large predatory fish.

103.2 Razorbelly   *Hilsa kelee*

IDENTIFICATION: D 16–19, A 17–23. Deep-bodied and silvery with a black spot behind the gill cover. Belly compressed into a keel. Tail deeply forked. Small specimens have several black spots on their sides. SIZE: 35 cm. BIOLOGY: An Indo-Pacific coastal species; extends southwards to KZN. Shoals are often recorded in Durban harbour. Filter-feeds on plankton and is an important component of the diet of larger predatory fish. Spawns in summer.

103.3 Goldstripe sardinelle   *Sardinella gibbosa*

IDENTIFICATION: D 17–20, A 17–21. A small silvery fish distinguished by a thin gold-coloured stripe along sides. At least four easily confused species occur in South African waters. SIZE: 20 cm. BIOLOGY: A tropical Indo-Pacific species that frequents coastal waters including estuaries. Strays occur as far south as Eastern Cape.

103.4 Pilchard or sardine   *Sardinops sagax*

IDENTIFICATION: D 17–20, A 18–22. Silver and blue, with 10–15 black spots along sides. The single dorsal fin lacks spines. SIZE: 28 cm. BIOLOGY: Pelagic; forms large shoals which are harvested commercially on the West Coast. Huge tonnages were caught in the 1960s, followed by a collapse of the stock. During the famous Natal winter 'sardine run' they are driven ashore by predatory game fish. Filter-feeds, and is an important prey item for large fish, birds and dolphins. Previously called *Sardinops ocellatus*. SIMILAR SPECIES: The round herring, *Etrumeus whiteheadi* (Walvis Bay to Natal), is another commercially important clupeid. It has 11–14 anal fin rays, lacks lateral spots, and its eyes are noticeably red after capture.

103.5 Cape anchovy   *Engraulis encrasicolis*

IDENTIFICATION: D 14–17, A 15–21. Small and silvery. Snout projects past lower jaw. Belly lacks enlarged, sharp scales (scutes). SIZE: 13 cm. BIOLOGY: Indo-Pacific; a coastal pelagic shoaling species. Filter-feeds on plankton and is an important food-source for many fishes, birds and mammals. Spawns inshore on the Agulhas Bank; juveniles recruit westwards and are a major component of the purse-seine fishery, usually being processed into fish-meal and fish-oil.

103.6 Glassnose or bony   *Thryssa vitrirostris*

IDENTIFICATION: D 13–14, A 34–43. Silvery, with a black spot behind the gill cover. Snout blunt, translucent and overlaps the underslung mouth, the interior of which is orange. SIZE: 20 cm. BIOLOGY: A West Indian Ocean coastal shoaling species; common in bays and estuaries. Filter-feeds on plankton. Frequent in the diets of game fish and an excellent bait fish. Spawns in winter. RELATED SPECIES: *Thryssa setirostris* has a long posterior extension to the upper jaw.

03.1 ▲

103.2 ▲

3.3 ▲

4 ▲

103.5 ▼

103.6 ▼

Wolfherring, Ladyfish, Tarpon, Milkfish, Silversides

104.1 **Wolfherring** *Chirocentrus dorab*

IDENTIFICATION: D 16–19, A 29–36. An elongate and compressed silver fish with deeply forked tail, a large mouth and numerous fang-like teeth. Back usually marked with bands of green and blue. SIZE: 1 m. BIOLOGY: An Indo-Pacific pelagic species found in coastal waters, in particular near the surf zone. Carnivorous, preying on small fishes.

104.2 **Ladyfish or springer** *Elops machnata*

IDENTIFICATION: D 22–27, A 15–18. Elongate, cylindrical silver fish with spineless dusky-yellow fins. Tail fin strongly forked. Mouth large, reaching back past the eye. The eyes have fatty eyelids. SIZE: 1 m and 11 kg. BIOLOGY: An Indo-West Pacific predatory species that inhabits coastal waters and estuaries. Preys on small fish. Larvae resemble eel leptocephali, but have forked tails and recruit to East Coast estuaries in summer.

104.3 **Oxeye tarpon** *Megalops cyprinoides*

IDENTIFICATION: D 17–20, A 24–31. A compressed, silver fish with a projecting lower jaw and a distinct forked tail. Fatty tissue occurs over the eyes. SIZE: 1 m. BIOLOGY: A tropical Indo-West Pacific species that occurs in coastal waters, estuaries and even freshwater. Usually solitary, it preys on small fishes. Larvae resemble eel leptocephali, but have forked tails.

104.4 **Milkfish** *Chanos chanos*

IDENTIFICATION: D 13–17, A 8–10. A large, silver fish with a small toothless mouth and large, powerful scissor-like tail. Eyes are covered by fatty tissue. SIZE: 1.8 m and 18 kg. BIOLOGY: An Indo-Pacific species inhabiting coastal waters, estuaries and bays. Feeds on detritus and small invertebrates on sand- and mud-banks. Spawning occurs at sea but ribbon-like larvae recruit to estuaries, where they metamorphose into juveniles.

104.5 **Hardyhead silverside** *Atherinomorus lacunosus*

IDENTIFICATION: DIV–VII + I 8–11, AI 12–17. A small fish with greenish dorsal surface and a distinct silver stripe along its sides. Upper surface speckled with darker colour-bearing cells. Two dorsal fins. SIZE: 15 cm. BIOLOGY: An Indo-West Pacific Ocean species that occurs in large schools in coastal waters. Feeds on zooplankton.

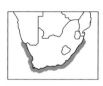

104.6 **Cape silverside** *Atherina breviceps*

IDENTIFICATION: DV–VIII + I 11–15, AI 15–18. A small fish with a distinct silver stripe along the sides. Two dorsal fins and scales with small black dots. SIZE: 11 cm. BIOLOGY: A shoaling species restricted to southern Africa. Occurs in surf zones and lower reaches of estuaries. Eggs have long sticky filaments for attachment to sand or vegetation.

104.1 ▲

104.2 ▲

04.3 ▲

104.4 ▲

.5 ▲

104.6 ▼

Catfish, Kingklip, Hake & Anglerfish

105.1 Natal seacatfish *Galeichthys* sp.

IDENTIFICATION: DII 7, A17–21. Robust, with brown dorsal surface and flanks. Scaleless body covered in mucus. Head compressed; six barbels on lower jaw. Dorsal fin and pectoral fins each bear a serrated spine. The Natal catfish is a newly recognised, unnamed species. SIZE: 45 cm. BIOLOGY: Occurs in shallow coastal waters. The male carries the fertilised eggs in his mouth until they hatch and lose their yolksacs. The spines inflict painful injuries. RELATED SPECIES: The white seacatfish, *Galeichthys feliceps* (common in Eastern Cape estuaries), is smaller, paler, and has a narrower head. The black seacatfish, *Galeichthys ater*, has brown spots on the belly. It occurs on the South Coast but not in estuaries.

105.2 Striped eel-catfish *Plotosus lineatus*

IDENTIFICATION: DI 4 + 80–90, A 58–82. Eel-shaped. Black, with 2–3 yellowish stripes from snout to tail. The second dorsal fin and the anal fin are joined to the pointed tail fin. Mouth surrounded by eight barbels. Dorsal fin and pectoral fins each bear a serrated venomous spine. SIZE: 32 cm. BIOLOGY: Indo-West Pacific, frequenting shallow coastal waters. Congregates in dense schools. Eats invertebrates. RELATED SPECIES: The eel-catfish, *Plotosus nkunga*, lacks horizontal stripes.

105.3 Kingklip *Genypterus capensis*

IDENTIFICATION: Elongate and eel-like, mottled pink and brown. Dorsal, tail and anal fins joined; pelvic fin reduced to a pair of rays. SIZE: 1.6 m. BIOLOGY: A southern African endemic, trawled on offshore banks deeper than 50 m, and is a favoured table fish that commands a high price. Bottom-dwelling; carnivorous on fish and invertebrates. Becomes sexually mature at 45–75 cm (4–6 years).

105.4 Shallow-water hake *Merluccius capensis*

IDENTIFICATION: D 10–12 + 38–43, A 37–41. Elongate, silvery-grey; white belly. Fins lack spines. Tail fin truncate. Eyes large; teeth sharply pointed. SIZE: 1.4 m. BIOLOGY: Bottom-living; migrates to the surface at night to feed on fish and crustaceans; cannibalistic on small hakes. Spawns in spring. An extremely important commercial species trawled on the West Coast. RELATED SPECIES: The deep-water hake, *Merluccius paradoxus*, is almost identical but has more vertebrae and dark spots on the gills.

105.5 Sargassum fish *Histrio histrio*

IDENTIFICATION: DI + I + I 11–13, A 7–8. Mottled brown, with a smooth skin. Many skin flaps and filaments on head and body. Pectoral fins have limb-like appearance. SIZE: 20 cm. BIOLOGY: Circumglobal, inhabiting oceanic waters and inshore reefs. It is often associated with clumps of floating seaweed.

105.6 Painted angler *Antennarius pictus*

IDENTIFICATION: DI + I + I 11–13, A 7–8. Colour variable; generally brown with irregular dark and white patches. The skin has tiny prickles. Many skin flaps and filaments on head and body. SIZE: 24 cm. BIOLOGY: One of at least five angler fish species in South Africa. Inhabits shallow reefs on the East Coast. Uses a lure-like filament on its head to attract small fishes, which it then engulfs with its large mouth.

5.1 ▲   105.2 ▼

105.3 ▲   105.4 ▼

.5 ▼   105.6 ▼

Scorpionfish, Gurnards & Lizardfish

106.1 **Raggy scorpionfish** *Scorpaenopsis venosa*

IDENTIFICATION: DXII 9, AIII 5. Skin flaps and tentacles decorate the head and body. There are straight bony ridges between the eyes. The upper spine on the gill cover is not split. SIZE: 18 cm. BIOLOGY: One of 18 Indo-Pacific species of scorpionfish inhabiting South African coral reefs and rocky areas. Preys on small fish and invertebrates.

106.2 **Popeyed scorpionfish** *Rhinopias eschmeyeri*

IDENTIFICATION: DXII 9, AIII 5. Compressed body that is highly variable in colour. The eyes are set high up on head, and the snout is long and depressed before the eye. Large tentacle over the eye. SIZE: 23 cm. BIOLOGY: An Indo-Pacific species inhabiting coral reefs and rocky areas. Preys on small fish and invertebrates.

106.3 **Devil firefish** *Pterois miles*

IDENTIFICATION: DXIII 9–11, AIII 6. A striking fish; reddish body lined with many white crossbars; dark spots on soft dorsal, anal and tail fins. Pectoral fins and sharp dorsal spines extremely long. The latter are toxic and cause painful wounds. SIZE: 30 cm. BIOLOGY: Indian Ocean; associated with calm reefs and tidal pools. Preys on crabs, shrimps and fish, preventing them from escaping by extending the pectoral fins like a net before engulfing them. RELATED SPECIES: Three other firefish are also regularly seen in Natal. *Pterois russellii* and *Pterois radiata* do not have spotted dorsal, anal and tail fins. *P. radiata* has 5–6 broad dark bars on its body; *P. russellii* numerous thin bars. *Pterois antennata* has 12 dorsal spines and a pair of long tentacles with dark bands above the eye.

106.4 **Bluefin gurnard** *Chelidonichthys kumu*

IDENTIFICATION: DIX–X + 15–16, A 14–16. Elongate red fish with large head encased in bony plates. Pectoral fins large, strikingly blue-edged, and have three rays that are not united by a web. SIZE: 60 cm. BIOLOGY: An Indo-West Pacific bottom-dweller of importance for the inshore trawl fishery. Pectoral rays act as feelers in the search for prey. RELATED SPECIES: The endemic Cape gurnard, *Chelidonichthys capensis*, is similar but does not have blue pectorals.

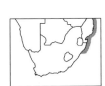

106.5 **Stonefish** *Synanceia verrucosa*

IDENTIFICATION: DXII–XIII 5–7, AIII 5–6. Ugly fish with a squat body, large head, and a row of sharp, grooved dorsal spines with poison sacs located at their bases. SIZE: 35 cm. BIOLOGY: A solitary, sedentary predator occurring on shallow reefs and sandy areas in the tropical Indo-Pacific. Wounds from this fish are potentially fatal and must be immersed in extremely hot water to break down the protein poison.

106.6 **Bartail flathead** *Platycephalus indicus*

IDENTIFICATION: DX + 13, A 13. Brown fish with a conspicuously flattened head that has smooth bony ridges. Mouth large and lower jaw longer than the upper one. Tail yellow with black crossbars. SIZE: 1 m. BIOLOGY: Indo-West Pacific, occurring principally in estuaries and shallow water. Lies camouflaged in sandy and muddy areas, where it snatches passing prey. Sexual maturity reached at 50 cm, and spawning occurs in winter and spring.

106.7 **Redband lizardfish** *Synodus variegatus*

IDENTIFICATION: D 10–13, A 8–10. Slender, elongate fish. Mouth large; fins without spines. About six red-brown vertical bars intersect a red horizontal band on the lower flanks. No scales on tail fin. SIZE: 23 cm. BIOLOGY: An Indo-West Pacific bottom-dwelling reef-fish. Snaps up small fish and crustaceans as they unsuspectingly swim by. Several similar species occur in South Africa.

106.1 ▲

106.2 ▼

106.3 ▼

106.4 ▼

106.5 ▼

106.6 ▼

106.7 ▼

Halfbeaks, Pipefish, Seahorses & Garfish

Syngnathiform fishes include many highly specialised and interesting species. Amongst them are the pipefish and seahorses, which are protected by bony plates arranged in a series of rings that encase the body. Their snouts are tubular and their tails are often prehensile.

107.1 Tropical halfbeak   *Hyporhamphus affinis*

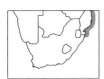

IDENTIFICATION: D 15–17, A 15–17. Elongate, silver fish with long needle-like lower jaw and scaly, short, triangular upper jaw. Dorsal and ventral fins are set far back and lack spines. Tail fin forked with lower lobe longer than upper lobe. SIZE: 26 cm. BIOLOGY: An Indo-West Pacific pelagic species that occurs around reefs. It feeds on plankton by skimming the open beak along the surface. Eggs have long filamentous strands for attachment to floating objects or seaweed. RELATED SPECIES: The endemic Cape halfbeak, *Hyporhamphus capensis*, which frequents estuaries, has a weakly forked tail. The spotted halfbeak, *Hemiramphus far*, has black spots along the flanks and no scales on the upper jaw.

107.2 Razorfish   *Aeoliscus punctulatus*

IDENTIFICATION: DIII + 10–11, A 12–13. Highly compressed body with bony plates, sharp ventral edge and long snout. Peculiar fin arrangement with dorsal fin situated at the end of the body and tail fin displaced ventrally. SIZE: 20 cm. BIOLOGY: A West Indian Ocean species that has a characteristic nose-down swimming position. Preys on planktonic crustaceans and is found along the edges of reefs.

107.3 Crowned seahorse   *Hippocampus cameleopardalis*

IDENTIFICATION: D 16–17, A 4. One of six species of seahorses found in South Africa, the crowned seahorse has a distinct crown-like structure on the head. Seahorses are small fishes with ring-like plates covering the body and a tapering prehensile tail. Single dorsal fin and no anal or tail fin. SIZE: 8 cm. BIOLOGY: Indo-Pacific and recorded offshore to a depth of 46 m. Male bears young in ventral pouch. RELATED SPECIES: The endemic Knysna seahorse, *Hippocampus capensis*, occurs in only a few South Coast estuaries. It lacks a dorsal crown.

107.4 Longsnout pipefish   *Syngnathus temminckii*

IDENTIFICATION: D 33–42, A 3. Slender, greenish-brown fish with 54–64 ring-like plates covering the body. Small mouth at end of long snout and small fin at end of tapering tail. SIZE: 30 cm. BIOLOGY: Occurs in estuaries, tidal pools and other coastal habitats often associated with plants such as eelgrass. Male carries eggs in an underslung pouch.

107.5 Trumpetfish   *Aulostomus chinensis*

IDENTIFICATION: DVIII–XIII + III 23–27, AIII 23–26. Brownish, elongate fish with a long, tube-like snout and a row of dorsal spines along the back. Dorsal and anal fins equal-sized and positioned well back on the body. SIZE: 50 cm. BIOLOGY: An Indo-West Pacific species which inhabits coral reefs where it feeds on small fish and crustaceans. Often floats head down in a vertical position. RELATED SPECIES: The closely related flutemouths, *Fistularia* species, do not have dorsal spines and possess a long filament on the tail fin.

107.6 Garfish or yellowfin needlefish   *Strongylura leiura*

IDENTIFICATION: D 18–21, A 22–26. Elongate, cylindrical body that is silvery-greenish with white undersides. Jaws elongate with rows of pointed teeth. SIZE: 1 m. BIOLOGY: An Indo-West Pacific species that occurs in coastal waters and estuaries. Usually seen at the surface, where it harries small fish.

107.1 ▲

7.2 ▲　　　　　107.3 ▲　　　　　107.4 ▲

5 ▲　　　　　107.6 ▼

Squirrelfish, Flagtails, Thornfish & Cardinals

108.1 Crown squirrelfish   *Sargocentron diadema*

IDENTIFICATION: DXI 12–14, AIV 8–10. Five species of *Sargocentron* have been recorded in South African waters. Most are small, red-coloured fish characterised by rough scales, prominent fin spines, large eyes and a long backward-pointing spine on the gill cover. The crown squirrelfish has numerous thin horizontal silvery-white stripes across the red body. The membrane between the dorsal fin spines is black. SIZE: 20 cm. BIOLOGY: Indo-Pacific, usually associated with caves in coral reefs in depths of 2–30 m. Nocturnal, feeding at night on crustaceans.

108.2 Blotcheye soldier   *Myripristis murdjan*

IDENTIFICATION: DXI 13–15, AIV 11–13. Seven species of *Myripristis* soldier fish have been recorded from South African waters. Most are small, oblong, red-coloured fish characterised by rough scales, prominent fin spines and large eyes. The blotcheye soldier has a blackish edge to the gill cover and dark red fins edged with white. SIZE: 30 cm. BIOLOGY: An Indo-West Pacific species usually associated with caves in coral reefs. Nocturnal, feeding at night on crustaceans and small fishes over reefs, but returning to daytime refuges before dawn.

108.3 Barred flagtail   *Kuhlia mugil*

IDENTIFICATION: DX 9–11, AIII 9–11. Small and silver, with dark stripes on the tail fin. Single dorsal fin, and two distinct spines on the gill cover. SIZE: 20 cm. BIOLOGY: Widespread in the Indo-Pacific, and common on reefs in KwaZulu-Natal. Feeds on crustaceans. Juveniles frequent tidal pools.

108.4 Thornfish   *Terapon jarbua*

IDENTIFICATION: DXI–XII 9–11, AIII 7–10. Small silvery fish with 3–4 curved brown stripes extending from the head across the body and onto the tail. Two prominent spines on each side of the gill cover. SIZE: 33 cm. BIOLOGY: Indo-West Pacific coastal species frequenting estuaries. Characteristically arches the body to expose spines when captured. Eats scales off live fish as part of its diet.

108.5 Crescent-tail bigeye   *Priacanthus hamrur*

IDENTIFICATION: DX 13–15, AIII 13–16. Compressed silvery-red fish with large eyes, upturned mouth, crescent-shaped tail and scales on the gill cover. Dorsal, anal and pelvic fins dusky. SIZE: 45 cm. BIOLOGY: An Indo-Pacific fish that inhabits coral and rocky reefs. Nocturnal in habit.

108.6 Bandtail cardinal   *Apogon aureus*

IDENTIFICATION: DVII + I 9, AII 8. More than 30 species of these small, robust fishes have been recorded from South African waters. They all have two dorsal fins, large eyes and a large mouth. The bandtail cardinal has a conspicuous dark band around the tail and three blue lines across the snout. SIZE: 8 cm. BIOLOGY: Indo-West Pacific, nocturnal and inhabiting coral reefs. The male carries the fertilised eggs in his mouth until they hatch.

108.7 Ninestripe cardinal   *Apogon taeniophorus*

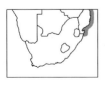

IDENTIFICATION: DVII + I 9, AII 8. Small, with two dorsal fins, large eyes and a large mouth. The ninestripe cardinal has horizontal black stripes and its tail spot is only apparent as a widening of the third stripe. SIZE: 8 cm. BIOLOGY: An Indian Ocean nocturnal species inhabiting coral reefs. The male carries the fertilised eggs in his mouth until they hatch. RELATED SPECIES: The broadstriped cardinal, *Apogon angustatus,* and the blackbanded cardinal, *Apogon cookii,* are similar but have a prominent black spot on the tail.

108.1 ▲   108.2 ▼   108.3 ▼

108.4 ▼   108.5 ▼

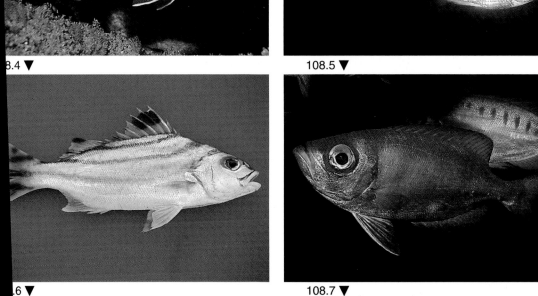

108.6 ▼   108.7 ▼

Rockcods

Fishes in the family Serranidae occur world-wide on tropical and subtropical reefs. They include the small, colourful anthiine basslets and the larger, solitary epinepheline rockcods, of which there are 30 species in South Africa.

109.1 Sea goldie *Pseudoanthias squamipinnis*

IDENTIFICATION: DX 15–17, AIII 6–7. Small. Clear differences between the sexes: females are orange-gold with an iridescent blue stripe below the eyes and males are reddish with an elongated third dorsal spine. SIZE: 10 cm. BIOLOGY: An Indo-Pacific species that is common on coral reefs. Capable of changing sex. Very popular with marine aquarists.

109.2 Coral rockcod *Cephalopholis miniata*

IDENTIFICATION: DIX 14–16, AIII 9. A striking, red-orange colour with blue spots over the body, head and tail, and the dorsal and anal fins. SIZE: 40 cm. BIOLOGY: An Indo-West Pacific species that inhabits coral and rock reefs. Occurs singly or in small groups and preys on fish and crustaceans.

109.3 Yellowbelly rockcod *Epinephelus marginatus*

IDENTIFICATION: DXI 15–16, AIII 8. Robust; mouth large, scales small. Brown with distinct yellow belly; irregular greenish-white blotches on body and dorsal fin. SIZE: 1.5 m and 30 kg. BIOLOGY: Widespread in the Atlantic, on rocky shores and reefs down to 200 m. Spawns in winter and spring; juveniles occur in tidal pools and gullies. Preys on small fishes and bottom-dwelling invertebrates.

109.4 Catface rockcod *Epinephelus andersoni*

IDENTIFICATION: DXI 13–15, AIII 8. Robust, elongate, and with a large mouth. Body is brown with dark spots, which extend over the dorsal and tail fins. Three distinctive black stripes across head. SIZE: 80 cm and 8 kg. BIOLOGY: Endemic, frequents rocky areas down to 50 m. Important inshore predator. Breeding occurs in spring in KwaZulu-Natal, and juveniles are found in tidal pools.

109.5 Potato bass *Epinephelus tukula*

IDENTIFICATION: DXI 14–15, AIII 7–9. Large, robust, grey-brown fish with dark ovate spots all over body and fins. Mouth huge. Three short spines on gill cover. Tail fin rounded. SIZE: 2 m and 100 kg. BIOLOGY: Indo-West Pacific species that frequents rock and coral reefs down to depths of 150 m. A solitary, territorial, top predator that can be aggressive to divers if provoked.

109.6 Lyretail or swallowtail rockcod *Variola louti*

IDENTIFICATION: DIX 13–15, AIII 8. Red with purple spots over body and fins. Tail fin lunate and all fins have a broad yellow margin. SIZE: 1 m. BIOLOGY: A magnificent Indo-Pacific predatory species that inhabits coral reefs and deeper offshore reefs to depths of 100 m. Preys on small reef fishes.

109.7 Koester *Acanthistius sebastoides*

IDENTIFICATION: DXI–XIII 15–17, AIII 7–8. Robust, with a large mouth and small scales. Beige body covered in brown blotches and small orange spots. Juveniles have a blue tail. SIZE: 35 cm. BIOLOGY: A southern African endemic, particularly common in the Eastern Cape. Lives in rocky areas from the shallow subtidal to about 30 m depth, where it preys on crabs and small fish.

109.8 Sixstripe soapfish *Grammistes sexlineatus*

IDENTIFICATION: DVII 13–14, AII 9. Similar in shape to rockcods but has fewer spines in fins. Dark-brown with horizontal yellow stripes across body. SIZE: 25 cm. BIOLOGY: Indo-Pacific. Solitary, inhabits reefs and tidal pools. Nocturnal in habit, and uses slimy toxic mucus produced by the skin to deter predators.

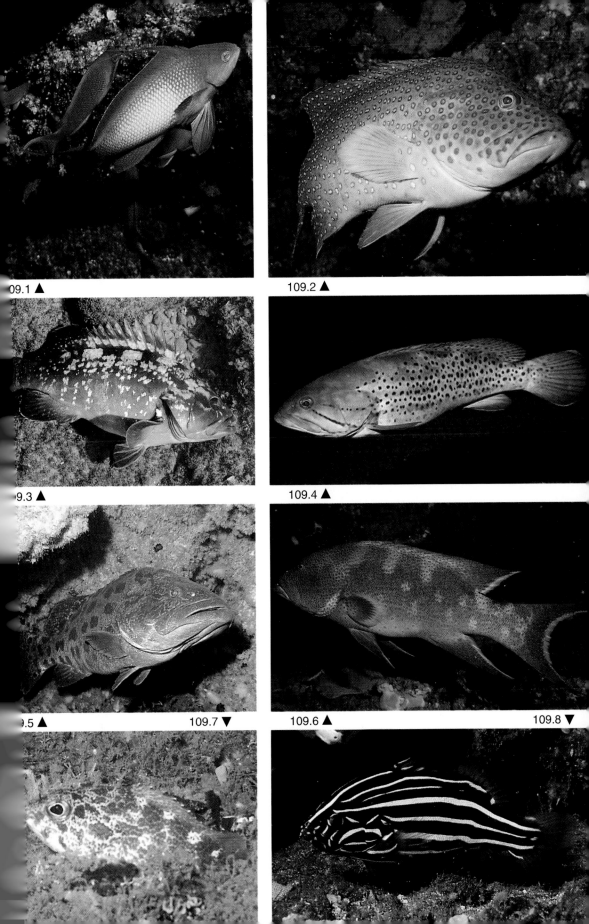

109.1 ▲

109.2 ▲

109.3 ▲

109.4 ▲

109.5 ▲   109.7 ▼

109.6 ▲   109.8 ▼

Rubberlips, Grunters & Cavebass

Nineteen species of the family Haemulidae have been recorded from reefs and estuaries in southern Africa. Grunters rub their pharyngeal teeth together after being captured, producing a characteristic sound, from which they acquire their common name. They are an important source of food for people in the tropics.

110.1 Whitebarred rubberlip *Plectorhinchus playfairi*
IDENTIFICATION: DXII 19–20, AIII 7. A dark fish characterised by four white vertical bars across the body. Lips pink and fleshy, particularly in older fishes. SIZE: 90 cm and 5 kg. BIOLOGY: A solitary West Indian Ocean species frequenting coral and rocky reefs from the surf zone to 80 m. Feeds on bottom-dwelling invertebrates.

110.2 Dusky rubberlip *Plectorhinchus chubbi*
IDENTIFICATION: DXI 16–17, AIII 7–8. Robust, bronze-grey with paler undersides. Eyes large, and lips become fleshy with age. SIZE: 75 cm and 8 kg. BIOLOGY: An Indian Ocean species frequenting reefs down to 80 m. Feeds on bottom-dwelling invertebrates and small fish. Groups of juveniles are often associated with floating seaweed.

110.3 Spotted grunter *Pomadasys commersonnii*
IDENTIFICATION: DXI 14–15, AIII 9–10. Elongate with a long sloping forehead, serrated gill covers, and small protrusible mouth. Sides and dorsal surface covered in dark spots. No spots on head. Small juveniles lack spots and also have only two anal spines. SIZE: 80 cm and 9 kg. BIOLOGY: An Indian Ocean species occurring in shallow sandy areas and estuaries. A sought-after angling species especially during summer 'grunter runs' in KwaZulu-Natal. Often seen with tails waving out of water on shallow banks where they feed on sand- or mud-prawns. Spawning occurs in KwaZulu-Natal during spring and early summer. Juveniles only occur in estuaries.

110.4 Javelin grunter *Pomadasys kaakan*
IDENTIFICATION: DXII 13–15, AIII 7–8. An oblong, silvery fish which becomes more robust with age. Juveniles have five dark vertical bars extending down flanks, a dark patch on the gill cover, and dark spots on dorsal fin membrane. SIZE: 50 cm and 6 kg. BIOLOGY: An Indo-West Pacific bottom-feeding fish that occurs in estuaries and offshore to 75 m in sandy or muddy areas. Spawning takes place in winter.

110.5 Pinky *Pomadasys olivaceum*
IDENTIFICATION: DXII 15–17, AIII 11–13. Silvery-olive with a small mouth and a distinct dark blotch on gill cover. SIZE: 30 cm. BIOLOGY: This West Indian Ocean species is one of the smallest grunters on our coast. It is particularly common in inshore waters, where it occurs in large shoals over sand and reef. Often caught for bait. An important prey item in the diets of large predators such as sharks and dolphins. Breeds all year round.

110.6 Striped grunter *Pomadasys striatum*
IDENTIFICATION: DXII 13–14, AIII 6–7. Silvery-brown with three distinct horizontal brown stripes, the lowest of which passes across the gill cover to the eye. SIZE: 22 cm. BIOLOGY: A West Indian Ocean species that occurs in small shoals in sandy areas between reefs. Feeds on bottom-dwelling invertebrates.

110.7 Cavebass *Dinoperca petersi*
IDENTIFICATION: DIX–XI 18–20, AIII 12–14. A dark-brown, oval fish with white specks on the body and a black band across the cheek. Anterior dorsal and anal rays markedly longer than posterior ones. Scales extend onto head and fins. SIZE: 60 cm. BIOLOGY: An Indian Ocean species that lives under rocky ledges or in caves. Can make loud drumming noises.

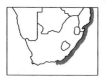

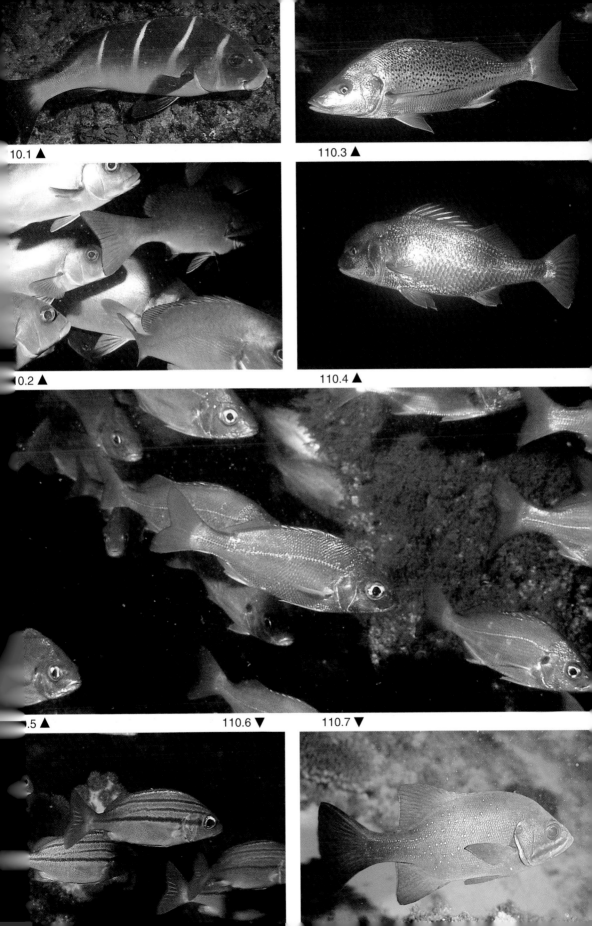

10.1 ▲

110.3 ▲

0.2 ▲

110.4 ▲

.5 ▲　　　　　　110.6 ▼　　　　110.7 ▼

Fusiliers & Snappers

The snapper family, Lutjanidae, is represented in southern Africa by 24 species. All are carnivorous, and many are vividly coloured and form shoals. They are important food fishes in the tropics.

111.1 Yellowback fusilier *Caesio xanthonota*

IDENTIFICATION: DX 15, AIII 12. Elongate blue and yellow fish with protrusible upper jaw and forked tail fin. Yellow dorsal surface extends onto head. SIZE: 30 cm. BIOLOGY: Occurs in vast shoals over Indo-Pacific coral reefs. A plankton-eater often sighted by divers at Sodwana Bay.

111.2 Blue-and-gold fusilier *Caesio caerulaurea*

IDENTIFICATION: DX 14–15, AIII 11–13. Elongate, with a protrusible upper jaw. Tail fin forked, with dark band on each lobe. Dorsal surface blue; conspicuous yellow horizontal stripe across flank to head. SIZE: 30 cm. BIOLOGY: Occurs in vast shoals over Indo-Pacific coral reefs. Planktivorous; often sighted by divers.

111.3 Kaakap or green jobfish *Aprion virescens*

IDENTIFICATION: DX 11, AIII 8. Sleek and elongate. Dark-green to bluish dorsal surface, paler underside. Tail forked. Snout blunt, grooved below nostrils. SIZE: 1.1 m and 11 kg. BIOLOGY: An Indo-Pacific predator usually found in the water column above coral or rocky reefs. Sexual maturity is reached at about 70 cm. Frequently caught off Zululand by ski-boat anglers and spearfishermen.

111.4 Twinspot snapper *Lutjanus bohar*

IDENTIFICATION: DX 13–14, AIII 8. Robust; reddish-purple with darker fins. Large mouth; prominent canines. Nostrils in a deep groove. Juveniles have two white dorsal spots. SIZE: 75 cm and 11 kg. BIOLOGY: Indo-Pacific. Usually solitary in caves down to 70 m. Preys on crustaceans, squid and fish. Sexual maturity attained at about 50 cm.

111.5 Bluebanded snapper *Lutjanus kasmira*

IDENTIFICATION: DX 14–15, AIII 7–8. Yellow. Four horizontal, black-edged blue stripes across body. Sometimes a dark blotch below the dorsal fin rays. Fins yellow. SIZE: 40 cm. BIOLOGY: Indo-West Pacific. Shoals over coral and rocky reefs to depths of 60 m. Preys on crustaceans and small fish. Sexual maturity occurs at 20 cm, and spawning takes place in late winter and spring.

111.6 Onespot snapper *Lutjanus monostigma*

IDENTIFICATION: DX 12–14, AIII 8. Yellow with a reddish head. Conspicuous black spot on the side below the soft dorsal fin. Fins yellow. SIZE: 40 cm. BIOLOGY: An Indo-West Pacific species that occurs on coral reefs and feeds on crustaceans. RELATED SPECIES: *L. russellii* has a larger spot above the lateral line and usually 7–8 narrow longitudinal stripes.

111.7 River snapper *Lutjanus argentimaculatus*

IDENTIFICATION: DX 13–14, AIII 7–8. Robust and reddish-brown with paler underside and red fins. A large mouth and prominent canines in both jaws. Juveniles have vertical bars. SIZE: 1 m and 12 kg. BIOLOGY: An Indo-West Pacific species found in estuaries and over reefs. Spawning occurs at sea, and juveniles recruit to estuaries where they frequent mangrove areas. Preys on fish and crabs.

111.8 Humpback snapper *Lutjanus gibbus*

IDENTIFICATION: DX 13–14, AIII 8. Deep-bodied, red with a steep forehead. All fins except pectorals are fringed with white. Tail fin markedly forked with rounded lobes. A distinct notch on the gill cover. SIZE: 50 cm. BIOLOGY: Indo-West Pacific; inhabits rocky and coral reefs down to depths of 120 m. Preys on invertebrates. Sexually mature at 25 cm. Juveniles occur in tropical estuaries.

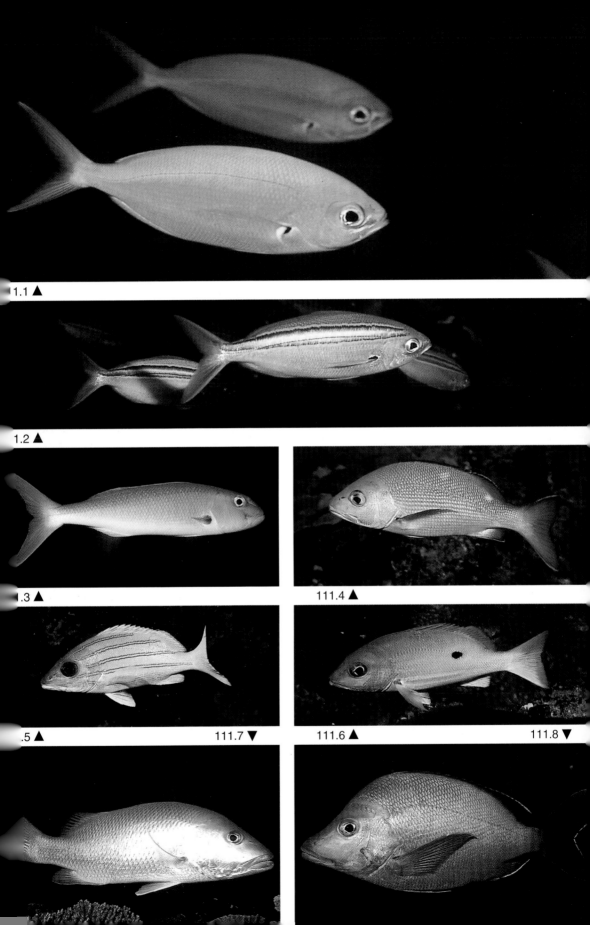

1.1 ▲

1.2 ▲

.3 ▲

111.4 ▲

.5 ▲   111.7 ▼   111.6 ▲   111.8 ▼

Sea Breams

The family Sparidae is particularly diverse in southern Africa, with 41 species occurring in our waters. Of these, 25 species are endemic. Sea breams are extremely important in the line-fishery. Sparids attain ages in excess of 30 years. Many are hermaphroditic: some are protandrous, changing sex from male to female, while others are protogynous, being female first and later changing sex and becoming male.

112.1 King soldierbream *Argyrops spinifer*

IDENTIFICATION: DXI–XII 10–11, AIII 8. Pink with blue fins and blue dots on scales. Blunt, angular head with the scaling between eyes ending in a point. First two dorsal spines very small but third to seventh spines are elongate. SIZE: 70 cm. BIOLOGY: Indo-West Pacific; occurs down to depths of 150 m.

112.2 River bream *Acanthopagrus berda*

IDENTIFICATION: DXI–XII 10–13, AIII 8–9. Deep-bodied, grey, with a steep forehead, pointed snout, forked tail, and yellow anal fin. Second anal spine large. SIZE: 75 cm and 3 kg. BIOLOGY: Widespread throughout the Indo-Pacific and usually found in estuaries. Feeds on bottom-dwelling invertebrates; spawns in estuary mouths in winter.

112.3 Fransmadam *Boopsoidea inornata*

IDENTIFICATION: DXI 10–11, AIII 11. A small, deep-bodied silvery fish with a bronze sheen, dark fins, a distinct lateral line, and large eyes. Snout pointed; mouth small and bears many fine, sharp teeth. Juveniles are bright orange in colour. SIZE: 40 cm. BIOLOGY: A shoaling fish usually found over rocky reefs to depths of 30 m. Omnivorous, with juveniles frequenting inshore seaweed beds particularly in the summer months. Regarded as a pest by anglers as it nibbles bait intended for larger fish.

112.4 Carpenter *Argyrozona argyrozona*

IDENTIFICATION: DXII 10, AIII 8. A moderately elongate, silvery-pink fish with reddish head and fins. Prominent canines but no molars. SIZE: 90 cm. BIOLOGY: A carnivore of commercial importance. Occurs offshore to depths of 200 m. Reaches sexual maturity at 22 cm, and spawns on the Agulhas Bank during spring and summer.

112.5 Santer *Cheimerius nufar*

IDENTIFICATION: DXI–XII 10–11, AIII 8. Silvery-pink fish, with an oval-shaped body, well-developed fins and strong canines. Juveniles have five vertical bars on their sides. Anal and pelvic fins have a bluish tinge, and the third to seventh dorsal spines are elongate. SIZE: 75 cm and 7 kg. BIOLOGY: An Indo-Pacific species found in shoals over reefs to a depth of 80 m. Preys on fish, squid and crustaceans. Sexual maturity is reached at 35 cm at an age of 4 years. Regularly caught by ski-boat fishermen.

112.6 Englishman *Chrysoblephus anglicus*

IDENTIFICATION: DXII 10, AIII 8. Deep-bodied, pink with red vertical bars and well-developed pink fins. Forehead blunt, almost vertical; mouth small, armed with canines and several rows of molars. SIZE: 1 m and 6 kg. BIOLOGY: Occurs offshore in depths of 20–100 m, often in association with slinger (113.4). Feeds on crustaceans, molluscs and fish. Sexual maturity is reached at about 40 cm.

112.1 ▲

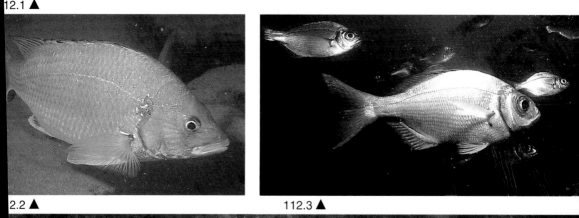

112.2 ▲ 112.3 ▲

112.4 ▲ 112.5 ▼ 112.6 ▼

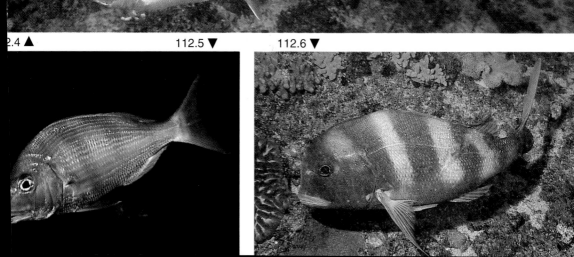

113.1 Dageraad   *Chrysoblephus cristiceps*
IDENTIFICATION: DXII 10, AIII 8. Deep-bodied, red fish with a steep, pointed head and moderately large mouth bearing enlarged canines and several rows of molars. Characteristic dark spot on the dorsal surface at the end of the dorsal fin, golden sheen over the gill cover, and blue mark under the eye. SIZE: 70 cm and 8 kg. BIOLOGY: Found over reefs to depths of 100 m. Preys on benthic invertebrates and small fish. Undergoes sex change: the small fish are females and change into males at about 40 cm. Spawning occurs in summer. An important component of the commercial and recreational ski-boat catch.

113.2 **Red stumpnose**   *Chrysoblephus gibbiceps*
IDENTIFICATION: DXI–XII 10–11, AIII 7–9. Silvery-pink, with 5–7 red vertical bars and numerous irregular black spots over the body. Steep forehead, which develops a conspicuous bump in old males. Pectoral fin long and tail fin strongly forked. SIZE: 75 cm and 8 kg. BIOLOGY: Frequents deep reefs down to 150 m. Preys on benthic invertebrates and small fish. Attains sexual maturity at 35 cm, and spawns in summer.

113.3 **Roman**   *Chrysoblephus laticeps*
IDENTIFICATION: DXI–XII 10–11, AIII 7–9. A robust scarlet-red fish with a prominent white saddle on its back, a white bar on the gill cover, and a blue line between the eyes. Conspicuous canines and several rows of molars in both upper and lower jaws. SIZE: 50 cm and 4 kg. BIOLOGY: A benthic predator that occurs on rocky reefs down to 100 m. Undergoes sex reversal from female to male at about 30 cm. Spawns in summer, and juveniles frequent shallow reefs. Sought after by ski-boat fishermen, commercial line-boats and spearfishermen.

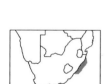

113.4 **Slinger**   *Chrysoblephus puniceus*
IDENTIFICATION: DXII 10, AIII 8. Deep-bodied, silvery-pink, with a steep forehead and small mouth. Conspicuous blue bar below eyes and iridescent blue spots on the body. SIZE: 85 cm and 4 kg. BIOLOGY: A deep-water shoaling reef-fish that is extremely important in the line-fisheries of Natal and Moçambique. A benthic carnivore that undergoes sex reversal from female to male. Spawning occurs in spring. Sex ratio varies, depending on the level of exploitation, with up to 100 females per male in heavily fished areas.

113.5 **False Englishman**   *Chrysoblephus lophus*
IDENTIFICATION: DXI 10, AIII 9. Deep-bodied and silvery-pink with about six vertical red bars. Steep forehead with a pitted bony area between the eyes. Pectoral fins long, tail fin deeply forked, and third to sixth dorsal spines markedly elongated. SIZE: 50 cm. BIOLOGY: Associated with offshore reefs down to 150 m in KwaZulu-Natal. A benthic carnivore caught by ski-boat fishermen.

113.6 **White karanteen**   *Crenidens crenidens*
IDENTIFICATION: DXI 11, AIII 10. A small, greyish fish with several faint horizontal stripes. Mouth small with many multi-pointed incisors. Anal fin often yellow. SIZE: 30 cm. BIOLOGY: A shoaling West Indian Ocean species generally found in shallow water in bays and estuaries. Omnivorous, feeding on seaweeds and invertebrates.

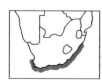

113.7 **Poenskop**   *Cymatoceps nasutus*
IDENTIFICATION: DXII 10, AIII 8. Large, robust, greyish-black; snout rounded and becoming white and fleshy with age. There are four prominent canines in the upper jaw and six in the lower jaw; two rows of rounded molars in each jaw. Juveniles are greenish-brown with white spots. SIZE: 1 m and 38 kg. BIOLOGY: Occurs on both shallow and deep rocky reefs. Carnivorous, feeding on echinoderms, molluscs and crustaceans. Spawns in winter; juveniles frequent shallow weedy areas. Much sought after by shore and ski-boat fishermen.

3.1 ▲

113.2 ▲

.3 ▲

113.4 ▲

.5 ▲

113.6 ▼

113.7 ▼

114.1 Zebra *Diplodus hottentotus*

IDENTIFICATION: DXI 12–13, AIII 11. An oval-shaped, golden-yellow fish with six distinct vertical black bars, of which the front one runs through the eye. The mouth is surrounded by fleshy lips, and a single row of sharp incisors and several rows of molars occur in each jaw. Juveniles have black pelvic fins. SIZE: 60 cm and 6 kg. BIOLOGY: Found in rocky areas down to 60 m, with juveniles in tidal pools. Omnivorous. Spawns in spring and summer. Caught on rocky shores by light-tackle anglers.

114.2 Blacktail *Diplodus capensis*

IDENTIFICATION: DXII 14–15, AIII 13–14. Oval-shaped silver fish with a marked black patch on the tail peduncle. Juveniles have several thin vertical crossbars. Mouth small with eight incisors and several rows of molars in each jaw. SIZE: 45 cm and 3 kg. BIOLOGY: A ubiquitous fish; this subspecies occurs all around our coast in a range of habitats. Juveniles are common in tidal pools and the lower reaches of estuaries. Omnivorous. Spawns throughout the year with a peak in late winter and spring. A fine light-tackle angling species.

114.3 Janbruin *Gymnocrotaphus curvidens*

IDENTIFICATION: DX 11–12, AIII 9–10. Robust and brown with conspicuous blue eyes. Protruding curved incisors and smaller conical teeth in each jaw. SIZE: 50 cm. BIOLOGY: Found on shallow reefs to depths of 30 m. Nowhere abundant and usually solitary, this species appears to be omnivorous.

114.4 White steenbras *Lithognathus lithognathus*

IDENTIFICATION: DXI 10, AIII 8. Large, elongate silvery fish with long sloping forehead and pig-like snout. Head longer than the length of the pectoral fin. Dark bars along flanks. SIZE: 1 m and 30 kg. BIOLOGY: A prized angling species that is caught along sandy beaches and in estuaries. Adults undertake spawning migrations up the East Coast during winter, and juveniles recruit to estuarine nursery areas in spring. Preys on sand-dwelling benthic invertebrates. RELATED SPECIES: The West Coast steenbras, *Lithognathus aureti*, occurs in large shoals in sandy areas, particularly north of Walvis Bay. The head is shorter than the length of the pectoral fin.

114.5 Sand steenbras *Lithognathus mormyrus*

IDENTIFICATION: DXI 12–13, AIII 10–11. A small, elongate silver fish with 12–14 brown vertical bars and brown fins. SIZE: 30 cm. BIOLOGY: A widely-spread sand-dwelling species that occurs around most of the African coast and in the Mediterranean. Abundant along the South Coast in sandy bays to depths of 30 m, where it preys on benthic invertebrates. Spawns in summer.

114.6 Red tjor-tjor or sand soldier *Pagellus bellottii natalensis*

IDENTIFICATION: DXII 10–11, AIII 10. A small pink fish with red fins and a red spot on the operculum. Mouth small with sharp conical incisors and two series of small molars. SIZE: 35 cm. BIOLOGY: A common shoaling fish over sandy areas, with juveniles particularly common in shallow South Coast bays. Preys on small benthic invertebrates and is important in the diet of larger coastal predators. Spawns in winter; larvae are common along the East Coast.

114.7 Dane *Porcostoma dentata*

IDENTIFICATION: DXIII 10–11, AIII 8–9. Plump orange-red fish with a purple band joining the eyes and a prominent red blotch at the base of the pectoral fin. Tail and anal fin fringed with white. Jaws with 4–6 projecting canines. SIZE: 30 cm. BIOLOGY: Forms small shoals over reefs and pinnacles in depths of 20–100 m. Preys on reef invertebrates. Spawning occurs in spring.

4.1 ▲

114.2 ▲

.3 ▲

114.4 ▲

5 ▲

114.6 ▲

114.7 ▼

115.1 Hottentot *Pachymetopon blochii*

IDENTIFICATION: DX–XI 11–12, AIII 10. A plump dark-grey fish with dark fins. Mouth small with five rows of incisors in each jaw but no molars. SIZE: 50 cm and 3 kg. BIOLOGY: This species supports a commercial line-fishery in the Western Cape. Generally found in small shoals in kelp beds. Omnivorous. Can attain an age of 12 years. Sexual maturity is reached at 25 cm, and spawning occurs throughout the year with peaks in summer and autumn.

115.2 Blue hottentot *Pachymetopon aeneum*

IDENTIFICATION: DX–XI 11–13, AIII 10. A robust silvery-bronze fish with a blue head. Mouth small with five rows of incisors in each jaw but no molars. SIZE: 60 cm. BIOLOGY: Frequents reefs from 20 to 80 m depth, though juveniles are common in shallower areas. Sexual maturity is reached at 25 cm. Spawns in summer. Preys on sessile invertebrates. Rarely caught from the shore.

115.3 Bronze bream *Pachymetopon grande*

IDENTIFICATION: DXI 11, AIII 10–11. Plump, oval-shaped and brown with a smallish head and a bump over the eyes. Five rows of incisors in each jaw but no molars. No scales on front of gill cover. SIZE: 70 cm and 5 kg. BIOLOGY: Occurs along shallow, often turbulent, rocky shores where it feeds on seaweeds and associated invertebrates. Spawning occurs in late summer and autumn. SIMILAR SPECIES:

115.4 *Polyamblydon germanum*, the German, is easily confused with the bronze bream, which is similar in appearance, but it has molars, only a single row of incisors, and scales on the gill cover. It occurs on deeper reefs from East London to Maputo.

115.5 Red steenbras *Petrus rupestris*

IDENTIFICATION: DXI 10–11, AIII 8. A large, robust reddish-to-copper-coloured fish with prominent canines in both jaws. Fins are dark red. Juveniles are orange with red spot on tail peduncle. SIZE: 2 m and 70 kg. BIOLOGY: A famous angling species, but its liver is poisonous due to the high vitamin A content. This powerful predator is found over deep reefs down to 160 m, where it preys on octopus and fishes. Sexual maturity is attained at 60 cm, and spawning occurs in spring. Juveniles occur over shallow reefs. Attains 30 years in age.

115.6 Scotsman *Polysteganus praeorbitalis*

IDENTIFICATION: DXII 10, AIII 8. Large, pinkish-red fish with a steep forehead and deep body tapering towards the tail. There are numerous blue dots on the sides and blue lines encircling the small eyes. The large mouth has 4–6 canines in each jaw and several rows of smaller conical teeth. SIZE: 90 cm and 10 kg. BIOLOGY: Lives on offshore reefs from 20 to 120 m. This powerful, solitary predator is caught mainly by skiboaters, but the stock has been severely depleted.

115.7 Seventy-four *Polysteganus undulosus*

IDENTIFICATION: DXII 10, AIII 8–9. A pinkish-red fish with 4–6 iridescent horizontal blue stripes along its sides and a marked black blotch above the pectoral fin. The head is convex with 4–6 canines and many small teeth in each jaw. SIZE: 1 m and 14 kg. BIOLOGY: Occurs on deep reefs to 200 m depth. Carnivorous on fish and squid. The stock is severely depleted due to overfishing on spawning aggregations.

115.8 Panga *Pterogymnus laniarius*

IDENTIFICATION: DXII 10, AIII 8. Pink, with several blue horizontal lines along its sides. Fleshy, furry lips with 4–6 canines in each jaw. SIZE: 45 cm. BIOLOGY: Common off the South Coast where it occurs over reefs and sand to depths of 140 m. An important component of the line- and trawl-fisheries. It is carnivorous on crustaceans, squid and small fishes, and spawns during the summer months.

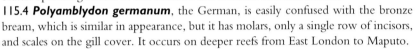

5.1 ▲

115.2 ▲

.3 ▲

115.4 ▲

5 ▲ 115.7 ▼

115.6 ▲ 115.8 ▼

116.1 Musselcracker   *Sparodon durbanensis*

IDENTIFICATION: DXI 11–12, AIII 10. Large, robust, silver-grey fish with a large head, blunt snout and powerful jaws. Four prominent incisors and rows of crushing molars. Juveniles have bright-orange fins. SIZE: 1.1 m and 23 kg. BIOLOGY: Found along rocky shores and shallow reefs where it preys on molluscs, red-bait and crabs. Spawning occurs in summer, and juveniles are common in Eastern Cape tidal pools. Initially, juveniles feed on algae but as molars develop they begin to take hard-shelled prey. This is a highly sought-after rocky-shore angling species.

116.2 Steentjie   *Spondyliosoma emarginatum*

IDENTIFICATION: DXI 11–13, AIII 10. Small, grey-blue, with dark fins and some pale yellow horizontal lines along upper sides. Jaws with 4–6 rows of pointed incisors and a single row of molars. SIZE: 45 cm. BIOLOGY: Occurs in large shoals above reefs to 30 m depth. The male constructs a nest in sand next to reefs and guards the eggs until they hatch. Feeds on benthic invertebrates. Steentjies are a favoured bait species.

116.3 Strepie   *Sarpa salpa*

IDENTIFICATION: DXI 14–16, AIII 13–15. Small, elongate, silver-green, with 8–10 horizontal yellow stripes running from head to tail. Upper incisors notched, lower incisors pointed; no molars. SIZE: 40 cm. BIOLOGY: An eastern Atlantic species that also occurs in the Mediterranean. Shoals are common in rocky and sandy areas, and it is also found in the lower reaches of estuaries. Spawns in KZN in winter and spring. Herbivorous and an excellent bait fish. Juveniles feed on small crustaceans.

116.4 Natal stumpnose   *Rhabdosargus sarba*

IDENTIFICATION: DXI 12–13, AIII 11. A silver-grey fish with a yellow streak on the belly. Yellow dots on the scales form faint horizontal lines over the body. Jaws have 6–8 compressed teeth in the front and rounded molars at the back. SIZE: 80 cm and 7 kg. BIOLOGY: An Indo-West Pacific species that frequents estuaries, shallow reefs and sandy areas to depths of 50 m. Preys on benthic invertebrates and spawns in winter and spring. Juveniles recruit to estuarine nursery areas.

116.5 Cape stumpnose   *Rhabdosargus holubi*

IDENTIFICATION: DXI 12–13, AIII 10–11. A silver fish with a blunt head and a conspicuous mid-lateral yellow stripe from head to tail. Jaws have 6–8 compressed incisors and rounded molars. In juveniles each incisor has three cusps. SIZE: 40 cm and 2 kg. BIOLOGY: Adults are found in estuaries and over inshore reefs. Juveniles occur chiefly in estuarine nursery areas where they graze in eelgrass beds. Adults feed on benthic invertebrates, and spawn in winter and spring.

116.6 White stumpnose   *Rhabdosargus globiceps*

IDENTIFICATION: DXI 11–13, AIII 10–11. Silvery, with 6–7 dark vertical crossbars and a blunt head. Jaws have 4–8 enlarged incisors and several rows of molars. SIZE: 50 cm and 3 kg. BIOLOGY: Found over sandy and rocky areas to 80 m depth. An omnivore which spawns during summer, with juveniles recruiting to estuarine nursery areas. Caught by line and seine-net fishermen in the Western Cape.

116.7 Bigeye stumpnose   *Rhabdosargus thorpei*

IDENTIFICATION: DXI 13, AIII 12. A silver fish with large eyes, a broad yellow band on the underside, and yellow anal and pelvic fins. There are several thin yellow stripes along the sides as well. Jaws have six compressed teeth in the front and four rows of molars at the back. SIZE: 40 cm and 1 kg. BIOLOGY: Occurs around coral and rocky reefs, where it feeds on molluscs and crustaceans. Juveniles are common in northern KwaZulu-Natal estuaries.

6.1 ▲

116.2 ▲

6.3 ▲

116.4 ▲

5 ▲

116.6 ▲

116.7 ▼

Galjoens & Knifejaws

117.1 Galjoen *Dichistius capensis*

IDENTIFICATION: DX 18–19, AIII 13–14. A deep-bodied, robust fish that varies from grey to black. Fin spines prominent; mouth small with curved incisors. Small scales cover the entire fish including the fins. SIZE: 80 cm and 6 kg. BIOLOGY: Southern African endemic. Frequents turbulent water off rocky and sandy shores. Omnivorous. Spawns in summer. Tagging has shown that individuals can remain in the same area for several years. South Africa's national fish and a popular Cape angling species. Previously named *Coracinus capensis*.

117.2 Banded galjoen *Dichistius multifasciatus*

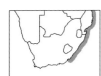

IDENTIFICATION: DX 21–23, AIII 13–14. Deep-bodied and robust; grey with about 14 alternating wide and narrow vertical dark bands. Small scales cover the entire body including the fins. Fin spines prominent. Mouth small, incisors curved, lips fleshy. SIZE: 30 cm. BIOLOGY: South-east African endemic; inhabits turbulent waters off rocky and sandy shores. Eats benthic invertebrates. Spawns off KZN in winter and spring. Juveniles occur in tidal pools and gullies. Formerly named *Coracinus multifasciatus*.

117.3 Stonebream *Neoscorpis lithophilus*

IDENTIFICATION: DVI–VIII 20–25, AIII 23–26. Silvery-grey, kite-shaped, with short anal and dorsal spines. Small scales cover the entire body. Mouth small, with rows of fine teeth. SIZE: 50 cm. BIOLOGY: Endemic; occurs in turbulent water along rocky and sandy shores. Feeds on seaweeds. Sexually mature at 35 cm. In KZN, spawning occurs in spring and summer. Juveniles occur in tidal pools and shallow gullies.

117.4 Grey chub *Kyphosus bigibbus*

IDENTIFICATION: DXI 12, AIII 11. Robust, grey, oval-shaped, with a forked tail and short pectoral fins. Mouth small with curved incisors. Brown band from corner of the mouth to the gill cover. SIZE: 70 cm and 10 kg. BIOLOGY: An Indo-West Pacific species that occurs over subtropical reefs. Herbivorous, feeding on seaweeds. Spawns during spring. Juveniles are often associated with drifting seaweed and other flotsam. RELATED SPECIES: The soft dorsal rays of the blue chub, *Kyphosus cinerascens*, are longer than the dorsal spines. The brassy chub, *Kyphosus vaigiensis*, has 14 dorsal rays and bronze lines.

117.5 Cape knifejaw *Oplegnathus conwayi*

IDENTIFICATION: DXII 11–14, AIII 11–13. A dark, oblong fish with teeth that are fused to form a parrot-like beak. Small scales cover the whole of the body. Juveniles are bright yellow with vertical black bands across the head and in front of the tail. SIZE: 90 cm and 6 kg. BIOLOGY: A South African endemic which, though rarely caught by anglers, is frequently seen by divers and shot by spearfishermen. It is found over inshore reefs, often in territorial pairs. Omnivorous; feeds on seaweeds, sponges and red-bait.

117.6 Natal knifejaw *Oplegnathus robinsoni*

IDENTIFICATION: DXI 20–24, AIII 14–17. Oblong, grey. Teeth fused to form parrot-like beak. Similar to previous species but its dorsal and anal rays are more numerous and its tail is more forked. Juveniles (117.6j) are bright yellow with five vertical black bars across the body. SIZE: 60 cm. BIOLOGY: Endemic. Occurs on reefs down to 100 m from Durban northwards. Feeds on reef-dwelling invertebrates. Rarely taken by line but frequently seen by divers.

117.7 Whitespotted rabbitfish *Siganus sutor*

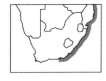

IDENTIFICATION: DXIV 10, AVII 9. Oval-shaped, with prominent fin spines. Mouth small with tiny teeth. Colour variable; live fish often lack the white spots or marbled patterns evident on capture. SIZE: 30 cm. BIOLOGY: A herbivorous, shoaling, West Indian Ocean species, associated with shallow weedy areas including estuaries. Spines extremely sharp and coated with toxic mucus.

117.1 ▲

117.2 ▲

.3 ▲

117.4 ▲

5 ▲

117.6j ▼

117.7 ▼

Smelts, Glassies, Kobs, Soapies & Pursemouths

118.1 Silver smelt   *Sillago sihama*

IDENTIFICATION: DXI + I 20–23, AII 21–23. Small, silvery, cylindrical fish with long snout. The scales are easily shed. SIZE: 30 cm. BIOLOGY: An Indo-Pacific species frequenting estuaries and shallow sandy shores. Often partially buries itself in the sand. Feeds on sand-dwelling benthic invertebrates.

118.2 Glassy   *Ambassis gymnocephalus*

IDENTIFICATION: DVI–VIII 8–10, AIII 8–10. A small, brownish, translucent fish with a silver band along the sides and a deeply notched dorsal fin. There is a spine in front of the eye. SIZE: 10 cm. BIOLOGY: An Indo-Pacific species frequenting estuaries, where it completes its life cycle. Occurs in small shoals and filter-feeds on plankton.

118.3 Squaretail kob *Argyrosomus thorpei*

IDENTIFICATION: DX + I 26–29, AII 7. Silvery, with a row of white spots on its sides. Scales present behind pectoral fin, distinguishing it from other species of kob. SIZE: 1 m and 12 kg. BIOLOGY: Predatory; offshore. RELATED SPECIES: The dusky kob *Argyrosomus japonicus* and the silver kob *A. inodorus* are very similar but have a dark blotch at the base of the pectoral fin that lacks scales. *A. japonicus* (1.9 m, 75 kg) predominates in warmer waters, nearshore and in estuaries, and has a deeper, shorter tail-fin peduncle. *A. inodorus* is smaller (1.4 m, 36 kg) and occurs west of Kei River in deeper waters; tail-fin peduncle narrower and longer. Both were previously named *A. hololepidotus*. All three species are important commercial and recreational linefish. The small kob *Johnius dussumieri*, plentiful in KZN estuaries, has a pointed tail. The snapper kob *Otolithes ruber* (KZN) has two prominent canines in each jaw.

118.4 Geelbek   *Atractoscion aequidens*

IDENTIFICATION: DX + I 27–31, AII 8–9. Large, elongate and silvery with a blue-grey dorsal surface; tail fin slightly concave. Interior of mouth and gill cover yellow. SIZE: 1.2 m and 23 kg. BIOLOGY: A predator that occurs near reefs and in the water column. Shoals of reproductively ripe fish migrate from the Cape to KZN in winter with the sardine run. Sexually mature at 60 cm. Caught by ski-boat and line fishermen.

118.5 Baardman   *Umbrina canariensis*

IDENTIFICATION: DX + I 25–29, AII 7. Grey-brown, with a curved dorsal surface and a short, thick barbel on chin. Juveniles have narrow oblique dark bars. SIZE: 55 cm and 10 kg. BIOLOGY: Occurs over sandy and reef areas on the West, South and East coasts of Africa. Preys on benthic invertebrates. Juveniles are usually found in shallow water. RELATED SPECIES: The slender baardman, *Umbrina ronchus*, occurs in KZN. Difficult to distinguish underwater, but more slender and has fewer dorsal rays.

118.6 Slender soapy   *Secutor insidiator*

IDENTIFICATION: DVIII–IX 15–17, AIII 14. A silvery, highly compressed oval fish with a protrusible mouth, which points upwards. The upper flanks have a series of black dots. The fins are dusky. SIZE: 15 cm. BIOLOGY: Occurs in shoals in shallow coastal waters of the Indo-West Pacific. Spawning occurs along the KZN coast in summer, and juveniles are common in estuaries, where they feed on plankton. RELATED SPECIES: *Leiognathus equula* is similar but has a downward-pointing mouth. A thick layer of mucus covers the body, and there are faint vertical bars and a dark saddle on the tail peduncle.

118.7 Smallscale pursemouth   *Gerres acinaces*

IDENTIFICATION: DIX–X 9–11, AIII 7. Small, silvery, with a series of grey-brown spots along sides. Snout pointed and mouth protrusible. Tail fin dark, deeply forked. SIZE: 25 cm. BIOLOGY: Occurs in shoals in estuaries of the tropical and subtropical Indo-Pacific. Preys on benthic invertebrates and spawns at sea.

118.1 ▲

118.2 ▲

.3 ▲

4 ▲

118.6 ▼

118.5 ▲

118.7 ▼

Batfish & Angelfish

119.1 Orbicular batfish *Platax orbicularis*

IDENTIFICATION: DV 36–37, AIII 26–27. Deep, compressed body with elongate dorsal, anal and pelvic fin rays. Juveniles yellowish with dark vertical bars that fade with age. SIZE: 30 cm. BIOLOGY: An Indo-West Pacific species that frequents coral reefs, seaweed beds, shipwrecks and jetties. Preys on benthic invertebrates.

119.2 Cape moony *Monodactylus falciformis*

IDENTIFICATION: DVIII 25–30, AIII 25–29. A compressed, kite-shaped, silver fish with rudimentary pelvic fins. Eyes large; mouth small with rows of tiny needle-like teeth. Lobes of dorsal and anal fins dark. Juveniles have about a dozen dark vertical bars that fade with age. SIZE: 30 cm. BIOLOGY: A shoaling West Indian Ocean species that frequents shallow reefs, bays and estuaries. Feeds on plankton; usually seen in mid-water. Spawns off estuary mouths and beaches in summer; juveniles often found in estuarine weed-beds. RELATED SPECIES: The Natal moony, *Monodactylus argenteus,* is very similar but the teeth are flattened and tricuspid. Dorsal and anal fins are generally yellow but have dark anterior lobes. Juveniles have only two dark stripes over the head.

119.3 Spadefish *Tripterodon orbis*

IDENTIFICATION: DIX 19–21, AIII 15–17. Deep, compressed, oblong body with long dorsal spines. Head rounded, lips thick, mouth small. Juveniles have several vertical dark bands that fade with age. SIZE: 75 cm and 7 kg. BIOLOGY: A West Indian Ocean species that frequents inshore reefs. Preys on planktonic and bottom-dwelling invertebrates.

119.4 Old woman *Pomacanthus rhomboides*

IDENTIFICATION: DXI–XII 22–25, AIII 21–23. Oval-shaped, compressed fish with large anal and dorsal fins. Adults are dull brown but the hind part of the body is paler. Strong spine on the edge of the gill cover. Juveniles are black with 15–20 vertical blue-white bars. SIZE: 45 cm. BIOLOGY: Occurs in the West Indian Ocean on coral reefs, where it feeds on encrusting invertebrates. Juveniles are found in tidal pools.

119.5 Emperor angelfish *Pomacanthus imperator*

IDENTIFICATION: DXIV 19–21, AIII 19–20. Adults are oval-shaped and compressed, with almost horizontal yellow stripes, a yellow tail, a blue-edged dark band over the eyes, and a white mouth. There is a conspicuous spine on the gill cover. Juveniles (119.5j) are strikingly different, with concentric blue and white lines on an almost black body. SIZE: 40 cm. BIOLOGY: Indo-West Pacific; occurs on coral reefs, where it feeds on sponges and other invertebrates. Juveniles often found in tidal pools particularly in the late summer months.

119.6 Jumping bean *Centropyge acanthops*

IDENTIFICATION: DXIV 16–17, AIII 16–18. A small, brilliant-blue fish with golden head and dorsal surface. SIZE: 8 cm. BIOLOGY: A West Indian Ocean species frequenting coral reefs and often found hiding amongst coral rubble. Feeds on small invertebrates. A popular aquarium fish.

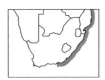

119.7 Semicircle angelfish *Pomacanthus semicirculatus*

IDENTIFICATION: DXIII 21–23, AIII 20–22. Oval-shaped and compressed with elongate filaments on the dorsal and anal fins. Adults are greenish-brown with numerous blue spots on the body and fins. The gill cover has a blue margin and a conspicuous spine. Juveniles are black with alternating blue-and-white semicircular lines, and pale blue edges to the dorsal and anal fins. SIZE: 40 cm. BIOLOGY: Indo-West Pacific, inhabiting coral reefs. Feeds on algae and benthic organisms. Juveniles frequent tidal pools and are sometimes found far south due to dispersal of larvae by the Agulhas Current.

19.1 ▲   119.2 ▲

9.3 ▲   119.4 ▲

5 ▲   119.6 ▼   119.5j ▲   119.7 ▼

Butterflyfish

Butterflyfish of the family Chaetodontidae are conspicuous inhabitants of coral reefs, and there are 24 species of these colourful, disc-shaped fish in southern Africa. They are usually active by day and seek shelter close to the reef at night. Several have conspicuous 'eye-spots' on the body which may serve to confuse would-be predators.

120.1 **Threadfin butterflyfish** *Chaetodon auriga*

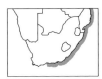

IDENTIFICATION: DXIII 23–24, AIII 20–21. Small, whitish fish. Tail and anal and dorsal fins yellow. Black bar through eye, and a characteristic black spot on soft dorsal fin; grey chevron markings on flanks. In adults, the soft dorsal fin has a trailing filament. SIZE: 20 cm. BIOLOGY: Indo-Pacific; frequents coral reefs where it is usually seen in pairs. Feeds on small reef invertebrates. Juveniles inhabit tidal pools on the East Coast.

120.2 **Blackedged butterflyfish** *Chaetodon dolosus*

IDENTIFICATION: DXII–XIII 21–22, AIII 18–19. Grey, with small brown spots, hind part of the body black, and a dark vertical stripe through the eye. Edge of dorsal and anal fins white. Tail fin yellow. SIZE: 15 cm. BIOLOGY: Known only from the East Coast of Africa and Mauritius, where it is found on coral reefs.

120.3 **Brownburnie** *Chaetodon blackburnii*

IDENTIFICATION: DXVI–XVII 21–23, AIII 16–18. In contrast to other butterflyfishes, this species is relatively drab, with most of the body and the dorsal, anal and tail fins dark-brown in colour. The front of the body is yellow with a dark band extending through the eye. SIZE: 12 cm. BIOLOGY: A West Indian Ocean species which occurs on coral and rocky reefs. Preys on small invertebrates.

120.4 **Gorgeous gussy** *Chaetodon guttatissimus*

IDENTIFICATION: DXIII 21–23, AIII 16–18. Pale-yellow body with numerous small black spots extending onto and coalescing on the dorsal and anal fins. A black band extends through the eye and the tail fin. The dorsal and anal fins are yellow-edged. SIZE: 12 cm. BIOLOGY: Widely distributed in the Indian Ocean; occurs in small groups on coral reefs. Consumes small invertebrates.

120.5 **Whitespotted butterflyfish** *Chaetodon kleinii*

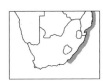

IDENTIFICATION: DXIII 20–23, AIII 18–19. Body yellowish, with a white spot at the centre of each scale. Jaws black; dorsal and anal fins yellow. Black band through the eye and running onto the chest. SIZE: 15 cm. BIOLOGY: Indo-West Pacific; occurs singly or in small groups on coral reefs down to 40 m. Feeds on small invertebrates and seaweeds.

120.6 **Racoon butterflyfish** *Chaetodon lunula*

IDENTIFICATION: DXII–XIII 23–25, AIII 17–19. A yellow fish with darker flanks. A black band through the eye precedes a characteristic broad white band across the forehead. Juvenile has a black spot on the soft dorsal fin and at the base of the tail fin. SIZE: 20 cm. BIOLOGY: Indo-Pacific; found in pairs on coral and rocky reefs down to 30 m. Juveniles frequently encountered in tidal pools. Eats small invertebrates and seaweeds.

120.7 **Pearly butterflyfish** *Chaetodon madagaskariensis*

IDENTIFICATION: DXIII–XIV 20–21, AIII 16–17. Greyish-white, with broad orange bands across the rear of the body and the tail fin. About six black chevron marks on sides, a black band through eyes, and a black bar in front of the dorsal fin. SIZE: 15 cm. BIOLOGY: An Indian Ocean species that frequents coral and rocky reefs down to 40 m. Often found on seaward or outer reefs. Eats invertebrates and seaweeds.

20.1 ▲

120.2 ▲

0.3 ▲

120.4 ▲

120.7 ▼

.5 ▲

120.6 ▼

121.1 **Doublesash butterflyfish** *Chaetodon marleyi*

IDENTIFICATION: DXI 22–25, AIII 18–19. Silvery-white, with a yellowish-brown vertical bar through the eye and two broader vertical bars extending across the flanks. Dorsal, anal and pelvic fins yellow or orange; brown vertical bars through the tail fin and base. Juveniles characterised by a black spot in the soft dorsal fin. SIZE: 15 cm. BIOLOGY: Endemic to South Africa; generally seen in pairs on rocky reefs, though occasionally found in estuaries. Juveniles frequent tidal pools and shallow reefs. Feeds on invertebrates and also grazes on seaweeds.

121.2 **Maypole butterflyfish** *Chaetodon meyeri*

IDENTIFICATION: DXII 23–24, AIII 18–20. A striking, bluish-white fish with about ten characteristic curved black lines radiating across the body and fins. Fins and body margins are yellow. SIZE: 20 cm. BIOLOGY: Indo-Pacific; occurs in pairs or small groups on coral reefs. Apparently preys exclusively on coral polyps.

121.3 **Limespot butterflyfish** *Chaetodon interruptus*

IDENTIFICATION: DXIII 21–23, AIII 17–20. A bright-yellow fish with a conspicuous black spot on its flanks. A black bar through the eye, and a thin black line through the tail fin base and along the trailing edge of the soft dorsal and anal fins. SIZE: 20 cm. BIOLOGY: Pairs of this Indo-Pacific species occur on coral reefs, where they feed on small invertebrates.

121.4 **Vagabond butterflyfish** *Chaetodon vagabundus*

IDENTIFICATION: DXIII 23–26, AIII 19–21. White, with hind part of the body yellow. Black vertical bars through the eye and across the hind part of the body and the tail fin. About eight thin diagonal lines run towards the head across the front of the body and a further 14 thin lines run diagonally towards the tail across the flanks. SIZE: 23 cm. BIOLOGY: Widespread in the Indo-Pacific on coral and rocky reefs down to 30 m. Usually occurs in pairs. Feeds on seaweeds and small invertebrates.

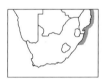

121.5 **Longnose butterflyfish** *Forcipiger flavissimus*

IDENTIFICATION: DXII 22–24, AIII 17–18. Bright-yellow fish with a black and white head, small black spot on anal fin, and distinctive elongate snout. SIZE: 15 cm. BIOLOGY: Widespread throughout the Indo-Pacific; frequents coral and rocky reefs. Specialised mouth enables it to probe small holes and crevices for invertebrate prey.

121.6 **Schooling coachman** *Heniochus diphreutes*

IDENTIFICATION: DXII 23–25, AIII 17–18. White, with two characteristic broad black bars, which cross the body diagonally. Fourth dorsal spine elongated into a filament which extends past the tail. Soft dorsal, tail and pectoral fins are yellow. SIZE: 25 cm. BIOLOGY: Indo-West Pacific species; occurs on coral and rocky reefs. Preys on small invertebrates. Juveniles sometimes remove parasites from other fish. RELATED SPECIES: Two other *Heniochus* species are also found on Zululand reefs. The coachman, *Heniochus acuminatus*, is very similar but adults develop a short, stout spine in front of each eye. The masked coachman, *Heniochus monoceros*, has a dark snout and a short horn in front of each eye.

121.7 **Moorish idol** *Zanclus cornutus*

IDENTIFICATION: DVI–VII 39–42, AIII 31–37. Black and white, with conspicuous yellow marking on the flanks. Fine blue vertical lines through the black parts of the body. The tail fin has a blue edge. The dorsal fin spines are extended into a long filament. Snout pronounced, with yellow saddle across the bridge. SIZE: 22 cm. BIOLOGY: Widespread in the Indo-Pacific; lives in groups on coral and rocky reefs, where it preys on invertebrates. Larvae are disc-like and often carried south by the Agulhas Current.

121.1 ▲

121.2 ▲

121.3 ▲   121.4 ▼

121.5 ▲

121.6 ▲   121.7 ▼

Kingfish

Some 50 species of the family Carangidae are found off southern Africa. Identification can be difficult and requires detailed examination of the scales and counts of lateral-line scutes (enlarged scales that bear spines or ridges).

122.1 Coastal kingfish *Carangoides caeruleopinnatus*

IDENTIFICATION: DVIII + I 20–23, AII + I 16–20. Silvery-blue. Deep-bodied, almost oval. Black spot on gill cover. Scaleless patch from gill cover to pectoral fin and down to pelvic fin; 20–38 small hard scutes on tail peduncle. SIZE: 40 cm and 4 kg. BIOLOGY: Indo-West Pacific; one of 12 *Carangoides* species in southern Africa. Shoals over deep coastal reefs. Rarely caught from the shore. Eats crustaceans and fish.

122.2 Blue kingfish *Carangoides ferdau*

IDENTIFICATION: DVIII + I 26–34, AII + I 21–26. Blue above and silvery-white below. Snout blunt; 21–37 small hard scutes near the tail. Scaleless patches from the gill cover to the pelvic-fin base and at the base of the long, sickle-shaped pectoral fin. Juveniles can have six vertical dusky bars on flanks. SIZE: 70 cm and 8 kg. BIOLOGY: Indo-West Pacific; inhabits coastal waters and offshore reefs. Preys on soft benthic invertebrates.

122.3 Yellowspotted kingfish *Carangoides fulvoguttatus*

IDENTIFICATION: DVIII + I 25–30, AII + I 21–26. Blue-green with silver flanks. Black blotch on gill cover and at least three large blotches on the lateral line. Outer edge of anal fin white; trailing edge of caudal fin black. Scaleless area in front of pelvic fin; 15–21 scutes on the tail peduncle. SIZE: 1.2 m and 18 kg. BIOLOGY: Coastal Indo-West Pacific; shoals over reefs down to 100 m. Feeds on fish and crustaceans.

122.4 Giant kingfish *Caranx ignobilis*

IDENTIFICATION: DVIII + I 18–21, AII + I 15–17. The largest of seven *Caranx* species in South Africa. Silvery-grey with small black spots on flanks. Forehead steep. A small scaleless patch between the gill cover and pelvic fin; another at the pectoral-fin base. Pectoral fin sickle-shaped, extends to anal fin; 26–38 prominent lateral-line scutes near the tail. SIZE: 1.6 m and 55 kg. BIOLOGY: A large, powerful Indo-Pacific predator. Preys mainly on reef fish. Sexually mature at 65 cm; spawns in summer. Juveniles found in Natal estuaries. A prized game fish, caught from ski-boats and the shore.

122.5 Bluefin kingfish *Caranx melampygus*

IDENTIFICATION: DVIII + I 21–24, AII + I 17–20. Elongate, silvery-green; black, blue and yellow spots on the flanks. Second dorsal, anal and tail fins brilliant blue; 27–42 strong lateral-line scutes near the tail. Breast completely scaled. Pectoral fin yellow, sickle-shaped. SIZE: 1 m and 8 kg. BIOLOGY: Indo-Pacific. Groups over coral and rocky reefs; preys on fish. Matures at 45 cm and spawns in summer. Juveniles occur in Kosi estuary.

122.6 Blacktip kingfish *Caranx heberi*

IDENTIFICATION: DVIII + I 19–21, AII + I 15–17. Silvery-green; fins yellow. Tail fin forked, dorsal tip black. Body elongates with age. The breast generally lacks scales except for a small patch in front of the pelvic fin; 30–40 lateral-line scutes near tail. SIZE: 1 m and 12 kg. BIOLOGY: Forms small shoals over coastal Indo-Pacific reefs. Most common during summer. Preys on small fish.

122.7 Bigeye kingfish *Caranx sexfasciatus*

IDENTIFICATION: DVIII + I 19–22, AII + I 14–17. Dusky-grey. Conspicuous large eyes, partially covered by fatty tissue. Black spot on top of gill cover. White tip on soft dorsal fin; 27–36 strong lateral-line scutes near the tail. Juveniles have six dusky vertical bars. SIZE: 85 cm and 8 kg. BIOLOGY: Indo-Pacific; common in Natal. Adults found on reefs; juveniles in estuaries in summer. Matures at 50 cm. Eats pelagic fish and invertebrates.

22.1 ▲

122.2 ▲

2.3 ▲

122.4 ▲

.5 ▲

122.6 ▲

122.7 ▼

123.1 Golden kingfish *Gnathanodon speciosus*

IDENTIFICATION: DVII + I 18–20, AII + I 15–17. Yellow, with 7–12 dark vertical bars. Fins yellow; tips of the forked tail fin are black. Adults acquire a few black blotches on the flanks. Mouth protactile and lacking teeth. Breast completely scaled. SIZE: 1.2 m and 14 kg. BIOLOGY: An Indo-Pacific coastal species; occurs in small groups near reefs. Bottom-dwelling invertebrates are sucked up through its extended mouth.

123.2 Leervis or garrick *Lichia amia*

IDENTIFICATION: DVII + I 19–21, AII + I 17–21. Elongate, silvery-green. Mouth large. Fins dark. The conspicuous curved lateral line lacks scutes. Tiny scales cover the body. Juveniles yellow with several black vertical bars. SIZE: 1.5 m and 32 kg. BIOLOGY: Swift and aggressive; hunts in the surf backline. Migrates north to KZN following the winter sardine run. Matures at 60 cm. Spawns off KZN in spring; juveniles occur in Cape estuaries. A highly rated game fish caught by shore-anglers and spearfishermen.

123.3 Talang queenfish *Scomberoides commersonnianus*

IDENTIFICATION: DVI–VII + I 19–21, AII + I 16–19. Elongate, silvery; 5–8 large black spots on the flanks. Mouth large, pectoral fins short, tail fin deeply forked. No scutes on lateral line. Oval scales occur on mid-body. SIZE: 1.2 m and 14 kg. BIOLOGY: Indo-West Pacific; coastal, usually near reefs. Eats fish and crustaceans. RELATED SPECIES: Two smaller species occur on the East Coast. The doublespotted queenfish, *Scomberoides lysan*, has two rows of spots on the flanks; the needlescaled queenfish, *Scomberoides tol*, has 5–8 black blotches on the flanks and needle-like scales mid-body.

123.4 Giant yellowtail *Seriola lalandi*

IDENTIFICATION: DVII + I 30–35, AII + I 19–22. Large and elongate. Blue above and silvery-white below, with a horizontal bronze stripe from the snout to the tail. Upper jaw narrow. Base of the tail has a moderate keel. Tail fin yellow. SIZE: 1.5 m and 35 kg. BIOLOGY: A circumglobal predator that shoals in coastal and offshore waters of the Cape, often associated with reef pinnacles. Migrates annually up the East Coast following the sardine run. A highly prized game fish, it is caught by spearfishermen, shore and ski-boat anglers. It is also an important component of commercial line-fish catches in the Cape.

123.5 Southern pompano *Trachinotus africanus*

IDENTIFICATION: DVI + I 21–23, AII + I 19–21. Robust and blunt-nosed. Silvery; lacks spots or scutes on the flanks. The fins are dusky yellow, and the tail fin strongly forked. SIZE: 90 cm and 25 kg. BIOLOGY: An Indian Ocean species; inhabits shallow coastal waters in the vicinity of reefs. Eats mussels, which it crushes with strong pharyngeal grinding plates. Often captured by shore anglers in the surf zone. RELATED SPECIES: **123.6 *Trachinotus botla***, the largespotted pompano, is silvery-blue; its flanks have large black spots. Fins blue; no lateral-line scutes. Occurs in KwaZulu-Natal.

123.7 Maasbanker *Trachurus trachurus*

IDENTIFICATION: DVIII + I 30–36, AII + I 24–32. Spindle-shaped, silvery; dorsal surface olive-green. Eye large, and has much adipose (fatty) tissue. Black spot on the gill cover. Lateral line has up to 78 scutes. SIZE: 70 cm. BIOLOGY: Coastal, pelagic; shoals are harvested commercially in the Cape. Preys on plankton; undertakes vertical migrations following its prey. Spawns in late winter and spring. Eaten by seals and dolphins.

123.8 Elf or shad *Pomatomus saltatrix*

IDENTIFICATION: DVII–VIII + I 23–28, AII 23–27. Silvery with a dorsal greenish sheen. Mouth large; single row of sharp teeth in each jaw. Scales small, easily shed. SIZE: 1.2 m and 10 kg. BIOLOGY: A powerful predator. Follows the sardine run to KZN and spawns there in spring. Matures at 25 cm. Juveniles common in South Coast bays. Once severely depleted, but a bag limit, minimum size and closed season have led to a recovery.

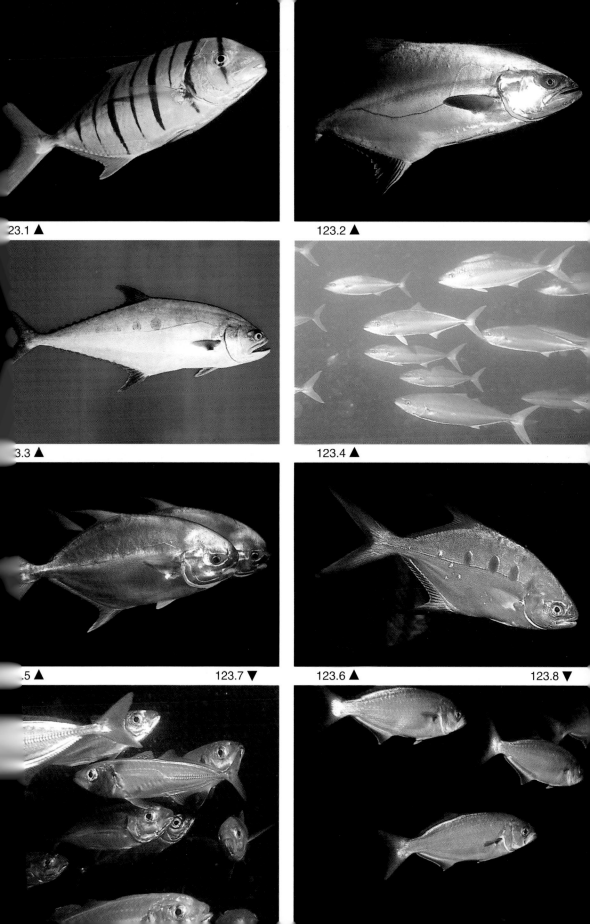

23.1 ▲

123.2 ▲

3.3 ▲

123.4 ▲

.5 ▲

123.7 ▼

123.6 ▲

123.8 ▼

Snoek, Tuna & Billfish

Tunas are wide-ranging, migratory species. World-wide, some 6 million tonnes are caught annually. Eighteen species occur around southern Africa. Billfish (swordfish, marlin and sailfish) are close relatives of tunas.

124.1 Snoek *Thyrsites atun*

IDENTIFICATION: DXVIII–XXI 10–12 + 5–7, AII–III 8–11 + 5–7. Elongate, silvery. Dorsal spines prominent; 5–7 small finlets preceding the tail. Mouth extends back to eye. Teeth large, sharp. Pelvic fins small. Tail fin large and forked. SIZE: 2 m and 8 kg. BIOLOGY: Predatory; forms migratory shoals on the West Coast. Matures at 55 cm; spawns in winter. Caught commercially on hand-lines; often smoked or dried.

124.2 Mackerel *Scomber japonicus*

IDENTIFICATION: DIX–X + 12 + 5 finlets, AI + 12 + 5 finlets. Torpedo-shaped. Narrow tail peduncle with two keels. Dorsal surface metallic green with wavy lines; belly silver-white with dark spots. SIZE: 50 cm. BIOLOGY: A cosmopolitan, shoaling plankton-feeder. Spawns in spring. Harvested off the Cape. SIMILAR SPECIES: *Rastrelliger kanagurta* (Durban northwards) has a black spot on the pectoral fin and golden stripes on the flanks.

124.3 Eastern little tuna *Euthynnus affinis*

IDENTIFICATION: DXV–XVII + 12–13 + 8 finlets, A 13–14 + 7 finlets. Robust, spindle-shaped, silvery. Dorsal surface dark blue; wavy dark lines on upper flanks. Distinct black spots below the pectoral fin. A row of tiny conical teeth in each jaw. SIZE: 1 m and 11 kg. BIOLOGY: Indo-Pacific; shoaling. Preys on small fish. Matures at 45 cm and larvae occur on the KwaZulu-Natal shelf during summer. Juveniles often used as bait. SIMILAR SPECIES: The frigate tuna, *Auxis thazard,* has no ventral black spots. It often schools with *E. affinis.*

124.4 Skipjack tuna *Katsuwonus pelamis*

IDENTIFICATION: DXV–XVI + 14–15 + 8 finlets, A 14–15 + 7 finlets. Spindle-shaped. Blue above; silvery belly with 4–6 longitudinal stripes. Mouth large; teeth small, conical. A keel on the tail peduncle. SIZE: 1 m and 18 kg. BIOLOGY: Cosmopolitan, schooling in the tropics. Abundant off the East Coast in summer. Preys on pelagic fish and squid. Matures at 45 cm. The spawning area is further north in the Indian Ocean.

124.5 King mackerel *Scomberomorus commerson*

IDENTIFICATION: DXV–XVIII + 16–18 + 9–10 finlets, A 17–19 + 9–10 finlets. Elongate and silvery; irregular dark vertical bars on the flanks. Mouth large, conspicuous teeth. Lateral line distinct. Tail fin large. Three distinct keels on tail peduncle. SIZE: Up to 2 m and 39 kg. BIOLOGY: A shoaling Indo-Pacific predator. Abundant off Natal and Transkei in summer and autumn. Preys on pelagic fish. Matures at 90 cm. Does not spawn off South Africa, and local fish appear to migrate south from Moçambique.

124.6 Queen mackerel *Scomberomorus plurilineatus*

IDENTIFICATION: DXV–XVI + 19–21 + 8–10 finlets, A 19–22 + 8–10 finlets. Silvery blue, elongate, with rows of dark dashes on the flanks. Tail strongly forked. Three keels on the tail peduncle. SIZE: 1.2 m and 10 kg. BIOLOGY: West Indian Ocean. Spawns in Moçambique; shoals migrate to KZN. Eats small fishes. Matures at 80 cm.

124.7 Sailfish *Istiophorus platypterus*

IDENTIFICATION: D 39–48 + 6–8, A 8–15 + 5–8. Metallic-blue; streamlined, with a long spear-like snout. First dorsal fin very high and brilliant blue; middle rays longest. Tail fin well developed; two keels on tail peduncle. SIZE: 3 m and 100 kg. BIOLOGY: Wide-ranging in all tropical oceans; abundant off KZN in summer. Preys on pelagic fish and squid. Grows extremely fast. Mot commonly caught billfish in South Africa.

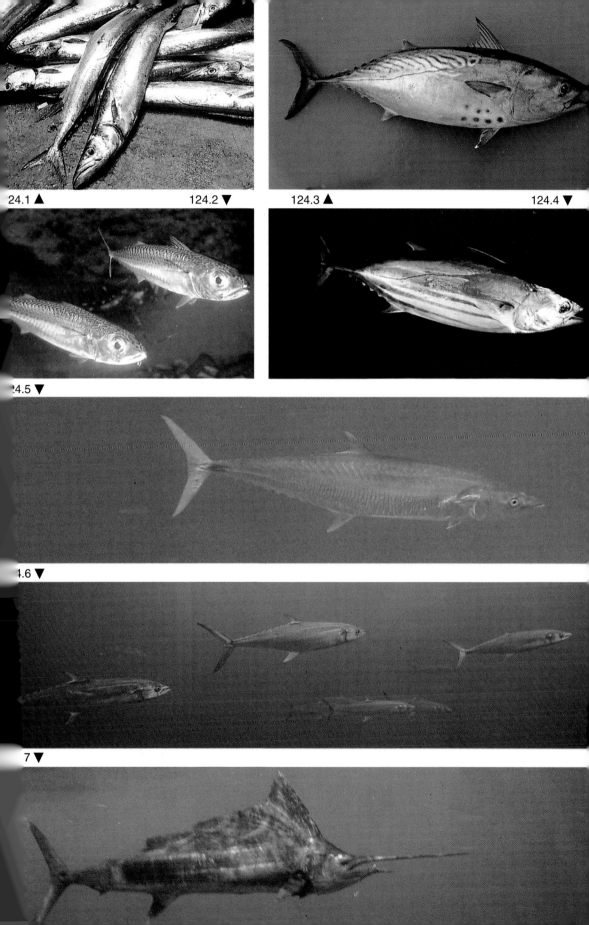

24.1 ▲   124.2 ▼   124.3 ▲   124.4 ▼

24.5 ▼

4.6 ▼

7 ▼

Remoras, Hawkfish, Fingerfins & Sweepers

125.1 Shark remora   *Echeneis naucrates*
IDENTIFICATION: D 34–42, A 32–38. A slender fish with a characteristically flattened head bearing a sucking disc with 21–28 grooves. White with a black horizontal stripe from tip of lower jaw to tail. SIZE: 1 m. BIOLOGY: Distributed world-wide in warm waters. Attached to sharks and manta rays by means of the sucking disc. Generally feeds on small pieces of food wasted by the shark.

125.2 Spotted hawkfish   *Cirrhitichthys oxycephalus*
IDENTIFICATION: DX 12, AIII 6. Small, reddish-brown mottled fish with enlarged pectoral fin rays and little tufts on the dorsal spines. One of seven species of hawkfishes found in South African waters. SIZE: 10 cm. BIOLOGY: Commonly occurs on Indo-Pacific coral reefs. Maintains territories. Usually well camouflaged; preys on small invertebrates.

125.3 Freckled hawkfish   *Paracirrhites forsteri*
IDENTIFICATION: DX 11, AIII 6. The dorsal surface is dark reddish-brown to black. Numerous red spots on head and pectoral fin base. White stripe along dorsal fin base. Ventral surface white to brown. SIZE: 22 cm. BIOLOGY: Associated with Indo-Pacific coral reefs. Rests on live coral and darts out to snap up passing crustaceans and small fishes.

125.4 Twotone fingerfin   *Chirodactylus brachydactylus*
IDENTIFICATION: DXVII–XVIII 28–31, AIII 8–10. Reddish-brown with paler undersides and distinct red pectoral fins with elongated rays which do not reach past the anus. A row of five bluish-white spots occurs on the flanks. Fleshy lips surround the small mouth. SIZE: 40 cm. BIOLOGY: A common endemic fish frequenting tidal pools, subtidal gullies and offshore reefs. Feeds on small invertebrates. Sexual maturity is attained at 25 cm. Small, compressed, silvery juveniles recruit to tidal pools in summer, and acquire the adult coloration at about 5 cm.

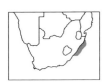

125.5 Natal fingerfin   *Chirodactylus jessicalenorum*
IDENTIFICATION: DXVII–XVIII 26–27, AIII 7–8. Red fish with a dark patch at the base of the pectoral fin. Longest pectoral-fin ray extends back to the anal fin. Lips fleshy. SIZE: 50 cm and 10 kg. BIOLOGY: An endemic species that is associated with rocky banks. Preys on benthic invertebrates. Frequently speared in KwaZulu-Natal, though rarely caught by anglers. RELATED SPECIES: Easily confused with the bank steenbras, *Chirodactylus grandis,* which is similar but dull brown, has shorter pectoral rays, fewer soft dorsal rays and occurs mainly in the Cape.

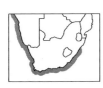

125.6 Redfingers   *Cheilodactylus fasciatus*
IDENTIFICATION: DXVII–XIX 23–25, AIII 9–11. Small fish with mottled-brown bars extending onto the dorsal, anal and tail fins. Pectoral fins red with lower rays thickened and elongated, extending back to the anal fin. SIZE: 30 cm. BIOLOGY: A southern African endemic found in tidal pools and subtidal gullies. Preys on small invertebrates. Juveniles are silvery and lack the extended pectoral rays. RELATED SPECIES: The barred fingerfin, *Cheilodactylus pixi* (shown in the lower left side of the photograph), is similar but the tail is translucent and the body markings form continuous vertical bars.

125.7 Dusky sweeper   *Pempheris adusta*
IDENTIFICATION: DVI 9, AIII 37–43. Small, silvery-bronze fish with an oblique mouth, large eye, short dorsal fin, and a very long, dusky anal fin. SIZE: 17 cm. BIOLOGY: Indo-West Pacific; forms shoals in caves and under overhangs on coral and rocky reefs. Nocturnal in habit, and preys on planktonic invertebrates.

25.1 ▲

25.2 ▲

125.3 ▲

5.4 ▲

125.6 ▼

125.5 ▲

125.7 ▼

Damselfish

126.1 Spot damsel *Abudefduf sordidus*

IDENTIFICATION: DXIII 14–16, AII 14–15. One of seven East Coast species of *Abudefduf*. A small, oval, grey fish with a black saddle on the tail peduncle, a black spot at the base of the pectoral fin, and five pale vertical crossbars. SIZE: 20 cm. BIOLOGY: Indo-Pacific; frequents coral and shallow rocky reefs. Juveniles are common in tidal pools in summer. Omnivorous. Adults defend home territories against intruders.

126.2 Sergeant major *Abudefduf vaigiensis*

IDENTIFICATION: DXIII 12–13, AII 11–13. Characterised by five vertical black bands and a yellow dorsal surface. SIZE: 20 cm. BIOLOGY: Indo-Pacific; common on coral and shallow rocky reefs. Juveniles abundant in tidal pools. Omnivorous. Adults defend home ranges and nesting sites. RELATED SPECIES: The fourbar damsel, *Abudefduf natalensis* has four broad dark bars; the one on the tail peduncle joins dark bands along the tail-fin lobes. The stripetail damsel, *Abudefduf sexfasciatus*, has five dark bars; the last is separate from the band along the ventral tail-fin lobe but joins one on the upper lobe.

126.3 Sash damsel *Plectroglyphidodon leucozonus*

IDENTIFICATION: DXII 15–17, AII 12–13. One of six *Plectroglyphidodon* species on the East Coast, the sash damsel is brown with a pale vertical bar across the mid-body. Juveniles have a black spot on the soft dorsal fin. SIZE: 15 cm. BIOLOGY: Indo-West Pacific, inhabiting coral reefs and shallow rocky areas. Aggressively guards territories. Juveniles frequent tidal pools.

126.4 Twobar clownfish *Amphiprion allardi*

IDENTIFICATION: DX–XI 15–17, AII 13–15. Small, brown, with orange-yellow fins and two bluish-white vertical bars. Juveniles have a third bar across the tail. SIZE: 15 cm. BIOLOGY: Endemic to East African coral reefs; usually shelters amongst the tentacles of large sea anemones. Omnivorous. Eggs are attached to the reef during summer and are cared for by the parents.

126.5 Nosestripe clownfish *Amphiprion akallopisos*

IDENTIFICATION: DVIII–X 17–20, AII 12–14. Pinkish, with a pale stripe along the dorsal surface from snout to tail peduncle. SIZE: 10 cm. BIOLOGY: An Indian Ocean coral-reef species usually associated with large sea anemones. Hermaphroditic: if a large dominant female is removed from its anemone, then one of the smaller males associated with her will develop into a female.

126.6 Chocolate dip *Chromis dimidiata*

IDENTIFICATION: DXII 12–13, AII 12–13. One of seven *Chromis* species recorded from South Africa. The front half of the body is distinctly brown and the back half white. SIZE: 7 cm. BIOLOGY: An Indian Ocean species that occurs on coral reefs down to 30 m, where it shelters amongst branching corals.

126.7 Domino *Dascyllus trimaculatus*

IDENTIFICATION: DXII 14–16, AII 14–15. Small, round, black; juveniles have white spots on the forehead and dorsal flank but these are lost in adults. One of three South African *Dascyllus* species. SIZE: 14 cm. BIOLOGY: An Indo-West Pacific species that shelters amongst branching corals. Juveniles sometimes associated with sea anemones or hidden amongst the spines of sea urchins.

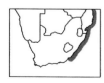

126.8 Blue Pete *Pomacentrus caeruleus*

IDENTIFICATION: DXIII 14–15, AII 15–16. One of four *Pomacentrus* species along the East Coast, this small fish is brilliant blue with a yellow ventral surface. Hind part of the dorsal fin, and the pectoral, anal and tail fins are yellow. SIZE: 10 cm. BIOLOGY: Skulks in West Indian Ocean coral reefs.

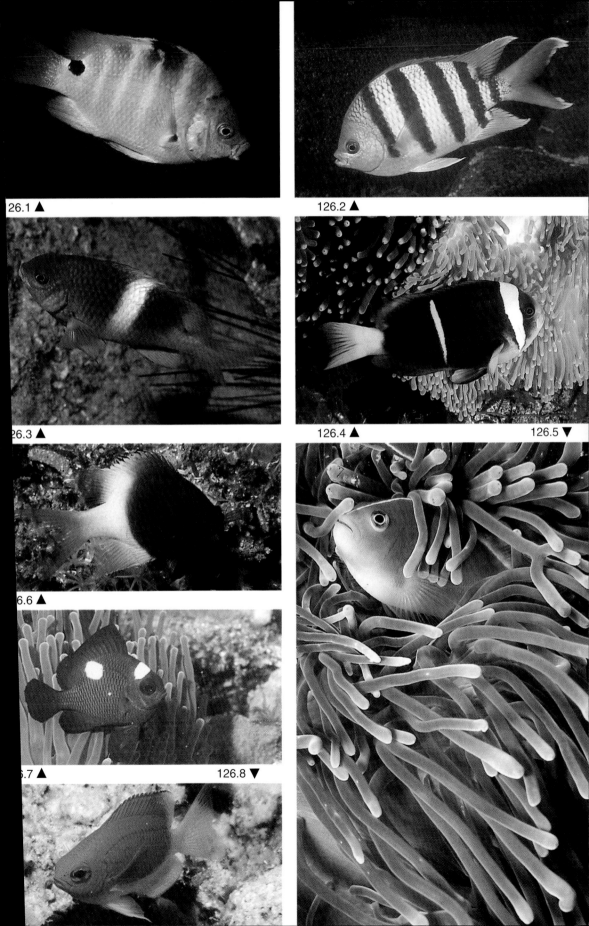

26.1 ▲

126.2 ▲

26.3 ▲

126.4 ▲ 126.5 ▼

6.6 ▲

.7 ▲ 126.8 ▼

Wrasses

Many wrasses undergo sex-change from female to male, and alter their colours as they mature, making identification of the 66 species extremely difficult. Most species have pharyngeal teeth with which they crush hard-shelled invertebrates.

127.1 Birdfish   *Gomphosus caeruleus*

IDENTIFICATION: DVIII 13, AIII 11. Although colour varies with age and sex, the elongate tubular snout with a terminal mouth is characteristic of this species. Juveniles and females are yellowish-grey but mature males are blue-green with yellow fin edges. SIZE: 30 cm. BIOLOGY: An Indian Ocean species that frequents coral reefs. Preys on small benthic invertebrates.

127.2 Saddleback hogfish   *Bodianus bilunulatus*

IDENTIFICATION: DXII 10, AIII 12–13. An elongate, robust, red fish with a yellow-ish belly and a large black saddle on the tail peduncle and below the soft dorsal fin. Two pairs of protruding canines. Juveniles with a large black area at the rear of the body. One of six *Bodianus* species on the East Coast. SIZE: 55 cm and 7 kg. BIOLOGY: Indo-West Pacific, inhabiting coral reefs.

127.3 Diana's hogfish   *Bodianus diana*

IDENTIFICATION: DXII 10, AIII 12. Elongate, with sloping forehead. Brownish-red, with black spot on tail and 3–4 yellow spots on upper flank. Scales on hind upper flank have black dots. Juveniles are dark with white blotches on the body and black spots on the fins. SIZE: 25 cm. BIOLOGY: Indo-West Pacific; solitary on coral and rocky reefs down to 80 m.

127.4 Picture wrasse   *Halichoeres nebulosus*

IDENTIFICATION: DIX II, AIII 11. Elongate, green. Females have irregular dark bars on sides, a large maroon patch on the belly, a black spot behind the eye and on the gill cover, and a yellow-edged black spot on the dorsal fin. Males are primarily green with a reduced maroon patch. SIZE: 10 cm. BIOLOGY: Indo-West Pacific species inhab-iting shallow inshore areas. Preys on benthic invertebrates. SIMILAR SPECIES: The rain-bow wrasse, *Thalassoma purpureum*, lacks the maroon patch on the belly. Pink lines radiate from behind its eyes.

127.5 Checkerboard wrasse   *Halichoeres hortulanus*

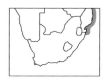

IDENTIFICATION: DIX 11, AIII 11. Males blue-green, head with orange lines. Scales black-edged, giving a checkered appearance. Yellow blotch at base of dorsal fin. Females paler, with a black patch behind the yellow blotch; tail yellow. Juveniles black and white with a large, central, yellow-edged spot on the dorsal fin. SIZE: 25 cm. BIOLOGY: One of seven South African *Halichoeres* species. Inhabits Indo-Pacific coral reefs.

127.6 Goldbar wrasse   *Thalassoma hebraicum*

IDENTIFICATION: DVIII 13, AIII 11. Elongate, greenish-blue; tail lunate. Conspicuous vertical yellow bar from front of dorsal fin to belly. Head yellow; several curved blue bands extend to the pectoral fin. Juveniles and females differ slightly, but also have the vertical yellow bar. SIZE: 25 cm. BIOLOGY: One of seven *Thalassoma* species found on East Coast reefs. Usually solitary; preys on hard-shelled invertebrates.

127.7 Bluestreak cleaner wrasse   *Labroides dimidiatus*

IDENTIFICATION: DIX 11, AIII 10. Small, elongate; white and blue with a black hori-zontal bar from the snout, widening towards the tail, and bars through the dorsal and anal fins. The tail is truncate. SIZE: 12 cm. BIOLOGY: Inhabits Indo-Pacific coral reefs. Males and attendant harems set up 'cleaner stations' and remove parasites from larger reef fishes, even venturing into their mouths and under their gill covers. SIMILAR SPECIES: A small blenny, *Aspidontus taeniatus*, mimics the cleaner wrasse but tears chunks of flesh from its unsuspecting 'clients'. It has an underslung mouth, unlike *Labroides*.

27.1 ▲

127.2 ▲

7.3 ▲

127.4 ▼

127.5 ▼

.6 ▼

127.7 ▼

Parrotfish & Emperors

128.1 Ember parrotfish   *Scarus rubroviolaceus*

IDENTIFICATION: DIX 10, AIII 9. One of 13 *Scarus* species off the East Coast. The ember parrotfish is a robust fish with a blunt snout. Teeth fused to form beak-like structure. Colour varies with age and sex. The mature male (128.1a) is bluish-green; front of body pinkish with blue band above mouth. Blue stripes on the lower lip and chin join and reach the eye. Tail fin lunate. The female (128.1b) is red and has greenish upper flanks with irregular black stripes. SIZE: 65 cm. BIOLOGY: Widespread in the Indo-Pacific, and dwells on coral reefs. Feeds on algae, which are scraped from the reef with the beak. Common at Sodwana Bay.

128.2 Blue humphead parrotfish   *Chlorurus cyanescens*

IDENTIFICATION: DIX 10, AIII 9. A striking fish with a distinct hump on the head. Body dark blue with the posterior half bright-green. Thin blue margins to dorsal, tail, anal and pelvic fins. SIZE: 50 cm. BIOLOGY: West Indian Ocean, inhabiting coral reefs, where it feeds on algae scraped from the reefs. Occasionally seen by divers at Sodwana Bay.

128.3 Glowfish   *Gnathodentex aureolineatus*

IDENTIFICATION: DX 10, AIII 8–9. A silvery-pink fish with several yellowish horizontal lines across the upper body. Conspicuous yellow patch at end of the dorsal fin. Canines flare outwards. Eyes large. SIZE: 30 cm. BIOLOGY: A Indo-West Pacific coral-reef inhabitant. Solitary or shoaling. Carnivorous.

128.4 Grey barenose   *Gymnocranius griseus*

IDENTIFICATION: DX 9–10, AIII 9–10. Silver-grey, with a vertical black bar through the eye. Juveniles have several other diffuse dark vertical bars which fade with age. Strong conical teeth. SIZE: 80 cm. BIOLOGY: Indo-West Pacific, living over reefs and sandy areas down to 80 m.

128.5 Redspot emperor   *Lethrinus lentjan*

IDENTIFICATION: DX 9, AIII 8. One of at least ten *Lethrinus* species found off the East Coast. Brown to green with blue tinges. Snout pointed. Few or no scales on pectoral base. Red spot on gill cover and pectoral base. SIZE: 40 cm. BIOLOGY: Distributed throughout the tropical Indo-West Pacific. Feeds on bottom-dwelling invertebrates.

128.6 Blue emperor   *Lethrinus nebulosus*

IDENTIFICATION: DX 9, AIII 8. The blue emperor has the long sloping forehead and scaleless cheeks typical of this genus. Though predominantly brownish in colour, each scale has a blue centre. There is dense scaling on inner base of pectoral fin. SIZE: 75 cm and 7 kg. BIOLOGY: Indo-West Pacific, occurring on coral and rocky reefs down to 50 m. Preys on benthic and planktonic invertebrates. Attains sexual maturity at 50 cm. Caught by ski-boat anglers in Natal.

128.7 Bigeye barenose   *Monotaxis granoculis*

IDENTIFICATION: DX 10, AIII 9. Silvery-grey; upper flank darker with two distinct white vertical bars. Head heavy, eye large; snout becoming increasingly blunt with age. SIZE: 60 cm. BIOLOGY: Inhabits Indo-Pacific coral reefs and adjacent rubble and sandy areas, down to a depth of 60 m.

128.1a ▲

128.1b ▲

28.2 ▲

128.3 ▲

8.4 ▲

128.6 ▼

128.5 ▲

128.7 ▼

Barracudas, Mullet & Goatfish

129.1 Pickhandle barracuda *Sphyraena jello*

IDENTIFICATION: DV + I 9, AII 8. The pickhandle barracuda is one of 10 species of *Sphyraena* in southern Africa. It is elongate and silvery with about 20 vertical bars. The large mouth extends to the eye. Teeth sharp. Dorsal fins dusky and widely separated. SIZE: 1.5 m and 16 kg. BIOLOGY: An Indian Ocean predator; frequents coastal reefs in small shoals. Preys on fish. Juveniles common off Durban. RELATED SPECIES: The great barracuda, *Sphyraena barracuda*, has irregular dark blotches on the sides and only 75–85 lateral-line scales, fewer than any other barracuda occurring off southern Africa.

129.2 Southern mullet *Liza richardsonii*

IDENTIFICATION: DIV + I 8–9, AIII 9. Elongate and silvery; dorsal surface dark. Yellow spot on the gill cover. Snout pointed, teeth feeble. When the pectoral fin is folded forward, it just reaches the eye. SIZE: 40 cm. BIOLOGY: Endemic; most common in the W. Cape; caught on sandy beaches by commercial treknetters. Feeds on phytoplankton and benthic diatoms. Spawns in spring and summer. Juveniles frequent estuaries.

129.3 Groovy mullet *Liza dumerilii*

IDENTIFICATION: DIV + I 8, AIII 9. Elongate greyish-silver, with a yellow spot on the gill cover. Snout pointed, teeth feeble. When pectoral fin is folded forward, it reaches past the eye. Scales have a characteristic groovy pattern, most easily seen on the dorsal surface of the head. SIZE: 40 cm. BIOLOGY: Endemic. Usually found in estuaries. Spawns at sea in spring and summer. Small juveniles (1–2 cm) recruit to estuarine nursery areas over an extended period. Feeds on benthic diatoms and minute animals. RELATED SPECIES: The striped mullet, *Liza tricuspidens*, is larger (up to 75 cm) and is characterised by its tricuspid teeth and about ten dark horizontal stripes along the sides. It has a yellow spot on the gill cover. Extends from Mossel Bay to northern Zululand.

129.4 Flathead mullet *Mugil cephalus*

IDENTIFICATION: DIV + I 8, AIII 8. Elongate, silvery; dorsal surface darker. Diffuse horizontal stripes on sides. Snout blunt, mouth toothless. Eyes almost entirely covered by transparent adipose (fatty) eyelids. When folded forward the pectoral fin does not reach the eye. SIZE: 80 cm. BIOLOGY: Cosmopolitan; shoals in coastal waters and estuaries and penetrates freshwater. Feeds on diatoms and detritus. Matures at 45 cm; spawns at sea in winter. Juveniles (1–2 cm) recruit to estuarine nursery areas in spring. Large pre-spawning aggregations form in estuary mouths. RELATED SPECIES: The bluetail mullet, *Valamugil buchanani* (Indo-West Pacific, south to Knysna), is also large but has a bright-blue tail and lacks adipose eyelids. The freshwater mullet, *Myxus capensis* (Knysna to Natal), is smaller, has a pointed snout, and flattened teeth in the upper jaw. Dorsal surface dark, ventral surface silvery. Frequents rivers and the upper parts of estuaries but spawns at sea.

129.5 Blacksaddle goatfish *Parupeneus rubescens*

IDENTIFICATION: DVIII + I 8, AI 6. Ten species of goatfish occur in South African waters, the blacksaddle goatfish being the most common. They all have elongate bodies with two dorsal fins, fleshy lips and two long chin barbels. The blacksaddle goatfish varies in colour but has a characteristic black marking in front of the tail. SIZE: 42 cm. BIOLOGY: Goatfish are usually found in the vicinity of reefs throughout the Indo-Pacific. They use their movable barbels to detect food on the seabed. The blacksaddle goatfish is the most common goatfish in South African waters.

129.6 Flame goatfish *Mulloides vanicolensis*

IDENTIFICATION: DVII + I 8, AI 6. Elongate, greyish-white body with yellowish dorsal area and conspicuous yellow horizontal stripe extending from eye to tail. Fins yellow. Lips fleshy. A pair of long chin barbels. SIZE: 33 cm. BIOLOGY: An Indo-West Pacific species occurring in shoals around coral reefs.

129.1 ▲

129.2 ▲　　　　　　　　　　　　129.3 ▲

129.4 ▲　　　　　　129.5 ▲　　　　　　129.6 ▼

Blennies & Triplefins

130.1 Snakelet   *Halidesmus scapularis*
IDENTIFICATION: DI 58–63, A 48–52. Small, elongate, eel-like brown fish with long dorsal and anal fins but no pelvic fins. Dark spot above gill cover. SIZE: 20 cm. BIOLOGY: An endemic intertidal and shallow subtidal species that lives in crevices. Feeds on amphipods and isopods.

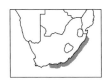

130.2 Horned rockskipper   *Antennablennius bifilum*
IDENTIFICATION: DXI–XIII 17–20, AII 17–21. Elongate, mottled, with about seven dark bands and many small blue dots along flanks. Pair of tentacles on nape, but no tentacles on eye. SIZE: 10 cm. BIOLOGY: West Indian Ocean species inhabiting tidal pools. Eggs are attached to rocks.

130.3 Ringneck blenny   *Parablennius pilicornis*
IDENTIFICATION: DXI–XII 18–24, AII 20–25. Small, dusky, with up to nine vertical bars that separate into spots ventrally. Two dark bands on underside of head. No tentacles on top of head; tentacle above eye consists of several filaments arising from a broad base. SIZE: 12 cm. BIOLOGY: An Atlantic species found in tidal pools. Feeds on seaweed and crustaceans. Eggs are attached to rocks and guarded by the male. RELATED SPECIES: The endemic horned blenny, *Parablennius cornutus*, is irregularly banded and its eye-tentacle consists of a central stalk with many side branches.

130.4 Maned blenny   *Scartella emarginata*
IDENTIFICATION: DXI–XIII 12–16, AII 14–18. Small, brown with variable dark vertical bands and small dark spots. Mane-like row of tentacles on top of head. Eye and nostril tentacles consist of several short filaments. Teeth in upper jaw not movable. SIZE: 10 cm. BIOLOGY: Occurs in tidal pools and shallow waters of the Atlantic and Indian oceans. Feeds on algae and benthic invertebrates. Breeds year-round; eggs are attached to rocks and shells, and guarded by the male.

130.5 Rippled rockskipper   *Istiblennius edentulus*
IDENTIFICATION: DXIII 20–21, AII 21–23. Small, dusky, with alternating dark vertical bars and pale stripes on its sides, and spots on the dorsal fin. Deep notch in middle of dorsal fin. Females paler than males, often with spots near tail and lacking the fleshy crest found on the male's head. Small tentacles on nostril, eye and nape. Upper jaw teeth freely movable. SIZE: 13 cm. BIOLOGY: Indo-Pacific; inhabits shallow reefs and tidal pools. Eggs are attached to rock surfaces. RELATED SPECIES:
130.6 *Istiblennius dussumieri*, the streaky rockskipper, extends south to Transkei. Distinguished by dusky vertical bands and branched eye-tentacles; lacks nape-tentacles.

130.7 Bandit blenny   *Omobranchus banditus*
IDENTIFICATION: DXI–XIII 19–21, AII 21–23. Small, with several dark vertical bands on head and body; a black spot above the gill opening. No tentacles on the eye or nostril. Enlarged curved canine on each side of lower jaw. A fleshy crest caps the head in males. SIZE: 6 cm. BIOLOGY: Endemic; abundant in tidal pools and subtidal gullies. Eggs are attached to the substrate. Larvae and juveniles have a characteristic large spine on the gill cover. RELATED SPECIES: The kappie blenny, *Omobranchus woodi*, which occurs in Eastern Cape estuaries, has irregular dark bands only on the front of the body.

130.8 Hotlips triplefin   *Helcogramma sp.*
IDENTIFICATION: DIII + XIII–XIV + 9–11, AI 17–20. Tiny fish with dorsal fin divided into three parts. Body has reddish-brown blotches. Adult male has brilliant blue and black throat. SIZE: 4 cm. BIOLOGY: Occurs in tidal pools and shallow subtidal areas in West Indian Ocean. RELATED SPECIES: **130.9 *Cremnochorites capensis***, the Cape triplefin, has four spines in the first dorsal fin and branched eye-tentacles. Subtidal, Cape Point to Port Alfred.

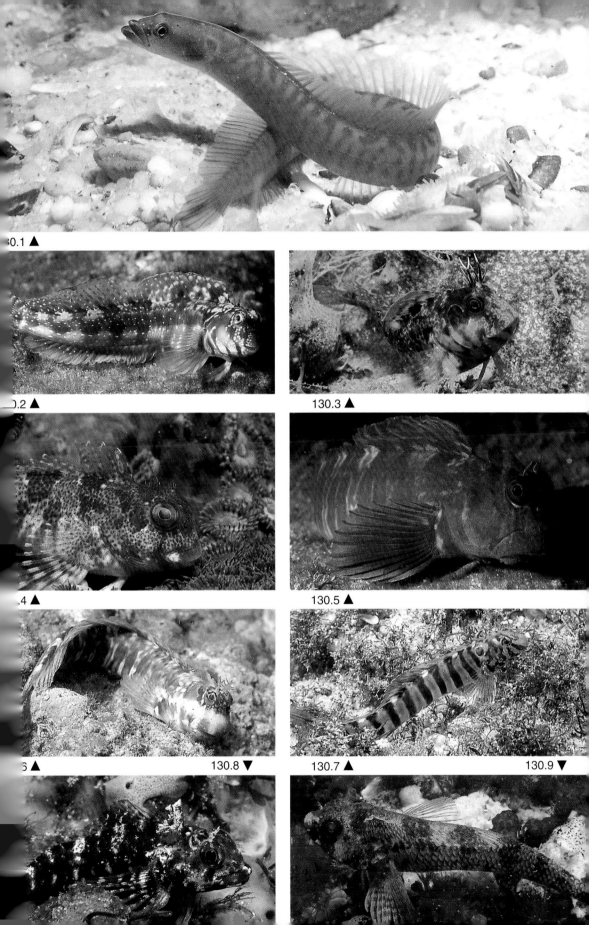

30.1 ▲

0.2 ▲

130.3 ▲

.4 ▲

130.5 ▲

6 ▲ 130.8 ▼

130.7 ▲ 130.9 ▼

Klipfishes

All 39 species of southern African klipfishes (family Clinidae) are endemic to the area. The females do not lay eggs, but undergo internal fertilisation and give birth to fully developed juveniles. Many species inhabit shallow rocky areas, but a few are characteristic of sandy habitats. Most feed on small invertebrates.

131.1 Agile klipfish   *Clinus agilis*

IDENTIFICATION: DXXXII–XXXVIII 2–4, AII 20–25. Small, grey-green with dark crossbars. Small tufts of cirri on the tips of the first 15 dorsal spines. Eye-tentacle flattened with short branches. Slight notch between third and fourth dorsal spines. SIZE: 10 cm. BIOLOGY: Hides among weeds in pools and estuaries.

131.2 Bluntnose klipfish   *Clinus cottoides*

IDENTIFICATION: DXXXI–XXXVI 4–6, AII 21–25. Small, mottled, with a black blotch on the gill cover. Dorsal and anal fins long, first three dorsal spines shorter than fifth or sixth, no notch in dorsal fin. Pelvic fin rays elongate. Bony ridge and small tentacles above the eye. SIZE: 15 cm. BIOLOGY: Hides in tidal pools. Breeds year-round.

131.3 Super klipfish   *Clinus superciliosus*

IDENTIFICATION: DXXXI–XLII 5–10, AII 21–30. Elongate, with a highly variable dark-green to red mottling. Lips fleshy. Pelvic rays long. First three dorsal spines bear clusters of cirri at the tips and are longer than the other dorsal spines, thus forming a crest. Distinct notch between third and fourth dorsal spines. Eye-tentacle flat and has a few short branches. SIZE: 30 cm. BIOLOGY: Abundant in tidal pools and subtidal gullies. Breeds throughout the year. Eats invertebrates; readily takes the bait of budding young anglers.

131.4 Speckled klipfish   *Clinus venustris*

IDENTIFICATION: DXXXVII–XLI 2–3, AII 23–28. A small, mottled, reddish-brown fish with elongate pelvic rays. No distinct notch in the dorsal fin although the second dorsal spine is longer than rest. Tentacle above eye with several short branches at tip. SIZE: 12 cm. BIOLOGY: Inhabits tidal pools; common on subtidal reefs down to 15 m.

131.5 Westcoast klipfish   *Clinus heterodon*

IDENTIFICATION: DXXX–XXXII 6–7, AII 20–22. Elongate with irregular crossbars. Tips of anal and pelvic fins red; tips of dorsal fin and tentacles above eyes pale. Yellow- or blue-edged spot on shoulder; two bands across cheek. SIZE: 13 cm. BIOLOGY: Lives in tidal pools; feeds on invertebrates.

131.6 Nosestripe klipfish   *Muraenoclinus dorsalis*

IDENTIFICATION: DXLI–XLVIII 1, AII 25–31. Eel-like, greenish-brown. A conspicuous white stripe from tip of snout to the origin of the dorsal fin. Elongate pelvic-fin rays, and flattened tentacle with fine cirri above the eye. SIZE: 10 cm. BIOLOGY: Lives under stones in tidal pools, often quite high up on the shore. Feeds on benthic invertebrates.

131.7 Grass klipfish   *Pavoclinus graminis*

IDENTIFICATION: DXXX–XXXV 4–6, AII 21–24. Elongate, mottled and highly variable in colour. Dorsal fin with a low anterior crest but usually no notch between third and fourth spines. No tentacle above eye. SIZE: 20 cm. BIOLOGY: Lives in tidal pools and subtidal gullies where it assumes the colour of the seaweeds with which it is associated.

131.8 Rippled klipfish   *Pavoclinus laurentii*

IDENTIFICATION: DXXIX–XXXIII 4–5, AII 20–22. Elongate, with variable mottled markings. First three dorsal spines form a crest. A distinct notch between the third and fourth spines. No tentacle above eye. SIZE: 13 cm. BIOLOGY: Hides amongst tide-pool seaweeds. Feeds on invertebrates.

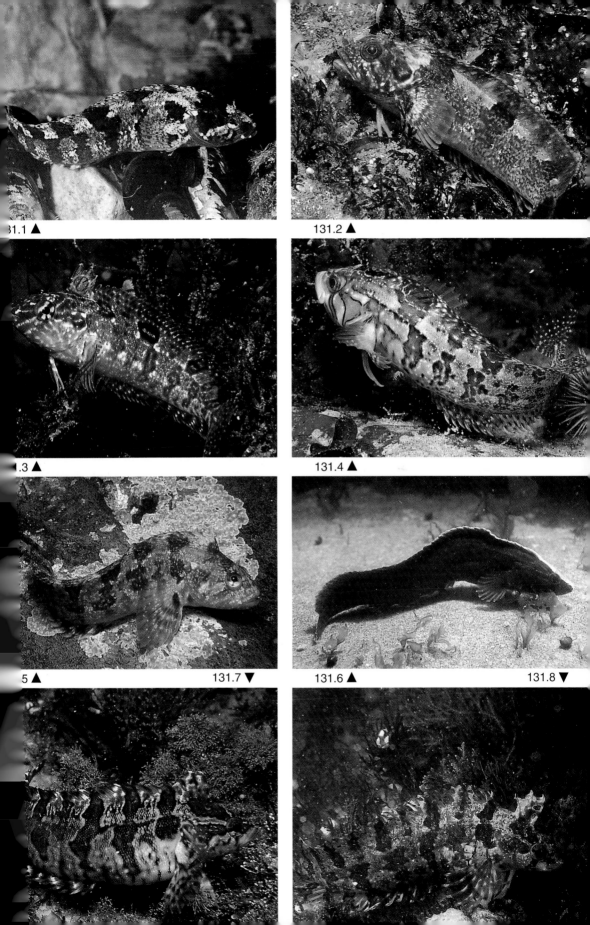

81.1 ▲

131.2 ▲

.3 ▲

131.4 ▲

5 ▲

131.7 ▼

131.6 ▲

131.8 ▼

Gobies, Suckerfish & Toadfish

Gobies are the most diverse group of fish in southern Africa. Most of the over 100 species are dull in colour, small and difficult to identify without microscopic examination, but several brightly-coloured species live on tropical coral reefs.

132.1 Barehead goby *Caffrogobius nudiceps*

IDENTIFICATION: DVI + I 11–12, AI 10–11. One of six *Caffrogobius* species in southern Africa, this small dusky, mottled fish has a rounded head, and a pale vertical bar at the base of the pectoral fin. The pelvic fin is disc-like, and the first of the two dorsal fins is striped. SIZE: 14 cm. BIOLOGY: A common endemic permanently resident in tidal pools. Breeds year-round; attaches eggs to stones and shells. Omnivorous, it eagerly takes the bait of junior 'bent-pin' anglers. RELATED SPECIES: Two similar species are common in tidal pools: the banded goby, *Caffrogobius caffer*, and prison goby, *Caffrogobius gilchristi*. The banded goby has 8–12 vertical bars on the sides and 13–14 dorsal rays in the second dorsal fin. The prison goby, which also occurs in estuaries, has 11–12 dorsal rays in the second dorsal fin and many irregular, thin, vertical dark bars which are usually broken in the middle, thereby forming two longitudinal bands.

132.2 Knysna sandgoby *Psammogobius knysnaensis*

IDENTIFICATION: DVI + I 9–10, AI 9–11. Small, grey-brown, with brown spots on dorsal surface and flanks, and pale undersides. Head broad; gill opening extends across throat. Pelvic fins disc-shaped. A black spot on rear of the first dorsal fin in males. SIZE: 7 cm. BIOLOGY: A South African endemic usually found on sandy banks in estuaries, where it often partly buries itself. Breeds all year round, and feeds on small benthic invertebrates.

132.3 Bigfin mudhopper *Periophthalmus sobrinus*

IDENTIFICATION: DXIV–XVII + I 10, AI 9–10. Small, greyish-brown, with a steep forehead and protruding pop-eyes. Pectoral fins large, pelvic fins connected. Dorsal fins have a white margin and black stripe. SIZE: 14 cm. BIOLOGY: West Indian Ocean. Associated with estuaries, particularly mangroves. Spends long periods out of water, retaining oxygenated water in the gill chamber. Captures invertebrates on land or in the water. Breeds in estuaries; lays eggs in mud nests constructed by the territorial males. RELATED SPECIES: The African mudhopper, *Periophthalmus koelreuteri*, has fewer dorsal spines and a wider gap between the two dorsal fins; occurs in the same area.

132.4 Fire goby *Nemateleotris magnifica*

IDENTIFICATION: DVI + I 28–32, AI 27–30. Small elongate fish with a compressed head; front part of body greyish, rear part red and tail fin black. First dorsal fin elongate and yellow. SIZE: 9 cm. BIOLOGY: An Indo-West Pacific species which inhabits coral heads. Feeds on zooplankton.

132.5 Rocksucker *Chorisochismus dentex*

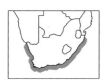

IDENTIFICATION: D 7–9, A 6–7. Mottled; pelvic fins form a prominent adhesive disc on underside of body. Head broad and compressed; mouth has long, conical teeth. Several other small species of suckerfish occur along the South African coast. SIZE: 30 cm. BIOLOGY: Endemic; occurs in tidal pools and shallow waters. Feeds on sea urchins, but more particularly on limpets, which it levers off the rocks with its strong teeth. Breeds throughout the year; eggs are attached to undersides of rocks.

132.6 Snakehead toadfish *Batrichthys apiatus*

IDENTIFICATION: DIII + 18–20, A 14–17. Small, brownish, with irregular, blotchy, dark-brown vertical bars extending onto the dorsal and anal fins. Skin loosely attached to head, body and fins. There is a network of fine ridges on the head. Four spines on gill cover, and small tentacles around mouth. SIZE: 10 cm. BIOLOGY: Endemic; found in tidal pools.

132.1 ▲

132.2 ▲

.3 ▲

132.5 ▼

4 ▲

132.6 ▼

Surgeonfish, Unicorns & Triggerfish

133.1 Pencilled surgeon *Acanthurus dussumieri*

IDENTIFICATION: DIX 25–27, AIII 24–26. One of a dozen *Acanthurus* species in South Africa. Compressed and oval. Brown, with thin, wavy, blue horizontal lines. Yellow band around the eye. A white spine in a black socket on the tail peduncle. Dorsal fin yellow; anal fin with blue margin; tail fin yellow with broad blue margin and black dots. SIZE: 54 cm. BIOLOGY: Indo-West Pacific. Inhabits coral reefs. Herbivorous. RELATED SPECIES: The brown surgeon, *Acanthurus nigrofuscus*, is brown with small orange spots on the head and a black spot at the end of the dorsal and anal fins. Tail spine and all fins are dark.

133.2 Powder-blue surgeonfish *Acanthurus leucosternon*

IDENTIFICATION: DIX 28–30, AIII 23–26. Compressed, oval; bright blue with a black and white head, yellow dorsal fin, and white anal and pelvic fins. Tail fin black and white, and the characteristic spine on each side of the tail peduncle is yellow. SIZE: 25 cm. BIOLOGY: An Indian Ocean coral-reef species; occurs singly or in small shoals. Feeds on seaweeds and invertebrates. Spine on tail can be raised and used for defence.

133.3 Convict surgeon *Acanthurus triostegus*

IDENTIFICATION: DIX 22–24, AIII 19–22. Compressed, oval, grey-green. Six vertical black bars on the body and a white ventral surface. Tail spine small. SIZE: 25 cm. BIOLOGY: A common shoaling Indo-Pacific species frequenting coral and rocky reefs. Feeds on algae. Matures at 10 cm; transparent juveniles recruit to tidal pools during the summer.

133.4 Bluebanded surgeon *Acanthurus lineatus*

IDENTIFICATION: DIX 27–29, AIII 25–28. Compressed, oval; head and upper flanks yellow, with horizontal black-edged blue stripes extending from the head to the tail peduncle. Undersides purple. Fins dusky and edged in blue. Tail spine long. SIZE: 38 cm. BIOLOGY: Indo-Pacific; frequents outer edges of coral reefs. Tail spine venomous.

133.5 Boomerang triggerfish *Sufflamen bursa*

IDENTIFICATION: DIII + 27–30, A 25–27. Greyish-brown, covered by tough scales. Two yellow curved bands run from the pectoral-fin base to the eye and the first dorsal fin. The first dorsal spine can be locked upright by the second spine. Pelvic fin reduced to a single spine. SIZE: 25 cm. BIOLOGY: One of three South African *Sufflamen* species. Indo-Pacific. Inhabits coral reefs. Feeds on invertebrates. Wedges itself in crevices at night.

133.6 Clown triggerfish *Balistoides conspicillum*

IDENTIFICATION: DIII + 25–27, A 21–22. Unmistakable, black with large round blotches on ventral part of body and an orange ring around the mouth. Body covered in tough scales. SIZE: 50 cm. BIOLOGY: Indo-West Pacific. Inhabits coral reefs and outer reef slopes. Preys on hard-shelled invertebrates, which it crushes with its strong teeth.

133.7 Redfang triggerfish *Odonus niger*

IDENTIFICATION: DIII + 33–36, A 28–30. Body deep and blue with a jutting lower jaw and red teeth. Tail lunate with extended lobes. SIZE: 50 cm. BIOLOGY: Found on coral-reef slopes in the Indo-West Pacific. Feeds on invertebrates.

133.8 Orange-spine unicorn *Naso lituratus*

IDENTIFICATION: DVI 27–30, AII 28–30. Oblong, compressed, olive-brown. Head angular. A pair of sharp bony plates on each side of the orange tail peduncle. Dorsal fin yellow with a blue margin. Tail fin emarginate. A curved yellow stripe runs from the eye to the mouth. SIZE: 45 cm. BIOLOGY: An Indo-Pacific coral-reef species. Herbivorous.

133.1 ▲ 133.2 ▲

133.3 ▲ 133.4 ▲

133.5 ▲ 133.7 ▼ 133.6 ▲ 133.8 ▼

Filefish, Tonguefish, Flounders & Soles

Flatfish have compressed, asymmetrical bodies, with both eyes on the same side of the head. They lie on one side of their bodies, the lower side of which is unpigmented. There are about 50 species in southern Africa.

134.1 Porky *Stephanolepis auratus*

IDENTIFICATION: DII + 28–34, A 30–34. A compressed, brownish-grey fish with irregular spots on sides and pale fins. First dorsal-fin spine is barbed and can be locked in an upright position by the second spine. Pelvic fin reduced to a single barbed spine. Body covered in small spinules. SIZE: 28 cm. BIOLOGY: Confined to East Africa; frequents shallow reefs and estuaries. Small juveniles often associated with flotsam.

134.2 Tropical flounder *Bothus mancus*

IDENTIFICATION: D 96–104, A 74–81. One of three *Bothus* species in South Africa, the tropical flounder is flattened and almost round in shape. The tail fin is distinct from the dorsal and anal fins. Eyes are on the left side of the head. The pectoral fin is elongated in males. Greyish with numerous small blue spots, a few black spots, and a large dark blotch on the middle of the body. SIZE: 42 cm. BIOLOGY: A widely distributed Indo-Pacific species inhabiting shallow sandy areas or flat reef areas. The characteristic transparent leaf-like larvae are abundant in offshore plankton samples on East Coast.

134.3 Cape sole *Heteromycteris capensis*

IDENTIFICATION: D 95–102, A 64–75. A small brown sole with pale and dark spots on the body and fins. Eyes on right-hand side of head. Dorsal and anal fins separate from the tail fin; no pectoral fins. Snout very hooked. SIZE: 15 cm. BIOLOGY: Southern African endemic inhabiting shallow sandy areas in estuaries. A peak in spawning occurs in summer, and juveniles of 1 cm recruit to estuaries.

134.4 Blackhand sole *Solea bleekeri*

IDENTIFICATION: D 61–74, A 46–59. Small, oval, brown, with dark spots on body and fins, and a black pectoral fin. Eyes on right side of head. Last dorsal and anal rays joined by a membrane to the tail-fin base. SIZE: 17 cm. BIOLOGY: A southern African endemic; occurs in estuaries and muddy and sandy coastal areas. Feeds on small benthic invertebrates. Sexual maturity is reached at 10 cm, and spawning occurs in spring. Juveniles recruit to estuaries at about 1 cm. RELATED SPECIES: The lemon sole, *Solea fulvomarginata* (False Bay to Transkei), is similar but has yellow fins.

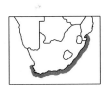

134.5 Sand tonguefish *Cynoglossus capensis*

IDENTIFICATION: D 103–110, A 81–88. One of eight species of the genus *Cynoglossus* in South Africa. Characterised by eyes on the left side of the head, snout hooked over lower jaw, continuous dorsal, caudal and anal fins, and absence of pectoral fins. Mottled grey to brown on upper surface. No lateral line on blind side of body but three lateral lines on eyed side. SIZE: 30 cm. BIOLOGY: A West African species confined to shallow muddy areas on the continental shelf. Feeds on benthic invertebrates. Larvae have conspicuous elongated anterior dorsal rays. SIMILAR SPECIES: The redspotted tonguefish, *Cynoglossus zanzibarensis* (Saldanha to Kenya), has red spots on the fins and more dorsal and anal rays.

134.6 East Coast sole *Austroglossus pectoralis*

IDENTIFICATION: D 90–110, A 80–95. An elongate, brown sole with eyes on the right side of the head. Right pectoral fin longer than head. Dorsal and anal fins continuous with tail fin. SIZE: 60 cm. BIOLOGY: Endemic; lives on muddy banks down to about 100 m. Preys on benthic invertebrates. Females grow larger than males. Spawning occurs on the Agulhas Bank. The most important commercial flatfish in South Africa. RELATED SPECIES: In the West Coast sole, *Austroglossus microlepis* (Cape to Natal), the right pectoral fin is shorter than its head. It is also commercially important.

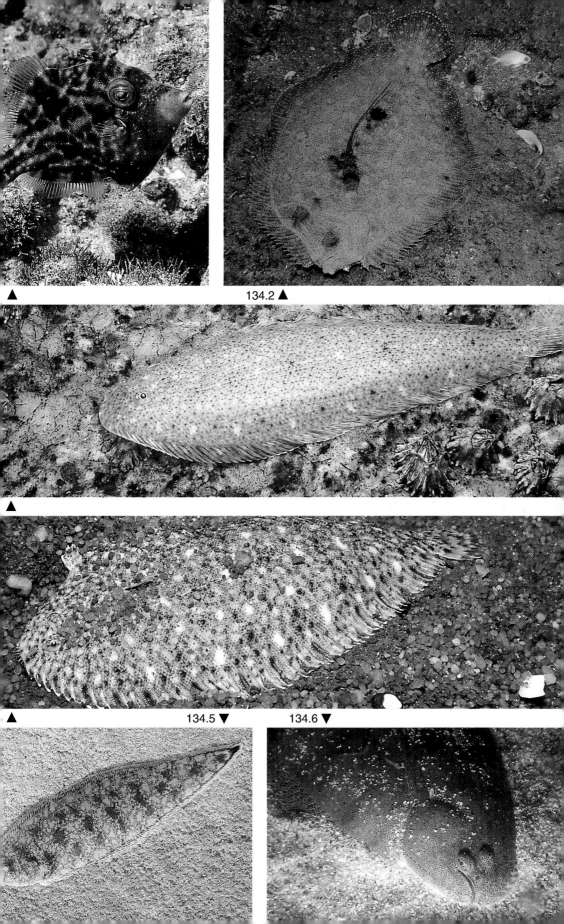

134.2 ▲

▲

134.5 ▼   134.6 ▼

▲

Boxfish, Pufferfish & Porcupinefish

The bodies of boxfish are enclosed by fused, bony plates. Puffers and porcupinefish are capable of inflating their bodies by swallowing water (or air). Their flesh is very toxic. All are feeble swimmers.

135.1 Longhorn cowfish   *Lactoria cornuta*

IDENTIFICATION: D 8–9, A 8–9. Body enclosed by an armour of fused bony plates. Two spines project forward from above the eyes; another two point backwards from the base of the anal fin. SIZE: 45 cm. BIOLOGY: Indo-Pacific, usually associated with coral-reef areas. Juveniles appear seasonally in KwaZulu-Natal. RELATED SPECIES: **135.2 *Lactoria diaphana***, the spiny cowfish, has two spines on the back and short spines over the eyes. Indo-West Pacific, but sometimes washed up on the Cape coast.

135.3 Boxy   *Ostracion cubicus*

IDENTIFICATION: D 8–9, A 9. Body box-like and enclosed by fused plates. Males yellow, with a bluish back and dark-edged white spots. Females olive with dark-edged white spots. Juveniles bright yellow with black spots. SIZE: 45 cm. BIOLOGY: An Indo-Pacific coral-reef fish. Juveniles are usually encountered in late summer along the East Coast and are sometimes cast ashore further south.

135.4 Spotted toby   *Canthigaster amboinensis*

IDENTIFICATION: D 11–12, A 11. One of eight *Canthigaster* species on the South African coast. Head more angular than most other pufferfish. Body dark and covered with tiny prickles when inflated. White spots over body and lines on head. SIZE: 15 cm. BIOLOGY: Indo-Pacific, usually associated with shallow rocky and coral reefs. Feeds on invertebrates.

135.5 Blackspotted blaasop   *Arothron nigropunctatus*

IDENTIFICATION: D 10–11, A 10–12. One of seven *Arothron* species on the South African coast. Grey, with black blotches on the plump body. Tail fin rounded with white margin. SIZE: 30 cm. BIOLOGY: This Indo-West Pacific species inhabits calm, shallow coastal waters.

135.6 Evileye blaasop   *Amblyrhynchotes honckenii*

IDENTIFICATION: D 9–10, A 8. Dorsal surface dark green with a yellow lateral band and white underside. Teeth fused to form a strong beak. Conspicuous green eyes. SIZE: 30 cm. BIOLOGY: An Indo-Pacific species that is found over reefs, sandy areas and in the lower reaches of estuaries. Can inflate itself when provoked. The flesh is extremely poisonous. Often buries itself in the sand where it awaits prey such as crabs and small fish. Regarded as a pest by anglers because of its bait-robbing habits.

135.7 Birdbeak burrfish   *Cyclichthys orbicularis*

IDENTIFICATION: D 11–13, A 10–12. Body inflatable and scales modified into non-movable spines. Pelvic fins absent. Brownish-grey but with black blotches on back and sides. SIZE: 30 cm. BIOLOGY: An Indo-Pacific species which inhabits coastal waters. Juveniles are pelagic and are often washed ashore along the Cape coast.

135.8 Shortspine porcupinefish   *Diodon liturosus*

IDENTIFICATION: D 14–16, A 14–16. Body inflatable and scales modified into short movable spines. Pelvic fins absent. Small black spots on body and fins. Dark band on throat. SIZE: 60 cm. BIOLOGY: Porcupinefishes are circumtropical and usually inhabit coral and rocky reefs. They inflate themselves into a round prickly ball when threatened. Juveniles are pelagic and often differently coloured.

135.1 ▲   135.2 ▼   135.3 ▼

135.4 ▼

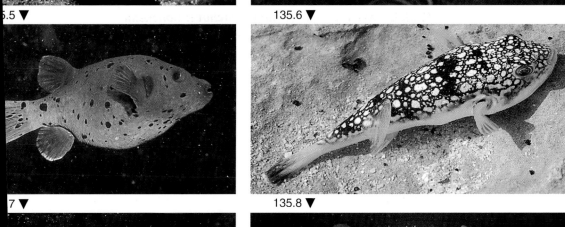

135.5 ▼   135.6 ▼

135.7 ▼   135.8 ▼

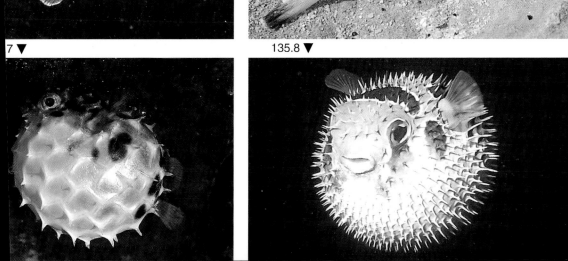

Reptilia : Snakes & Turtles

Very few snakes are completely marine, and only one of them is found in southern Africa. There are five species of turtles in the region. They spend virtually all their adult life at sea, and their legs are modified into flat flippers as an aid to swimming. Like tortoises, their bodies are enclosed in a hard, bony carapace. Turtles nest on sandy beaches, returning faithfully to the same beach year after year. In South Africa they nest only on a short stretch of coast in Zululand, between October and February. Temperature controls the sex of the offspring. Eggs kept at 20–24°C hatch as males; those above 29°C will all be female. Turtles are vulnerable to several human activities including disturbance at nesting beaches, capture for meat, strangulation in shark nets and death due to ingestion of tar-balls or plastic. Since protection was enforced for the Zululand nesting grounds, the populations there have slowly increased.

136.1 Green turtle *Chelonia mydas*

IDENTIFICATION: Like the loggerhead, the carapace consists of a series of non-overlapping plates. Distinguishable from the loggerhead because there are only four lateral plates in the row on either side of the central row, and the bill is not hooked. SIZE: 90 cm. BIOLOGY: Resident in southern Africa but never nests here. Breeds on the central Indian Ocean islands, where on several it has been hunted close to extinction. Juveniles eat small fish and molluscs, but adults subsist almost entirely on marine plants including seaweeds and seagrass. RELATED SPECIES: The olive ridley turtle, *Lepidochelys olivacea,* and the hawksbill turtle, *Eretmochelys imbricata,* are occasional visitors to our waters. The latter is recognised by the overlapping plates on its carapace and its 'tortoise-shell' patterning.

136.2 Leatherback *Dermochelys coriacea*

IDENTIFICATION: Easily recognised by its leathery back, which has seven longitudinal ridges. SIZE: With a maximum length of 2.5 m and a mass of 1.5 tonnes, the leatherback is the largest living reptile. BIOLOGY: Uncommon. About 60 breed in Zululand each year; only 6 nested in 1966 prior to conservation measures. Feeds largely on soft-bodied prey, particularly jellyfish and bluebottles.

136.3 Loggerhead turtle *Caretta caretta*

IDENTIFICATION: Upper surface of carapace broken up into a series of plates that do not overlap one another. On either side of the central row of plates there is a row of five plates. The beak is hooked. SIZE: 1 m. BIOLOGY: The most common turtle in southern Africa. Breeds in Zululand. Nesting holes 60–80 cm deep are dug at night above the high-water mark, and about 120 eggs deposited in them. Over a season, each female may lay almost 1000 eggs. Incubation lasts 60–70 days. The hatchlings (136.3j) emerge simultaneously at night and run the gauntlet to the sea, many falling prey to ghost crabs. Currents transport them south, and they spend the next 5–10 years in the gyres of the Agulhas Current, being recorded as far afield as Zanzibar, Madagascar and Cape Agulhas. Reproduction takes place at an estimated age of 12–15. Juveniles feed largely on floating, soft-bodied creatures, including bluebottles and bubble raft shells. Adults are more versatile and also consume mussels, rock lobsters, crabs, prawns and cuttlefish.

136.4 Yellow-bellied sea snake *Pelamis platurus*

IDENTIFICATION: Distinctively marked with a black upper surface and a yellow to yellow-brown lower surface. Tail flattened and yellow with black blotches. SIZE: 65 cm. BIOLOGY: Occurs widely in the Pacific and Indian oceans from the west coast of central America to the east coast of Africa. Found far out to sea and spends its entire life at sea. Gives birth to 3–8 young. Feeds on small fish. Has a potent venom that acts by paralysing muscles and is dangerous to humans, although its bite has never proved lethal.

36.1 ▲

6.2 ▲

136.3 ▲   136.3j ▼

.4 ▼

Coastal Birds

Many birds depend on the sea, but the focus here falls on birds associated with the open coast. Penguins (family Spheniscidae) occur only in the southern hemisphere; all are flightless but use their reduced wings to 'fly' underwater. Their feathers are modified to a scaly cover, and trap a layer of air when the birds submerge, helping to insulate the body. Apart from rare vagrants, there is only one species in southern Africa. Pelicans (family Pelecanidae) are characterised by the pouch that is suspended from the lower jaw, used to scoop up fish. Flamingoes (family Phoenicopteridae) have extraordinarily long legs and necks and a bent bill used to sieve food particles from the water. Gannets (family Sulidae) are closely related to cormorants, and have large bodies, heavy, sharply-pointed bills, and short, web-footed legs.

137.1 Jackass penguin *Spheniscus demersus*

IDENTIFICATION: Flightless, with flipper-like wings. Face and back black, separated by a white eye-stripe that extends along the side of the neck to meet the white chest. An inverted black 'U' is superimposed on the chest. Juveniles grey above, including the head. Chicks uniformly brown and fluffy (137.1j). SIZE: 60 cm. BIOLOGY: A common occupant of offshore islands, occasionally also colonising the mainland. Feeds largely on fish, notably pilchards, anchovy, horse mackerel and round herring, diving to depths of 30 m and circling underwater to concentrate the fish. Numbers have declined radically over the past 90 years, from an estimate of 2 million to about 150 000. Probable causes include oil spills, competition with commercial fisheries for food, harvesting of penguin eggs (which ceased in 1967) and collection of guano. The last activity prevents penguins from nesting in burrows dug in the guano, thus exposing their eggs to heat and predators.

137.2 Greater flamingo *Phoenicopterus ruber*

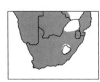

IDENTIFICATION: Overall colour white, tinged pink; open wings bright red with a black trailing edge. Bill pink with a black tip. SIZE: 130 cm. BIOLOGY: Commonly concentrates in coastal lagoons and estuaries and in shallow pans. Breeds most often in Botswana and northern Namibia. Feeds in shallow water, often rotating in a circle, stamping on the sediment to suspend small animals that are then filtered out by the bill. Tyre-shaped depressions are left in the mud. In lagoons where flamingoes are abundant, this 'treadling' profoundly affects the life of sandflats, decreasing the numbers of most invertebrates. RELATED SPECIES: The lesser flamingo, *Phoenicopterus minor*, is smaller, more obviously pink, and has a deep maroon bill. Feeds by filtering the water, never by treadling in circles.

137.3 Cape gannet *Morus capensis*

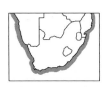

IDENTIFICATION: An elegant bird with a golden head, black stripe across the eyes and down the throat, and a broad black edging to the hind part of the wings. Pale blue ring around the eyes. Young birds brown, flecked with white. SIZE: 90 cm. BIOLOGY: Nests on offshore islands, forming huge colonies on flat areas. Incoming birds (137.3a) can recognise their mates in the dense colonies, and an elaborate greeting follows their return, including 'fencing' of bills (137.3b), head-bowing and mutual preening. Adults fly long distances to feed, plummeting from considerable heights to catch fish. Gannet guano is harvested and used as a fertiliser.

137.4 Eastern white pelican *Pelecanus onocrotalus*

IDENTIFICATION: White, tinged with pink during the breeding season. Trailing edge of wings black. Bill and facial skin yellow to orange. SIZE: 180 cm. BIOLOGY: Frequents bays, lagoons, estuaries, rivers and vleis; most common near the coast. Breeds on Dassen Island, at Lake St Lucia and at Walvis Bay. Feeds on fish, groups often co-operating to herd fish into shallow water. RELATED SPECIES: The pinkbacked pelican, *Pelecanus rufescens*, which is much less common and concentrated in KZN, is grey with a pink rump and back. In flight, the trailing edges of the wings are dark but never black.

37.1j ▲

137.1 ▲   137.2 ▼

37.3b ▼   137.3a ▼

.4 ▼

Cormorants, Herons & Egrets

Cormorants (family Phalacrocoracidae) are predominantly black, with short legs, webbed feet and slender bodies. Their feathers get wet when they enter the water, and cormorants are often seen spreading their wings out to dry them. They hunt underwater, swimming powerfully with their webbed feet. Herons and egrets belong to a closely-related family (Ardeidae) and are tall, slender birds with pointed bills and long, thin toes, and frequent shallow water, vleis, grass plains and ploughed land.

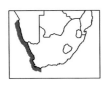

138.1 Whitebreasted cormorant   *Phalacrocorax carbo*

IDENTIFICATION: The largest of the cormorants, with a distinctive white chest, the rest of the body being brown, or glossy green-black when the birds are breeding. SIZE: 90 cm. BIOLOGY: Usually nests in colonies on coastal islands or cliffs, often on man-made structures, but does occur inland, wherever water is readily available. Feeds on fish.

138.2 Crowned cormorant   *Phalacrocorax coronatus*

IDENTIFICATION: Glossy green-black, with a red eye and an obvious tuft of upright feathers at the base of the bill. Gape of bill pale yellow. Tail proportionally longer than in other marine cormorants. SIZE: 55 cm. BIOLOGY: Nests in small colonies on rocky ledges and bushes on islands. Eats small fish and crustaceans, which it obtains by diving in shallow waters around islands.

138.3 Cape cormorant   *Phalacrocorax capensis*

IDENTIFICATION: Mature birds uniformly green-black. Head lacks a crest. Gape of bill bright orange-yellow; eye turquoise. Tail short. SIZE: 65 cm. BIOLOGY: The most common cormorant, but confined to the coast. Breeds between Namibia and Port Elizabeth. In Namibia it nests on artificial floating platforms, placed in the sea to accumulate guano. Often flies in long lines or V-shaped flocks (hence the Afrikaans name 'trekduiker'). This habit conserves energy, each bird gaining 'lift' from the one in front. Feeds on pelagic fish, particularly pilchards and anchovy. Flocks settle on the sea, and repeatedly dive to catch the fish. Other fish-eating birds often attracted to these feeding frenzies. Numbers dropped from about a million to 240 000 in the past 20 years.

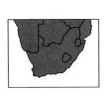

138.4 Bank cormorant   *Phalacrocorax neglectus*

IDENTIFICATION: Usually wholly black; white rump sometimes. Plumper than the Cape cormorant, and distinguished by its dark bill, dark eye, and a small crest on its head. SIZE: 75 cm. BIOLOGY: Makes seaweed nests on top of large boulders on islands. Dives shallowly; eats klipfish, gobies and invertebrates, including small rock lobsters. Total numbers low. Expanding seal populations have displaced it from some islands.

138.5 Grey heron   *Ardea cinerea*

IDENTIFICATION: Grey. Neck white, flecked black down its front. Black streak behind eyes. Cap white. Seen in flight, the lower surface of the wings is entirely grey. SIZE: 100 cm. BIOLOGY: Common in or near shallow vleis and dams, on rocky shores or in estuaries. Stands stock-still much of the time. Stalks and stabs fish, amphibians, crabs, insects and even other birds. RELATED SPECIES:
138.6 *Butorides striatus*, the greenbacked heron, has grey-green upper parts and a dark cap. Wing feathers sharply outlined with white. Legs yellow. Hunts crustaceans and fish on rocky shores and estuaries from Transkei northwards.

138.7 Little egret   *Egretta garzetta*

IDENTIFICATION: Body white, bill slender and black. Legs black; feet distinctively yellow (visible even when the bird is flying). SIZE: 65cm. BIOLOGY: Ubiquitous, but most common at marshes, dams and estuaries. Roosts in groups but usually feeds on its own. Wades in intertidal pools, and stabs fish and crabs. RELATED SPECIES: There are five other egrets in southern Africa, but none frequents rocky shores. The great white egret, with totally black legs and feet, forages in estuaries from the E. Cape northwards.

8.1 ▲ 138.2 ▲

.3 ▲ 138.5 ▼ 138.6 ▼ 138.4 ▲ 138.7 ▼

Waders & Wagtails

Five species of waders and the Cape wagtail are common on beaches and rocky shores. Many other waders (not dealt with here) are frequent on estuaries. All are associated with shallow water and eat small invertebrates. Several breed in the Arctic, migrating south for the (southern) summer. A curious correlation exists between the numbers of migrant waders in South Africa and the densities of lemmings in the Arctic. This is caused by Arctic foxes switching their attentions to waders when lemmings are scarce.

139.1 Curlew sandpiper   *Calidris ferruginea*

IDENTIFICATION: Small, predominantly pale grey, though breeding birds develop a rufous chest and face; bill slightly longer than head and curving gently downwards. In flight, upper parts mainly grey, with darker flight feathers and a narrow white wing-bar. Tail pointed; rump white. SIZE: 20 cm. BIOLOGY: An Arctic migrant, common everywhere except very arid areas. Huge flocks feed on pans, vleis and estuaries, probing the bill into wet mud to capture small molluscs, crustaceans and worms.

139.2 Turnstone   *Arenaria interpres*

IDENTIFICATION: Distinctive yellow-orange legs. Belly white, foreneck and chest mottled or black. Neck short, giving the bird a hunched appearance. Breeding plumage striking, with black bands on the white face; upper parts chestnut and black. Non-breeding plumage more drab: head mottled and wings brown with pale feather margins. In flight, recognised by the white rump and three parallel white lines down the back. SIZE: 22 cm. BIOLOGY: An Arctic migrant, most common here in summer. Small flocks feed gregariously on rocky shores or tidal sandflats, overturning pebbles and sea-weeds to capture small prey. Sometimes abundant on coastal islands, where it (and other waders) eliminate 75–90% of the small weed-dwelling invertebrates each year.

139.3 Sanderling   *Calidris alba*

IDENTIFICATION: White below, pale-grey above, with a dark shoulder-patch. Bill short and heavy. In flight, the white bar across the top of the wings is conspicuous. White rump-patch absent. SIZE: 20 cm. BIOLOGY: Flocks gather on sandy beaches and run up and down ahead of advancing waves. Feeds by lowering its head and ploughing through the surface of the sand as it runs, gathering small animals. Migrates over 12 000 km to breed in the Arctic.

139.4 Grey plover   *Pluvalis squatarola*

IDENTIFICATION: Relatively large. Grey, with mottled-brown upper wing and head. Breeding birds black below, spangled black and white above, with an intervening white line. Bill short, stout. Legs long and black. A distinctive black 'armpit' is visible in flight. SIZE: 30 cm. BIOLOGY: Common on tidal sandflats and sheltered shores. Feeds on invertebrates, especially worms.

139.5 Whitefronted plover   *Charadrius marginatus*

IDENTIFICATION: Small; neck very short, giving the bird a chubby appearance. Bill very short. Grey and brown above, but with a narrow white collar around the neck; white to pale fawn below. Head with a brown cap and a narrow dark line running through the eye. SIZE: 17 cm. BIOLOGY: A familiar sight on open-coast sandy beaches, singly or in small groups. Pecks worms, crustaceans and insect larvae from the surface of the sand. Nests in shallow depressions in the sand above the high-water mark.

139.6 Cape wagtail   *Motacilla capensis*

IDENTIFICATION: Grey with a long tail, a white throat, and a dark band across the upper breast. White brow above the eye. SIZE: 18 cm. BIOLOGY: Wagtails are not related to waders, but are similar in size and colour. The Cape wagtail occurs singly or in pairs, usually near water, and is often seen on rocky shores where it pecks insect larvae and crustaceans from seaweeds. The up-and-down wagging of its tail is a feature of all three southern African wagtails.

139.1 ▲   139.2 ▼   139.3 ▼

139.4 ▼   139.5 ▼   139.6 ▼

Gulls & Oystercatchers

Gulls (family Laridae) are raucous scavengers that are common around ships, harbours and rubbish dumps. They are moderately large birds with long, slender wings, webbed feet and robust, pointed beaks. The oystercatchers (family Haematopodidae) are the largest of the waders and have bills that are markedly flattened from side to side, an adaptation for prying open mussels. They have short legs and only three toes.

140.1 Hartlaub's gull *Larus hartlaubii*

IDENTIFICATION: White, with a grey back and upper surface to the wings. Wings tipped black and white. Head normally white, but sometimes pale grey during the breeding season. Eyes usually dark; legs and bill dark red. SIZE: 38 cm. BIOLOGY: Limited almost entirely to the coast, where it scavenges on offal and fish remains. Hovers around markets and rubbish dumps and even follows ploughs to capture displaced insects and worms. Preys on small invertebrates that it scuffles out of sandy beaches by treadling wet sand. Breeds on coastal islands, forming colonies on flat portions of the island and constructs a nest of vegetation, often in a natural depression. May breed in mixed colonies with swift terns. Has a curious habit of incorporating pebbles or snails' shells among the eggs in its clutch.

140.2 Kelp gull *Larus dominicanus*

IDENTIFICATION: White, with a black back and wing; trailing edge of wing white. Bill yellow with a red spot. Eye dark, ringed with red. Feet and legs olive-coloured. Immature individuals uniformly mottled-brown (140.2j). SIZE: 60 cm. BIOLOGY: A coastal bird that scavenges offal and animals cast up after storms. Collects stranded black mussels and smashes them by dropping them from the air onto rocks. Preys on white mussels (*Donax serra*) which it obtains by treadling in wet sand. Often eats the eggs or chicks of other island-breeding birds if they are disturbed and move off their nests. Its numbers have increased in recent years because it scavenges food from rubbish dumps. Nests on coastal islands or cliffs. RELATED SPECIES: The lesser blackbacked gull, *Larus fuscus,* is very similar but less common and confined to Natal and Moçambique. It is smaller, has chrome-yellow legs and feet, and a pale eye.

140.3 Greyheaded gull *Larus cirrocephalus*

IDENTIFICATION: Body white; head, back, and most of the upper surface of its wings grey. Tips of wings predominantly black, separated from the grey by a white flash. Legs and bill red. Eyes pale, with a red rim. Juveniles have darker upper parts than those of Hartlaub's gull, and a dark band at the tip of the tail. SIZE: 40 cm. BIOLOGY: Widely distributed and most common on inland lakes and dams. Occurs around the entire coast, but most often seen in KwaZulu-Natal. Scavenges on dead fish and offal, and preys on insects and crustaceans.

140.4 African black oystercatcher *Haematopus moquini*

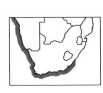

IDENTIFICATION: Unmistakable with its jet-black body, pink legs and bright orange-red bill and eyes. SIZE: 45 cm. BIOLOGY: Usually found in territorial pairs, but also roosts in flocks of 20–100 birds. Occurs on rocky and sandy shores and estuaries. Particularly abundant on islands on the West Coast. Feeds largely on mussels and limpets, but also on worms and whelks, scissoring the flesh from the shells of molluscs with its flattened bill. Where this species is abundant, such as on offshore islands, it has an important effect on the ecology of rocky shores: by substantially reducing the densities of limpets, it allows algal beds to develop. The algae support other invertebrates, which are eaten by smaller waders. Lays its eggs in a scraped hollow on the shore. The eggs and chicks are extremely well camouflaged, which, ironically, increases their chances of being accidentally damaged by off-road vehicles driven along beaches.

140.1 ▲   140.2j ▼   140.2 ▼

140.3 ▼   140.4 ▼

Terns

Terns (family Sternidae) have slender, pointed bills and more delicate bodies than gulls, and their tails are long and often forked. Most feed on fish, diving from considerable heights to capture them. Several of the species are difficult to tell apart: size and the colours of the bill and head are useful distinguishing features.

141.1 Caspian tern *Hydroprogne caspia*

IDENTIFICATION: Very large size, short tail and massive red bill make this tern easy to identify (141.1a). Body white with pale-grey back and upper wings. Black cap on head; tips of under-wing dusky, almost black (141.1b). SIZE: 55 cm. BIOLOGY: Widely distributed in Europe, Australasia and Africa. Locally common in the SW Cape, Eastern Cape and Zululand. Builds nests on the ground, scraping a shallow hollow.

141.2 Common tern *Sterna hirundo*

IDENTIFICATION: A medium-sized tern which is easily confused with several others. Distinguished by its relatively dark-grey upper parts and a grey rump and tail (visible in flight). When in breeding plumage, it has a black-tipped red bill. The bill is relatively long (approximately equal to the head length). The legs are dull red. Rump and tail greyish. SIZE: 35 cm. BIOLOGY: Strictly coastal and most abundant during spring to late summer. Often roosts in large groups, regularly numbering tens of thousands, around estuaries or on beaches. RELATED SPECIES: There are three similar species.

141.3 *Sterna vittata*, the Antarctic tern (Namaqualand to East London), is a winter visitor. Distinctly grey under-belly; grey neck separated from black cap by a white stripe across the cheek. Tail deeply forked and totally white. Legs shorter than in the common tern. In non-breeding plumage, the grizzled crown is distinctive.

The Arctic tern, *Sterna paradisaea* (Namibia to Maputo), has a uniformly red, much shorter bill (although the bill, like that of the common tern, is black in non-breeding plumage). Rump white. Most often seen at sea.

The roseate tern, *Sterna dougallii* (just north of Saldanha to Maputo, but largely confined to Algoa Bay), has a very pale back and its chest is tinged pink during the breeding season. Legs bright orange-red. Bill like that of the common tern – long, thin and dark – but its tail feathers are much longer, and pure white.

141.4 Sandwich tern *Sterna sandvicensis*

IDENTIFICATION: Relatively easy to recognise because of its long thin bill which is black, tipped with yellow. The back of the head has a slight crest. SIZE: 40 cm. BIOLOGY: Restricted to the coast, and seldom ventures far to sea. Roosts on beaches in large numbers. A summer visitor to most of the coast except Moçambique.

141.5 Swift tern *Sterna bergii*

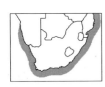

IDENTIFICATION: Smaller than the Caspian but bigger than the sandwich tern, this species is characterised by its large, uniformly yellow bill and a white forehead. SIZE: 50 cm. BIOLOGY: Resident in southern Africa year-round; nests communally, usually on coastal islands, forming a shallowly-scraped nest on the ground. Often breeds in association with Hautlaub's gull. RELATED SPECIES: The lesser crested tern, *Sterna bengalensis* (KZN northwards), is smaller, with paler underparts and a rich orange bill. Forehead black in breeding plumage.

141.6 Damara tern *Sterna balaenarum*

IDENTIFICATION: In breeding plumage, a black cap covers the entire head above the eyes. Small size, a pitch-black bill, and a grey rump and tail all distinguish this species. SIZE: 23 cm. BIOLOGY: Resident in southern Africa, but never abundant. Nests in dune-slacks and gravel plains up to 1 km from the sea. Vulnerable to disturbance by off-road vehicles. RELATED SPECIES: The little tern, *Sterna albifrons* (whole coast), is also small, but has a white rump and tail; even in breeding plumage, it has a white forehead.

41.1a ▲

141.1b ▲

1.2 ▲

1.3 ▲   141.5 ▼

141.4 ▲   141.6 ▼

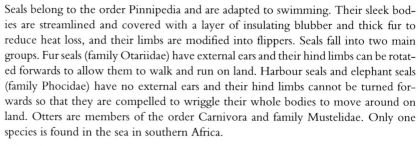

Mammalia: 1. Seals & Otters

Seals belong to the order Pinnipedia and are adapted to swimming. Their sleek bodies are streamlined and covered with a layer of insulating blubber and thick fur to reduce heat loss, and their limbs are modified into flippers. Seals fall into two main groups. Fur seals (family Otariidae) have external ears and their hind limbs can be rotated forwards to allow them to walk and run on land. Harbour seals and elephant seals (family Phocidae) have no external ears and their hind limbs cannot be turned forwards so that they are compelled to wriggle their whole bodies to move around on land. Otters are members of the order Carnivora and family Mustelidae. Only one species is found in the sea in southern Africa.

142.1 South African fur seal   *Arctocephalus pusillus pusillus*

IDENTIFICATION: Adults a uniform rich chocolate-brown colour, although darker when wet. Pups black. Teeth lying behind the canine teeth are cusped, unlike those of other fur seals. SIZE: Bulls 2.5 m, 300 kg; females 1.6 m, 75 kg. BIOLOGY: Fur seals spend much of their life at sea, and feed mainly on pelagic fish such as maasbanker, anchovies and pilchard as well as octopus, squid and small amounts of rock lobsters. Adult bulls 'haul out' in October and establish territories on islands or the mainland, defending them continually for the next 4–8 weeks, not even taking time to feed. In November, cows join them and the bulls establish harems of up to 50 cows. The cows give birth almost immediately and mate shortly after. Implantation of the embryo into the uterus is delayed for about 4 months so that the 8-month development leads to birth one year later when the cows haul out again. Fur seals were seriously depleted by sealing in the 1800s. Harvesting was subsequently controlled and quotas imposed. About 2.7 million seals were harvested in the 1900s, yet the population has recovered dramatically: it now approaches 2 million and is increasing by 3.7% per annum. Seals do compete with the fishing industry. They consume about 2 million tonnes of food each year, interrupt fishing operations and damage gear. However, culling is unlikely to reduce these effects significantly. Seals also adversely affect island-breeding sea-birds and have displaced some species from several islands. RELATED SPECIES: Sub-Antarctic fur seals, *Arctocephalus tropicalis,* occasionally visit S. Africa. They have a cream to burnt-orange chest and face, and the bulls have a prominent head-crest.

142.2 Southern elephant seal   *Mirounga leonina*

IDENTIFICATION: Enormous size and the swollen, proboscis-like snout of the male (142.2a) make this animal unmistakable. Females are fat and chubby, uniformly drab brown, and lack the proboscis (142.2b). Both sexes have only one pair of incisor teeth in the lower jaw, unlike all other seals in the family Phocidae. SIZE: Bulls 4.6 m and 3.5 tonnes; females 2.9 m, 350 kg. BIOLOGY: Common on sub-Antarctic islands; bulls haul onto boulder beaches to moult in November, followed shortly by females. Feeds mainly on squid and fish. Once heavily hunted for the oil their bodies yield, and totally eliminated from several islands during the late 1800s, although most populations have now recovered. A few stragglers find their way to S. Africa each year, and may join fur seal colonies. Isolated bulls sometimes flatten fur-seal females while attempting to mate with them.

142.3 Cape clawless otter   *Aonyx capensis*

IDENTIFICATION: Slender, sleek and seal-like in outline, with a robust, tapering tail. Upper part of body light to dark brown; lower parts paler, with a white belly. Chin and throat white and never spotted (unlike the spotted-necked otter, which does not occur in the sea). Hind feet partly webbed. Front legs have mobile, grasping fingers and are characteristically clawless. SIZE: 1.5 m, 18 kg. BIOLOGY: Lives throughout the wetter parts of southern Africa (usually near water) but also occurs in the sea, where it consumes crabs, fish and octopus. The crab *Plagusia chabrus* (42.1) is its main prey. Usually active during the early morning and evening.

2.1 ▲

142.2a ▲

142.2b ▼

142.3 ▼

Mammalia: 2. Whales & Dolphins

Highly specialised, whales and dolphins (Cetacea) have lost their hind limbs and modified their fore limbs into flippers for stability and steering; their horizontal tail-flukes provide propulsion. There are two groups: the toothed Odontoceti, and the Mysticeti which lack teeth but have comb-like baleen plates in their mouths, through which they strain water to extract plankton. Larger species were once hunted for their meat and oil, and their numbers were severely depleted. Strict protection has led to a recovery, but the numbers of most previously exploited species are still small. All cetaceans are air-breathers and must surface to breathe, expelling a spout of air and water vapour as they do so. Thirty-seven species occur in southern Africa, including the largest mammal that has ever existed, the blue whale, *Balaenoptera musculus,* which attains 33 m and 172 tonnes. Dolphins navigate by echolocation, emitting high-frequency clicks that bounce back to them off solid objects.

143.1 Southern right whale   *Balaena glacialis*

IDENTIFICATION: Lacks both a dorsal fin and throat-grooves. Lower jaw strongly curved to meet the arched snout. Wart-like white callosities bedeck the jaws, forehead and chin. Usually black-blue above, paler below. SIZE: 14–17 m; up to 47 tonnes. BIOLOGY: The baleen whale most frequently seen from the shore. Widespread in the southern hemisphere. Migrates south to sub-Antarctic waters in summer, where it feeds predominantly on copepods. Once hunted off Namibia and Moçambique but not reported there recently. Adult females visit our coast June to December, give birth and raise their young. Courting males compete for access to a female by pushing and shoving one another. Heavily exploited in the 1800s, and reduced to less than 10% of their original numbers. Protected from 1935 and now recovering at a rate of about 7% per annum in southern Africa. Named the 'right' whale because it was easy to hunt and floated after being killed: most other whales sink and are difficult to recover.

143.2 Humpback whale   *Megaptera novaeangliae*

IDENTIFICATION: One of the larger baleen whales. Recognised by its long, narrow flippers, almost one-third the body length. About 30 longitudinal grooves on the throat. Dorsal fin small and positioned far back. The tail-end of the body has several small humps on the upper surface. Totally or partially white individuals not unusual (143.2a). Jaws narrow in side view; upper jaw decorated with small knobs along its outer edge (143.2b). SIZE: 14–15 m. BIOLOGY: World-wide; spends summer feeding on krill in Antarctica. Circles underwater and releases air to form a 'bubble curtain' that concentrates its prey. Migrates north to overwinter on tropical and subtropical breeding grounds. Seldom feeds during winter, relying on blubber reserves to tide it over. Males produce melodious 'songs' that communicate over long distances underwater; these slowly change from year to year. Substantially overfished: only about 10% of the southern hemisphere population remained when it was protected in 1963.

Several other baleen whales occur off southern Africa, but seldom near the coast.

Bryde's whale (*Balaenoptera edeni*) is the most distinctive, having three longitudinal ridges on its head and about 45 throat-grooves that extend back to the navel. Up to 1967 it was hunted off Cape waters. Eats small shoaling fish.

The fin whale (*Balaenoptera physalus*) has a single longitudinal ridge on the snout; its lower jaw and chin are white on the right and black on the left. The dorsal fin is small, sharply pointed and directed backwards. Its baleen plates are grey with yellow streaks. Historically it was caught off Saldanha and Durban during its winter migrations to breeding grounds to the north.

The minke whale (*Balaenoptera acutorostrata*) is the second smallest of the baleen whales (10 m long) and the most abundant. It is spindle-shaped, with a pointed snout and about 55 throat-grooves that do not extend back to the navel. Its baleen plates are pale cream, edged with black.

143.1 ▲ 143.2a ▼ 143.2b ▼

144.1 Sperm whale   *Physeter macrocephalus*

IDENTIFICATION: The largest of the toothed whales, recognised by its enormous, blunt head. Lower jaw narrow and slung under the head, armed with heavy conical teeth. Dorsal fin represented by a triangular fleshy hump. Produces a single spout when it 'blows', characteristically directed forwards and to the left. SIZE: 15–18 m (males), 10–12 m (females); 60 tonnes. BIOLOGY: Occurs world-wide but seldom seen near the coast. Females tend to remain in warmer areas (40°S–40°N); adult males range far into cold waters. Feeds mainly on squid, especially deep-water species. Capable of prodigious dives: over 90 minutes and deeper than a kilometre. Produces clicks that travel at least 11 km underwater. Ambergris, formerly used by the perfume trade, forms in its rectum. Its head houses spermaceti, assumed to maintain hydrostatic equilibrium during deep dives, which was once used as a high-quality industrial lubricant and for cosmetic and pharmaceutical products.

144.2 Killer whale   *Orcinus orca*

IDENTIFICATION: Strikingly marked, with a black upper surface and white belly, a grey or white saddle behind the dorsal fin, and a distinctive oval white patch behind the eye. Dorsal fin very large, jutting vertically upwards in adult males but curving backwards in young males and females. SIZE: 7–9 m; 8 tonnes. BIOLOGY: Most common in the Antarctic and sub-Antarctic, but periodically recorded off South Africa. Carnivorous: eats fish, squid, birds, seals, dolphins and even attacks large whales. Hunts in packs. Never known to attack humans. RELATED SPECIES: The false killer whale, *Pseudorca crassidens,* is more slender, has a shorter dorsal fin, a small, streamlined head, no beak, and is uniformly black. There are six records of mass strandings of this species in South Africa.

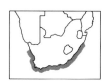

144.3 Common dolphin   *Delphinus delphis*

IDENTIFICATION: The head has a long, narrow, pointed beak. Body characteristically marked with a 'criss-cross' figure-of-eight pattern on the sides: a buff patch in the lower half of the body between the eye and the dorsal fin, and a pale-grey patch on the upper flank behind the dorsal fin. Dark stripe from flipper to lower jaw. SIZE: 2.5 m; 160 kg. BIOLOGY: One of the commonest warm-temperate and tropical species. Very gregarious, forming schools of 1000–5000. Feeds on shoaling pelagic fish. Most abundant near the coast; the more thinly-spread offshore population may be a separate species.

144.4 Bottlenosed dolphin   *Tursiops truncatus*

IDENTIFICATION: Body robust. Snout of moderate length, abruptly narrower than the head, which is thus roughly 'bottle-shaped'. Both jaws have 21–29 pairs of teeth. Upper half of body grey, with a darker grey 'cape' on the back, surrounding and including the dorsal fin. Belly off-white, often speckled with grey spots. SIZE: 2.6–3.3 m. BIOLOGY: Three populations appear to exist: a large form (3.3 m) close inshore in Namibia, a similarly large form occurring off the shelf around the entire coast, and a small (2.6 m) inshore form on the East Coast as far west as Cape Point (which may constitute a separate species, *T. aduncus*). Often joins bathers, and comes to the aid of new-born or injured dolphins. Forms schools of 20–50. Feeds mainly on fish but also squid, driving prey in a 'spearhead' formation before encircling them. Becomes entangled in shark nets. Concern has been expressed that in some Natal populations their rate of depletion is not matched by their birth rate. Chlorinated hydrocarbon levels are very high in the tissues of inshore animals in Natal, and the females 'offload' these toxins in their milk, with unknown effects on the suckling young.

44.1 ▲

44.2 ▲

4.3 ▲   144.4 ▼

145.1 Heaviside's dolphin *Cephalorhynchus heavisidii*

IDENTIFICATION: Distinctively shaped and patterned. Body thick-set, dorsal fin low and triangular, never curving backwards as it does in most dolphins. Front of body pale grey except for a dark line from the blow-hole to the dorsal fin. Remainder of upper surface black. Belly white, with a distinctive white lobe pointing obliquely backwards towards the tail. SIZE: 1.7 m. BIOLOGY: Endemic to the West Coast of southern Africa, where it occurs in small schools.

145.2 Dusky dolphin *Lagenorhynchus obscurus*

IDENTIFICATION: Beak very short and black-tipped. Fins, flippers and upper parts of body black, with two dark mid-body 'brush-marks' extending backwards and downwards onto the pale grey-white lower parts. The two-tone dorsal fin is also a useful means of identifying the dusky dolphin. SIZE: 1.9 m. BIOLOGY: Occurs in the coastal waters of continents and oceanic islands in southern-hemisphere temperate waters. Sometimes forms large schools and often rides the bow-waves of ships. Frequently leaps out of the water, and may even turn complete somersaults in the air.

145.3 Striped dolphin *Stenella coeruleoalba*

IDENTIFICATION: Dark blue-grey above; almost white below. Three distinct grey lines run backwards from the eye, one to the anus, a second to the base of the flipper, and a third between these which is short and tapering. A pale line flares upwards from the eye to the dorsal fin. SIZE: 2.5 m; 150 kg. BIOLOGY: Forms schools of 5–400. Occurs off the shelf edge, particularly on the South-east Coast. Feeds on fish and squid. RELATED SPECIES: The spotted dolphin, *Stenella attenuata*, has profuse spotting on the body. It occurs inshore on the Natal Coast.

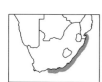

145.4 Humpback dolphin *Sousa plumbea*

IDENTIFICATION: Most easily recognised by the mid-dorsal elongate ridge on its back, which bears a small dorsal fin. Upper surface dark grey, shading to pale off-white below. SIZE: 2.8 m; 280 kg. BIOLOGY: Common inshore on the South-east Coast, forming schools of up to 25 (but usually much fewer). Total population in southern Africa small and in some areas possibly threatened by the effects of shark nets and pesticides, which are concentrated in its fatty tissues and then 'dumped' in the milk.

145.5 Southern right whale dolphin *Lissodelphis peronii*

IDENTIFICATION: One of the most striking and distinctive dolphins. Its lack of a dorsal fin and the contrast between its white face and belly and the black back make it instantly recognisable. SIZE: 2.3 m. BIOLOGY: Circumpolar in cold temperate waters of the southern hemisphere. Never abundant in southern Africa but periodically recorded off the Namibian coast.

145.1 ▲

145.2 ▲

145.3 ▲

145.4 ▲

145,5 ▼

Seaweeds

Algae include simple microscopic planktonic organisms and large macroalgae or sea-weeds. Identification of seaweeds is sometimes difficult because they vary in shape and colour, and because different phases of the life cycles of some species differ substantially. Texture and habitat are useful characters.

Seaweeds are separated into three divisions: the Chlorophyta or green algae, the Phaeophyta or brown algae, and the Rhodophyta or red algae. Members of the Chlorophyta are the easiest to distinguish, having a bright green colour. They contain the green pigments chlorophyll *a* and *b*, which are the same as those found in higher land plants. Chlorophyll *a* traps energy from sunlight and transfers it to chemical energy for the manufacture of organic compounds during photosynthesis.

The Chlorophyta have three basic body forms. The first two are simple: flat sheets one or two cells thick (order Ulotricales) or filaments of cells placed end to end (order Cladophorales). The rest (order Siphonales) consist of tubes grouped to form more complex plant bodies. The chloroplasts (tiny bodies housing the photosynthetic pigments) migrate within these tubes to make the most of light conditions, retreating from the surface to avoid destructive excess light. The life cycles of Chlorophyta are simple, alternating between a sporophyte and a gametophyte generation. The sporophytes produce spores which disperse, settle and develop into male and female sexual gametophytes. Gametes are released from the gametophytes and fuse to form new sporophytes. The two generations are similar in appearance.

The Phaeophyta have a wide variety of plant forms, including simple fans, membranous forking plants, cushions, and complex, tree-like forms such as the massive kelps. They are usually a yellowish-brown with grey or blackish tinges due to the pigments they contain, namely chlorophyll *a* and *c* plus additional pigments including fucoxanthin, which is brown. Many have a large, dominant sporophyte that alternates with a microscopic gametophyte. One group, the Fucales, lacks a separate gametophyte, and produces sexual gametes in special fertile branches on the sporophytes.

The Rhodophyta includes the large majority of the seaweeds. They contain blue and red pigments called phycobylins as well as chlorophyll *a*, and most are red or purple-red in colour (although a few are green or brown, and are easily confused with the Chlorophyta or Phaeophyta). Intertidal forms range in colour from green to reddish brown, and are often tough, wiry and much-branched to withstand wave action. Deep-water forms are usually purplish-red: their phycobylins absorb the blue-green light that penetrates deepest in the water, and transfer the energy to the chlorophyll for photosynthesis. Most deep-water algae form flat sheets to allow them to absorb the maximum amount of light, but they are often rippled and flexible. Many delicately branching species grow epiphytically on other algae and are held up to the light like the ferns and creepers in a forest. The life cycles of the Rhodophyta are extremely varied but usually have three distinct phases, further complicating their identification. The sporophyte is called a tetrasporophyte because its spores are produced in groups of four. The tetraspores develop into male and female gametophytes. Characteristic knobbles form on mature female gametophytes. The sporophyte and gametophyte generations are similar in some species. In others, such as *Gigartina* and *Notogenia erinacea* (Pl. 155), the tetrasporophytes are smooth and the groups of spores are visible as small dark spots, while the gametophytes are more branched with many surface papillae.

Seaweeds are of considerable economic importance. The kelps (Pl. 150) and *Gelidium* and *Gracilaria* (Pl. 157) are already commercially harvested, but many other species are potential targets for commercial exploitation.

Chlorophyta: 1. Simple Green Algae

Enteromorpha and *Ulva* form sheets one or two cells thick. *Chaetomorpha* species are simple unbranched filaments of cells placed end to end. In *Cladophora* the filaments are branched.

146.1 **Sea lettuce**   *Ulva* spp.

IDENTIFICATION: Membranous vivid-green blades, two cells thick, attached by a disc-shaped holdfast. SIZE: About 15–20 cm. BIOLOGY: Several species occur in mid- to high-tide pools or estuaries, where they can tolerate wide temperature and salinity changes. On the rising tide when cold water enters pools, the spores are shed from the edges of the blades and form a green scum. An early coloniser of bare rocks. SIMILAR SPECIES: *Monostroma* spp. are similar but only one cell thick. *Ulva* and *Monostroma* are widely eaten in the Far East.

146.2 **Green sea intestines**   *Ulva intestinalis*

IDENTIFICATION: Membranous green tubes, made up of sheets one cell thick. SIZE: About 15 cm. BIOLOGY: Like other *Ulva* spp. it tolerates temperature and salinity fluctuations and lives in mid- to high-tide pools and estuaries. Flourishes around sewage outfalls. Several different tubular species, all previously placed in the genus *Enteromorpha*. SIMILAR SPECIES: Can be confused with *Scytosiphon simplicissima* (152.6).

146.3 **Robust hair-weed**   *Chaetomorpha robusta*

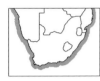

IDENTIFICATION: Stiff grass-green threads of a single row of large barrel-shaped cells up to 2 mm wide. The penultimate cells are spherical and contain the spores. Rows of transparent empty cells form at the tips when the spores are shed. SIZE: 100 mm x 2 mm. BIOLOGY: Grows on the sides of mid- to high-tide pools, just submerged. RELATED SPECIES: *Chaetomorpha crassa* (East London to Durban) has coarse unbranched green filaments of barrel-shaped cells, 2–3 mm wide, which coil around other algae and lack holdfasts. *Chaetomorpha aerea* (East London to Maputo) forms clumps of long, stiff, thread-like filaments, less than 0.2 mm wide, attached by a fibrous holdfast.

146.4 **Antenna weed**   *Chaetomorpha antennina*

IDENTIFICATION: Grass-green tufts of low-growing, unbranched, segmented filaments. The basal cell is very long and stiff, and the flexible terminal portion is made up of short cells. SIZE: Filaments about 20 mm long and less than 0.5 mm wide. BIOLOGY: Grows in high-tide pools in Natal.

146.5 **Blue whip cladophora**   *Cladophora flagelliformis*

IDENTIFICATION: Fine, wiry, blue-green, segmented filaments with untidy whorls of branches at the joints. Widens gradually to the tips, which are dark when filled with spores and transparent once the spores are shed. Holdfast is a small disc. SIZE: 15 cm; less than 0.8 mm wide. BIOLOGY: Untidy bunches on the sides of mid- to high-tide pools. RELATED SPECIES: *Cladophora mirabilis* (West Coast) also has a disc holdfast, but its coarse, dark-green filaments are up to 1 mm wide and fork repeatedly.

146.6 **Cape cladophora**   *Cladophora capensis*

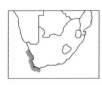

IDENTIFICATION: Fine filaments of a single row of cells, with many alternating and fairly short branches, stiff, grass-green. Holdfast fibrous. SIZE: Up to 30 cm, width 0.2 mm. BIOLOGY: Long flexible plants in subtidal gulleys, short untidy bushes in mid-tide pools. RELATED SPECIES: All with fibrous holdfasts. *Cladophora contexta* (Namibia to Cape Agulhas) forms grass-green compressed turf, up to 12 mm high. *Cladophora rugulosa* (Cape Agulhas to Durban) has long, coarse dark-green filaments. The lowest segments are long and truncheon-shaped with faint annular constrictions. *Cladophora radiosa* (Cape Point to East London) has similar coarse, dark-green branching filaments but the basal segments are long, smooth and cylindrical.

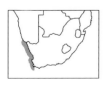

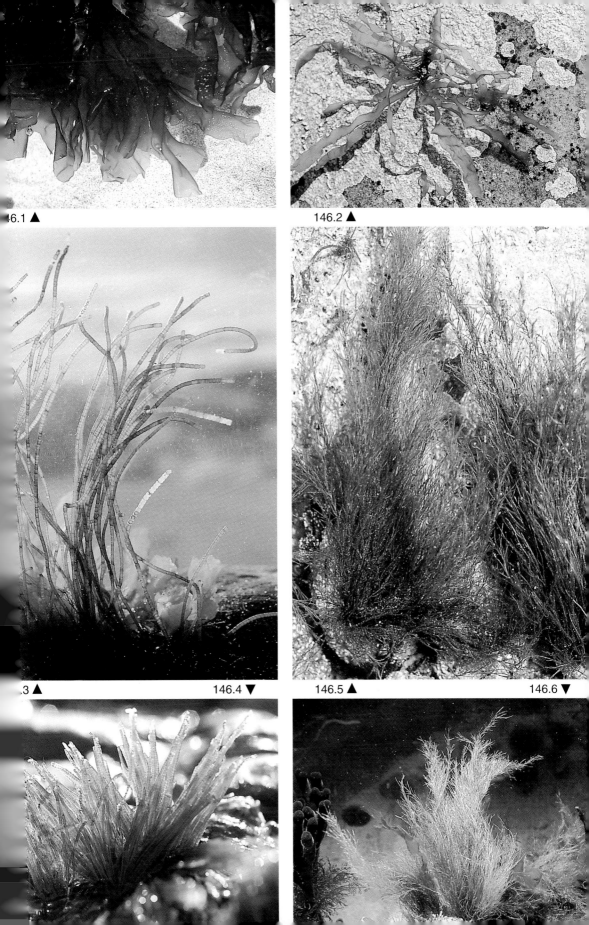

146.1 ▲

146.2 ▲

3 ▲

146.4 ▼

146.5 ▲

146.6 ▼

Chlorophyta: 2. Stalked Greens

These seaweeds (order Siphonales) are unusual in that they are made of continuous branching tubes with numerous nuclei and few or no cross-walls, even though the plant may be quite complex in structure. Their chloroplasts are able to make the best use of light conditions, by migrating through the tubes to the surface when it is dull, and away from the surface if the light is very bright. Many of them contain chemicals that act as deterrents to herbivores.

147.1 Sea moss *Bryopsis flanaganii*

IDENTIFICATION: Soft, dark-green cushions of feathery, unsegmented filaments with a bare stalk and short side-branches in the upper half. SIZE: Filaments about 50 mm high. RELATED SPECIES: There are several species with varying degrees of branching; all form soft, very dark green cushions of filaments.

147.2 Berry caulerpa *Caulerpa racemosa*

IDENTIFICATION: A mat of rhizomes giving rise to bunches of swollen light blue-green 'berries'. The shape of the berries varies and may be club-shaped, round or trumpet-shaped. SIZE: Bunches up to 20 mm high. BIOLOGY: Forming carpets on protected rocks at low-tide level.

147.3 Feathery caulerpa *Caulerpa holmesiana*

IDENTIFICATION: Stiff green feathers arising from a horizontal stem. The feathers have a short basal stem, which is annulated, and a long shaft with pinnate branches. SIZE: Feathers 30–50 mm high. BIOLOGY: Common on shallow, subtidal, sheltered rocks amid kelp, to a depth of 5 m.

147.4 Strap caulerpa *Caulerpa filiformis*

IDENTIFICATION: Tangled root-like rhizomes give rise to flattened, erect blades with cylindrical, annulated bases. Grass-green in colour, mottled with dark-green spots. SIZE: Blades about 20 cm x 5 mm. BIOLOGY: Form dense stands in sandy gulleys and can withstand being partially smothered by sand because the chloroplasts migrate to the exposed portions. The blades have special internal supports but are not divided into cells. RELATED SPECIES: *Caulerpa scalpelliformis* (northern KwaZulu-Natal and Moçambique) is a smaller species. The margins of the blades are coarsely serrated with large triangular teeth.

147.5 Wedge weed *Halimeda cuneata*

IDENTIFICATION: A series of flat, calcified discs linked together with flexible joints to form a distinctive, dark-green plant resembling a miniature prickly-pear cactus. SIZE: Plants 15 cm tall. BIOLOGY: Scattered subtidally on rocky ledges and pools and sandy gulleys on the South and East coasts. Survives periodic inundation by sand.

147.6 Sea brush *Chamaedoris delphinii*

IDENTIFICATION: Resembles dark-green paint brushes with a cylindrical annulated stalk and a terminal tuft of matted branched filaments. The base of the stalk is often calcified. Young plants consist of a transparent annulated tube with a bulbous tip. SIZE: 40 mm tall. BIOLOGY: Grows subtidally on rocky ledges and pools on East Coast.

147.7 Green fans *Udotea orientalis*

IDENTIFICATION: A tough, flat fan with a short stalk. Fibrous in texture, the fibres fanning outwards from the stalk. Colour green, often with concentric paler lines. SIZE: 30 mm. BIOLOGY: Inconspicuous; grows on submerged rocky ledges in Natal. Often found on the floors of pools that are periodically inundated by sand.

147.1 ▲

147.2 ▲

147.3 ▲

147.6 ▼

147.4 ▲

147.5 ▲

147.7 ▼

Chlorophyta: 3. Codiums and Valonia

Codiums have thick spongy branches made up of interwoven tubes, which end in specialised swellings (utricles) that cover the surface. When *Codium* is squeezed, the utricles separate. In *Pseudocodium* the cells adhere to one another. *Valonia* consists of interconnected balloon-like sacs.

148.1 False codium *Pseudocodium de-vriesei*

IDENTIFICATION: Plant erect; cylindrical branches with constrictions at irregular intervals. Surface cells are not separable by tearing or squashing, unlike the situation in *Codium* species. Grass-green. SIZE: 10 cm. BIOLOGY: Lives in low-tide rock pools.

148.2 Upright codium *Codium extricatum*

IDENTIFICATION: Plant upright; branches black-green, cylindrical, become progressively thinner; the tips have a halo of fine hairs. The cylindrical utricles separate easily from one another and bear several hairs (visible with a hand lens). SIZE: 15 cm. BIOLOGY: Inhabits sandy rock pools. RELATED SPECIES:

148.3 *Codium duthieae* (South Coast): branches often flattened; surface covered with a mixture of large and small utricles with rounded tips. *Codium fragile capense* (Namibia to Mossel Bay) has thick cylindrical branches and utricles with a small apical spine. *Codium isaacii* (West Coast) has uniformly small utricles with round tips (see 85.5). *Codium tenue* (East London to Durban): lower parts of plant flattened and the tips are cylindrical and translucent; utricles club-shaped with one or two hairs.

148.4 Flat-lobed codium *Codium platylobium*

IDENTIFICATION: Large with a basal attachment and broad, strap-shaped, flattened spongy lobes, deep black-green in colour. SIZE: Up to 50 cm long, 70 mm wide and 5 mm thick. BIOLOGY: Occurs in sandy-bottomed rock pools. A red encrusting seaweed, *Placophora binderi,* often grows on *Codium* species.

148.5 Stephens' codium *Codium stephensiae*

IDENTIFICATION: An encrusting codium consisting of thick, flattened lobes that are almost free from the substrate and stick up like masses of contorted ears. SIZE: Individual lobes 10–30 cm x 5 mm thick. BIOLOGY: Common subtidally and often forms extensive growths in rocky areas.

148.6 Lucas' codium *Codium lucasii capense*

IDENTIFICATION: Irregular cushions closely attached to the substrate and only free at the margins. The utricles have knob-like tips with a constricted neck. SIZE: Lobes 50 mm x 5 mm thick. BIOLOGY: Forms extensive lumpy sheets on sand-covered shady rocks in the mid-tide zone. RELATED SPECIES: *Codium pelliculare* (South Coast) forms thin, skin-like, hollow sacs about 20 mm wide and 2 mm thick. *Codium spongiosum* (East Coast) is made up of thick spongy cushions (50 mm x 15 mm thick) that disintegrate easily. *Codium prostratum* (East Coast) has creeping, closely-forking branches.

148.7 Golf-ball codium *Codium megalophysum*

IDENTIFICATION: Plants form green balls with a small attachment. Utricles are enormous, clearly visible. SIZE: Plants 20–50 mm in diameter. BIOLOGY: Small groups on the shady vertical sides of mid-tide pools. RELATED SPECIES: *Codium papenfussii* (South Coast) consists of flattened balls with a small attachment. The utricles are small and compacted into a continuous surface.

148.8 Green balloons *Valonia macrophysa*

IDENTIFICATION: Large, glistening, black-green, balloon-like sacs loosely joined to one another. SIZE: 50 mm x 20 mm. BIOLOGY: Occurs on low-shore, vertical rocky walls. RELATED SPECIES: *Valonia aegagropila* forms dark-green, shiny, irregular bubbly cushions, built up from a network of bulbous tubes. Size 80 mm x 2–3 mm. It occurs in rocky crevices near low-tide level in KwaZulu-Natal.

148.1 ▲

148.2 ▲

148.3 ▲

148.5 ▲

148.6 ▼

148.4 ▲

148.7 ▼

148.8 ▼

Phaeophyta: 1. Simple Brown Algae

Brown algae that are fan-shaped, or have flat branches with fanned tips, grow from a row of dividing cells (meristem) along the curved rim, e.g. *Padina, Zonaria* and *Stypopodium. Dictyota* grows from an apical cell which divides, forming regular forked (dichotomous) branching. *Dictyopteris* is similar but the fronds have a mid-rib.

149.1 Multi-fanned zonaria　*Zonaria harveyana*

IDENTIFICATION: Short overlapping fans with a few branches, yellow-brown with pale margins. Texture smooth, thin but fairly tough and pliable. SIZE: Fan about 30 mm high. BIOLOGY: Grows low on the shore to depths of about 2 m. Previously called *Zonaria multifidus* and *Homoeostrichus multifidus*.

149.2 Turkey-tail　*Padina boryana*

IDENTIFICATION: Delicate light-brown fans, with concentric light and dark bands, encrusted with lime. The margin is rolled under to protect the meristem. SIZE: Fans about 40 mm. BIOLOGY: Grows in clusters in mid-tide pools, especially very shallow pools on rocky platforms with a thin covering of sand.

149.3 Articulated zonaria　*Zonaria subarticulata*

IDENTIFICATION: Flat and branched; dark yellow-brown with a greyish tinge and dis-coloured dark spots; pale fan-shaped tips. Lateral margins irregular, interrupted at intervals by dark joints. Texture thin and leathery. Holdfast hairy. SIZE: Plants 20 cm, branches 10 mm wide. BIOLOGY: Subtidal on rocky shores. RELATED SPECIES: *Zonaria tournefortii* (Port Elizabeth to Moçambique) is similar but membranous, undulating, and uniformly yellow-brown with broader, overlapping fanned tips.

149.4 Zoned stypopodium　*Stypopodium zonale*

IDENTIFICATION: Large irregularly-split fans; light yellow-brown with a greenish iri-descence and dark concentric bands when young. Older plants dark and corrugated. SIZE: 25 cm. BIOLOGY: Subtidal on sand-covered rocks. Found as deep as 20 m on coral reefs. Produces a chemical that repels herbivorous fish.

149.5 Spotted dictyota　*Dictyota naevosa*

IDENTIFICATION: Thin forked fronds, expanding in width towards the tip. Light-brown with a blue-green sheen and dark spots. SIZE: 15 cm long, blades 5 mm wide. BIOLOGY: Grows in pools and gulleys at and below low-tide level. RELATED SPECIES: Two unnamed low-growing species are common in mid- to low-tide pools in KZN:
149.6 *Dictyota* sp. 1 is greenish-brown with a pale yellow margin. The forked fronds are 30–40 mm tall and uniformly 3 mm wide.
149.7 *Dictyota humifusa* has bright-blue, forking, overlapping fronds 20-30 mm long x 5 mm wide. *Dictyota dichotoma* (Cape Point to Zululand) is uniformly pale yellow-brown, profusely forked, and narrows towards the tips. Size 60 mm long, width 10 mm at the base, 3 mm at the tips. Occurs in rock pools below mid-tide; often grows on other algae. *Dictyota liturata* (Cape Point-Mossel Bay) is less profusely branched.

149.8 Smooth-tongued dictyopteris　*Dictyopteris ligulata*

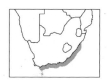

IDENTIFICATION: Elongate, thin, leathery blades, forked regularly at intervals of about 50 mm. Yellow-brown. Mid-rib distinct; margin smooth. SIZE: 20 cm long; 20 mm wide. BIOLOGY: Grows in calm, deep, mid-shore pools. RELATED SPECIES: Three species occur between East London and Zululand. *Dictyopteris delicatula* resembles the delicate, membranous fronds of *Dictyota dichotoma* but has a thin mid-rib. Fronds 2–3 mm wide, branching at 10 mm intervals.
149.9 *Dictyopteris macrocarpa* has blades with discoloured streaks and irregularly split margins. Small leaflets arise from the mid-rib. Size 20 cm by 15 mm. Fertile blades have 2–4 rows of dark, oval, spore-bearing patches.
149.10 *Dictyopteris serrata* has delicately veined blades with serrated margins and dark spots where the spores are borne.

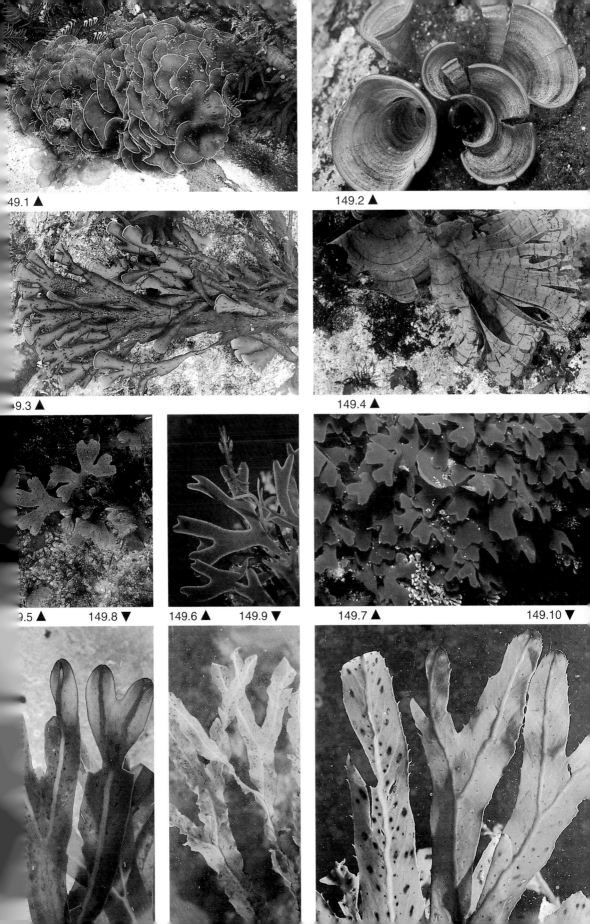

49.1 ▲

149.2 ▲

9.3 ▲

149.4 ▲

0.5 ▲   149.8 ▼

149.6 ▲   149.9 ▼

149.7 ▲   149.10 ▼

Phaeophyta: 2. Kelps

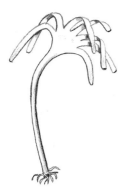

Kelps are the largest and fastest-growing algae, growing as much as 13 mm per day. The huge sporophyte plants are long-lived and form extensive underwater forests along the West Coast. They have sturdy root-like holdfasts and long stalks or stipes which support the blades. Unlike most algae, which have simple cells, the tissues of the kelps are complex with specialised cells for reproduction, photosynthesis, support and the transport of nutrients. Countless numbers of tiny spores are released from the blades and grow into microscopic male and female gametophyte plants. Kelp is harvested for extraction of alginic acid; this has many uses as a gel in food products, toothpaste, paint and ink. It is also used to stabilise earth embankments and to waterproof cement. It is an important fertiliser with a natural fungicidal action, good water-holding properties, and is rich in certain mineral salts. Fresh kelp contains growth hormones that dramatically increase the yield of commercial crops such as wheat. Kelp forests create a unique ecosystem. They break the force of the waves and provide a sheltered habitat; many animals also rely on kelps for food, either grazing the plants or consuming the fine soup of particles that continually erode from the tips of the fronds.

150.1 Split-fan kelp   *Laminaria pallida*

IDENTIFICATION: Stipe solid and stiff, terminating in a large fan-shaped blade, which is irregularly split into fronds. Spores are borne in patches on the blade. SIZE: Up to 5 m tall. BIOLOGY: *Laminaria* grows under the canopy of *Ecklonia maxima* to depths of 15 m but replaces it in deeper water, extending down to 30 m. The fronds curve over and touch the ocean floor, sweeping away animals like sea urchins and sea cucumbers that feast on kelp debris, spores and microscopic gametophyte plants. Young plants tend to grow only in the cleared areas around adult plants and so clumps develop. *Laminaria pallida* becomes more dominant than *Ecklonia maxima* as one moves further north along the West Coast.

150.2 Sea bamboo   *Ecklonia maxima*

IDENTIFICATION: The largest of local kelps; has a hollow gas-filled stipe expanding into a bulb at the top, which floats and holds the strap-shaped fronds at the surface of the water. The blades have thickened knobs along their edges and arise from a flat, tongue-like base. Spores are borne in extensive slightly-raised patches on the blades. SIZE: Up to 12 m. BIOLOGY: Small plants occur in the shallow water, larger individuals further offshore, down to depths of 10–15 m. Beneath their canopy is a host of other algae and animals such as sea urchins, mussels, rock lobsters and sea cucumbers. *Ecklonia maxima* is the dominant kelp on the southern West Coast but gives way to *Laminaria pallida* further north. Three species of red algae grow epiphytically on *Ecklonia maxima*: *Carpoblepharis flaccida* (160.1), *Suhria vittata* (160.2) and *Polysiphonia virgata* (160.3). The limpet *Patella compressa* (63.5) also lives on it.

150.3 Spined kelp   *Ecklonia radiata*

IDENTIFICATION: A smaller species of kelp with a short solid stipe and rather irregular fronds, which are usually prickly. SIZE: Up to 1 m long. BIOLOGY: Occurs in deep pools and shallow gulleys. Seldom forms solid stands like the other kelp species, and is the only southern African kelp that does not occur on the cold West Coast.

150.4 Bladder kelp   *Macrocystis angustifolia*

IDENTIFICATION: A slender flowing plant with a long flexible stipe that bears blades at regular intervals along its length. Each blade has a gas-filled bladder at its base, which expands into a rippled strap with a spined margin and pointed tip. At the tip of the plant the new blades are fused together and gradually separate as they grow. The spore-bearing tissue is confined to a few smooth blades at the base of the plant. SIZE: Plant up to 12 m long. BIOLOGY: This is the least common of the West Coast kelps and shelters inshore of the kelp forest, where there are large rock-rimmed lagoons.

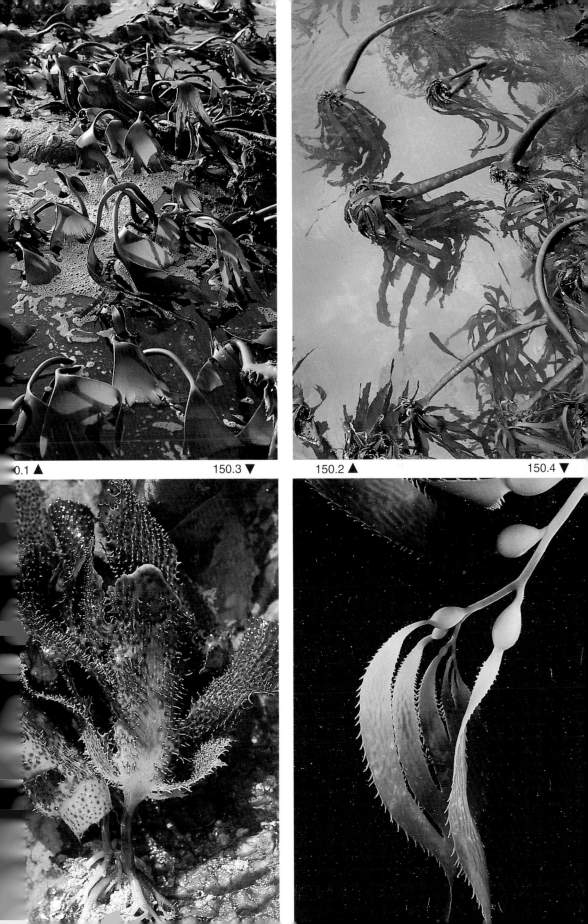

Phaeophyta: 3. Sargassum-like Algae

The Fucales is an order of brown algae that lacks an independent sexual gametophyte. The sporophyte plants produce male and female gametes directly: these fuse and form new sporophytes. Special fertile branches bear the gametes; the extruded eggs form tufts attached to the reproductive branches by mucous threads. They are not washed away: an advantage considering that most species live in heavy surf. *Desmarestia* belongs to a different order, Desmarestiales, with a microscopic gametophyte.

151.1 Different-leafed sargassum *Sargassum heterophyllum*

IDENTIFICATION: Bushy yellow-brown plants with a triangular stem, small air-floats and two types of blades. The lower 'leaves' are oval and marginally toothed; upper fronds small and spear-shaped with smooth margins. Fertile branches form inconspicuous small cylindrical tufts. SIZE: 25 cm long. BIOLOGY: Common in deep gulleys and pools; thrives in high-tide pools warmed by the sun. RELATED SPECIES: *Sargassum elegans* (South Coast) is similar but the stem is cylindrical in section.

151.2 Long-leafed sargassum *Anthophycus longifolius*

IDENTIFICATION: Tall plants with only one type of 'leaf', which is long and spear-shaped with marginal teeth and spurs. The stem is spirally twisted and knobbled at the base and flattened, with leafy wings and a strong mid-rib in the distal part. Bladders in the axils of the leaves are usually topped by a leaflet. The fertile branches are delicate and elongate. SIZE: Plant up to 50 cm, blades 20 mm wide. BIOLOGY: Occurs in deep gulleys and sublittorally on rocky shores.

151.3 Ornamented turbin-weed *Turbinaria ornata*

IDENTIFICATION: Small tufted bushes with a branching holdfast and a tough stem with spined, slightly flattened branches. The unusual fertile branches in the centre are thick and trumpet-shaped with spined margins. Yellow-brown in colour with a greyish tinge and dark spots on the fertile branches. SIZE: Plants 30 mm high. BIOLOGY: Clustered in shallow rock pools from mid-tide level down.

151.4 Acid weed *Desmarestia firma*

IDENTIFICATION: Elongate, flat, thin, leafy fronds with a serrated margin, a narrow mid-rib, and regular angular lateral veins that lead into pairs of lateral fronds. Translucent yellow-brown when young, darker when old. SIZE: Plants up to 1 m long, blades 30 mm wide. BIOLOGY: Grows subtidally beneath kelp. The plants contain sulphuric acid as a deterrent to herbivores, and soon disintegrate and lose their colour when stored.

151.5 Hanging wrack *Bifurcaria brassicaeformis*

IDENTIFICATION: Swathes of long, tough, cylindrical axes hang from creeping holdfasts. Colour yellow-brown. Fertile branches arise on either side of mature axes. They are flattened, spear-shaped, and bear two rows of lateral cavities that release gametes along the edges. SIZE: Axes 20 cm x 3–4 mm. BIOLOGY: Favours wave-pounded areas; forms glistening carpets on low-shore rocks.

151.6 Upright wrack *Bifurcariopsis capensis*

IDENTIFICATION: A conical disc-shaped holdfast gives rise to a tough, upright bushy plant. Yellow-brown in colour with grey tinges. The fertile branches are cylindrical, elongate and pitted with scattered openings. SIZE: Plants up to 20 cm tall. BIOLOGY: Grows on the floors of deep rock pools and gulleys.

151.7 Constricted axils *Axillariella constricta*

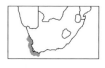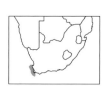

IDENTIFICATION: Axes thick, leathery, with flattened wing-like expansions, which are irregularly constricted, forming serial triangular shapes. Yellow-brown when young and blackish when old. The oval fertile branches are borne along the axes on the upper corners of the triangles. SIZE: Axes 25 cm long. BIOLOGY: Grows on shallow subtidal rocks among kelp.

151.1 ▲

151.2 ▲

151.3 ▲

151.4 ▲

151.7 ▼

151.5 ▲

151.6 ▼

Phaeophyta: 4. Bladders & Strings

152.1 Oyster thief   *Colpomenia sinuosa*

IDENTIFICATION: Yellowish, crinkled balls with a thin smooth skin and large internal cavity. SIZE: Cushion about 30 mm in diameter. BIOLOGY: Often grows epiphytically on other algae in low-tide pools or on sheltered rocks. During photosynthesis they fill with oxygen and may float to the surface. They are a nuisance in the oyster farms in Australia because they grow on the oysters and float them away – hence the name 'oyster thief'. SIMILAR SPECIES:

152.2 Brown brains   *Leathesia difformis* (Lüderitz to Zululand) forms a thick-walled spongy yellow-brown cushion, with a small central cavity. Diameter 25 mm. Grows in mid-tide rock pools and subtidally, often epiphytic on other algae or on the reef-worm *Gunnarea* (22.3).

152.3 Starred cushion   *Iyengaria stellata* (Cape Point to Port St Johns) has knobbled yellowish cushions composed of hollow, compacted branching tubes. Diameter 50 mm. Favours sheltered rock platforms and shallow pools at mid-tide level.

152.4 Dead-man's fingers   *Splachnidium rugosum*

IDENTIFICATION: Swollen yellow-brown, finger-like bladders, filled with slimy mucilage, turgid when wet and becoming withered and wrinkled when dry. The spores are released from pores in the surface skin. SIZE: Branches 80 mm x 10 mm. BIOLOGY: On rocks exposed at low tide. Its mucus protects it against desiccation during low tide.

152.5 Furry slime-strings   *Chordariaceae*

IDENTIFICATION: Cylindrical, furry, brown axes that are soft and slippery with a few branches. SIZE: About 20 cm x 5–10 mm. BIOLOGY: Seasonally common in mid-tide pools and gulleys. RELATED SPECIES: Members of the family Chordariaceae are very difficult to tell apart without microscopic examination. Genera include *Myriogloea*, *Myriocladia*, *Levringia* and *Papenfussiella*. They all occur on the West Coast to Cape Agulhas; *Levringia* extends to Durban.

152.6 Sausage skins   *Scytosiphon simplicissima*

IDENTIFICATION: Hollow brown tubes constricted at intervals, several cells thick, with a gas-filled cavity in the centre. Yellow-brown in colour. SIZE: About 15 cm. BIOLOGY: Attached to rocks in shallow pools and subtidally. SIMILAR SPECIES: Compare with the green alga *Enteromorpha intestinalis*, which consists of green tubes only one cell thick (146.2).

152.7 Cape cord-weed   *Chordariopsis capensis*

IDENTIFICATION: Stringy plants with a narrow cylindrical central axis and many short side-branches that stick out at right-angles in all directions. Colour ranges from yellow-brown to blackish. SIZE: Plants about 15 cm long; strings 1 mm wide. BIOLOGY: Lies flaccidly on the floors of mid-tide rock pools.

152.8 Bristle-tips   *Phloiocaulon suhrii*

IDENTIFICATION: Axes slender and tough with delicate, bristly, feather-like appendages alternating up the branches. Yellow-brown to blackish. SIZE: Axes about 40 mm tall and 1 mm wide. BIOLOGY: Occurs on the sides of high-shore pools. SIMILAR SPECIES: *Halopteris funicularis* (West Coast to Cape Agulhas) is a small tufted plant with numerous small, stiff, bristle-like branches coming off in all directions but pointing towards the tip, like a 'besom' reed-brush. The bristles are dark brown with a dark terminal cell in fertile plants. Plants 20–30 mm tall. It occurs among short turf in the intertidal, but is most often encountered in the shallow subtidal zone.

52.1 ▲

152.2 ▲

2.3 ▲

152.4 ▲

.5 ▲

152.7 ▼

152.6 ▲

152.8 ▼

Rhodophyta: I. Flat Red Algae

153.1 Purple laver *Porphyra capensis*

IDENTIFICATION: Thin, membranous blades. Slippery when wet, becoming like crumpled black plastic when dry. Dark purplish to purple-green with a yellow margin in male and a pink margin in female gametophytes. Sporophytes are microscopic. SIZE: Blades about 15 cm long. BIOLOGY: Occurs on the highest rocks in the Littorina Zone and is able to withstand drying as water is lost from the mucilaginous layer between the cells and not the cells themselves. It is confined to the high-shore because grazers eliminate it lower on the shore. Grows extremely fast. Edible; a related species is widely cultivated in the Far East and commands a considerable price. It is also a traditional dish in Wales and the Hebrides where it is mixed with oatmeal and fried as 'laver bread'.

153.2 Spotted iridaea *Mazzaella capensis*

IDENTIFICATION: Strap-shaped blades, brown with dark spots and a rough texture when fertile, not slippery. The blade is channelled and curled inwards where it meets the short stipe. Holdfast tiny and simple. SIZE: Blades about 30 cm long. BIOLOGY: Common intertidally on rocky shores on the West Coast, associated with *Aeodes orbitosa* (153.3). Previously called *Iridaea capensis*.

153.3 Slippery orbits *Aeodes orbitosa*

IDENTIFICATION: Blades broadly lobed, tough and extremely slippery; margin simple to finely toothed. Stipe absent; holdfast a thickened disc. Olive-brown or yellowish. SIZE: Blades about 30 cm long. BIOLOGY: Common on rocks in the mid- to low-tide level on the West Coast. Adult plants apparently unpalatable to grazers.

153.4 Orange sheets *Schizymenia obovata*

IDENTIFICATION: Flat, irregularly-lobed blades; holdfast with short branches. Texture fleshy, very finely corrugated surface, horny when dry. Colour characteristic brownish-orange and may have black patches caused by a microscopic epiphyte. SIZE: 30 x 15 cm. BIOLOGY: Occurs on the floors of mid-tide pools and gulleys.

153.5 Rippled ribbon-weed *Grateloupia longifolia*

IDENTIFICATION: Flaccid, membranous, strap-shaped blades. Marginal zones longer than the central region so that the frond ripples and coils. Plum-red, shimmering blue underwater. SIZE: About 30 cm long and 30 mm wide. BIOLOGY: Found in sandy gulleys and intertidal pools, where it ripples in the runnels. SIMILAR SPECIES: *Sarcodia dentata* (Cape Point to East London) is purplish-red; the blade is haphazardly forked, with a thickened, toothed margin. Texture fleshy.

153.6 Red rubber-weed *Pachymenia carnosa*

IDENTIFICATION: Thick, flat, irregularly-lobed blade. Margin smooth, stipe flat or cylindrical, holdfast a disc. Texture like rubber inner-tubing, not slimy. Colour orange-red to blood-red. SIZE: Large, 50 cm. BIOLOGY: Grows at the sublittoral fringe near kelp. RELATED SPECIES: *Pachymenia cornea* (West Coast to East London) has circular, purplish-red blades similar in texture, but with irregular holes and double-lipped, toothed margin. Size about 20 cm.

153.1 ▲

153.2 ▲

153.3 ▲

153.5 ▼

153.4 ▲

153.6 ▼

Rhodophyta: 2. Membranous Red Algae

154.1 Broad wine-weed *Epymenia obtusa*

IDENTIFICATION: Flat, dichotomising blades with broadly-rounded tips and no veins. Texture thin, crisp, not slimy. Colour clear reddish-pink. Stipe often forked; holdfast disc-shaped. Fertile blades bear rosettes of tiny leaflets. SIZE: Blades about 90 x 15 mm. BIOLOGY: Grows subtidally, often associated with kelp holdfasts. Commonly encrusted with the white bryozoan *Membranipora tuberculata* (48.6), and a colonial ascidian, *Botryllus anomalus* (98.6).

154.2 Veined tongues *Neuroglossum binderianum*

IDENTIFICATION: Membranous leaf-like blades with a thick flattened stipe and mid-rib that stops near the tip without fanning into veins. The blades have an untidy appearance and bear lateral leaflets, which are spotted when fertile. Holdfast branching. Transparent pinkish-red. SIZE: About 20 cm x 20 mm. BIOLOGY: Common in the shallow waters below kelp, and favours sheltered, sand-covered rocks. SIMILAR SPECIES: **154.3 *Botryoglossum platycarpum*** (West Cöast) has the same colour and habitat but is smaller and neater in appearance. The delicate, membranous blades have a solid mid-rib at the base which fans out into veins in the tip. The blades are irregularly forked with rippled margins. Stipe long and narrow. Holdfast a disc.

154.4 Veined oil-weed *Hymenena venosa*

IDENTIFICATION: Delicate, membranous, irregularly forking, strap-like blades with several parallel veins and a short stipe. Margin rippled. Colour pink, iridescing like oil on water. SIZE: About 15 cm x 20 mm. BIOLOGY: Subtidal, common in kelp-beds and often grows amongst sponges.

154.5 Black spot *Botryocarpa prolifera*

IDENTIFICATION: Flat, thin leathery blades without a mid-rib. Colour a distinctive coppery-red with blackish discolorations particularly along the margins. Holdfast a small disc. The stipe is short and may branch. The blades are oval, often with a tattered margin – small blades grow from the surface and the margin. Fertile plants have scattered rosettes of tiny leaflets on the surface of the blades. SIZE: Blades 20 cm x 40 mm. BIOLOGY: Subtidal beneath kelp.

154.6 Split disc-weed *Thamnophyllis discigera*

IDENTIFICATION: Blades thin, translucent pinkish-red, arising directly from a disc holdfast without any stipe. The blade is circular in outline but is much divided and branched, with many smaller truncated lobes at the tips. SIZE: About 15 cm long. BIOLOGY: Occurs in sheltered gulleys on sand-covered rocks. SIMILAR SPECIES: There are three species of *Acrosorium* (Cape Peninsula to East London) consisting of tiny, membranous pink blades which grow epiphytically on other algae (see 157.4). *Acrosorium maculatum* is 20 x 4 mm and forks two or three times. *Acrosorium acrospermum* is similar with white dots on the surface, 20 x 5mm. *Acrosorium ciliolatum* has uniformly narrow blades with curling tips and short lateral branchlets, 20 x 1.5 mm.

154.7 Cape wine-weed *Epymenia capensis*

IDENTIFICATION: Narrow, flat, regularly-forking blades, without veins, arising from a disc holdfast and short stipe. Thin, crisp texture. Colour clear wine-red. SIZE: Plants 10 cm long, blades 3–6 mm wide. BIOLOGY: Grows on exposed rocks at the sublittoral fringe, associated with *Gelidium* and *Plocamium*. SIMILAR SPECIES: **154.8 *Rhodophyllis reptans*** (Cape Point to Port Elizabeth) is also narrow, flat and regularly forked but the blades are much finer (2 mm wide), more branched, and narrow towards the tips. Colour clear wine-red; shimmers iridescent-blue underwater. Occurs commonly at depths of about 3 m in the understory. *Rhodymenia natalensis* (Transkei to Maputo) has narrow, flat, forking red fronds on a long stalk. 10 cm tall x 2–3 mm. Grows on low-shore rocks.

154.1 ▲   154.2 ▲

154.3 ▲   154.6 ▼   154.4 ▲   154.7 ▼   154.5 ▲   154.8 ▼

Rhodophyta: 3. Balloon- & Tongue-like Red Algae

The different phases in the life cycle of these red algae, although similar in size, may differ considerably in the degree of branching and surface texture, complicating identification. In *Gigartina* and *Nothogenia erinacea* the tetrasporophyte is smoother and has fewer surface papillae than the gametophyte.

155.1 Lance-weed   *Nemastoma lanceolata*

IDENTIFICATION: Long, narrow, soft, fleshy blades, forking once or twice, with tapering tips. Clear reddish-pink in colour. SIZE: Blades 30 cm long and about 2 cm wide. BIOLOGY: Grows submerged at depths of 1–5 m; often seen growing on the floors of deep pools and gullies.

155.2 Corrugated red alga   *Phyllymenia belangeri*

IDENTIFICATION: Dark purplish-red blades with disc holdfast and irregular lobes. The surface is coarsely textured, becoming strongly corrugated, tough and pliable as it matures. SIZE: About 20 cm long. BIOLOGY: A deep-water species that grows in the understory beneath kelp.

155.3 Tongue-weed   *Gigartina polycarpa*

IDENTIFICATION: Tough, fleshy, oval blades with a small disc holdfast and a short stem with lateral blades. Blades rubbery and dark reddish-brown. The gametophyte is covered with knobbled papillae reminiscent of a rough tongue. The tetrasporophyte is more ear-like with a pointed tip, a smooth marginal region, and the surface has a network of ridges and grooves. SIZE: Blades 15 cm long, 20–30 mm wide. BIOLOGY: Common low on the shore in sheltered areas. Is a potential source of carageenan, used as a gel. Previously called *Gigartina radula*. RELATED SPECIES: **155.4 *Gigartina bracteata*** (West Coast) is deep red, and has thicker, tougher blades with a few surface and marginal papillae in the gametophyte. The tetrasporophyte is smooth but, when the spores are released, becomes a lacy mass of holes with very short papillae on the margins. The blades are irregularly lobed and about 25 cm long. A deep-water species growing beneath kelp.

155.5 Twisted gigartina   *Sarcothalia stiriata*

IDENTIFICATION: The two generations differ, but both have a rubbery, succulent texture, thickened margins, and are yellow to purplish-brown. Tetrasporophyte smooth; a narrow stipe expands into an irregularly forked blade. Dark spots mark spore-bearing tissue. Gametophyte contorted; fleshy leaflets form irregular marginal ranks. SIZE: 6–8 cm. BIOLOGY: Occurs on wave-washed shores. RELATED SPECIES: **155.6 *Gigartina paxillata*** (Knysna to Cape Agulhas) is similar but has thinner, softer fronds; margins not thickened. The tetrasporophyte has forked tips and a few large peg-like papillae which may branch and have black spots. Gametophytes fairly flat with many small papillae and few leaflets. Olive-green to yellow or purple-brown.

155.7 Hedgehog seaweed   *Nothogenia erinacea*

IDENTIFICATION: Elongate fronds, yellowish to dark-brown, thin and not succulent; become black and papery when dry. Gametophytes covered with many small, branched outgrowths. Tetrasporophyte usually forked and fairly smooth with a few papillae concentrated along the margins. SIZE: Fronds 8–12 cm long; 6 cm wide. BIOLOGY: Abundant in the Upper Balanoid Zone on flat, sheltered rocks. RELATED SPECIES: **155.8 *Nothogenia ovalis*,** the balloon-weed, forms groups of oval, gas-filled, balloon-like bladders. Abundant on rocks adjacent to sand or gravel.

155.9 Tattered-rag weed   *Grateloupia filicina*

IDENTIFICATION: Fronds narrow, elongate, flat, with many lateral branches of varying lengths. Thin, not succulent; papery when dry. Yellow to purple-brown; black when dry. SIZE: Fronds 70 mm x 4 mm. BIOLOGY: Common in mid-shore rock pools. Should be compared with *Chordariopsis capensis* (152.7).

155.1 ▲

155.3 ▼

155.2 ▲

155.4 ▼

155.5 ▼

155.6 ▼

7 ▼

155.8 ▼

155.9 ▼

Rhodophyta: 4. Fork-branched Red Algae

156.1 Flat galaxaura *Dichotomaria diessingiana*

IDENTIFICATION: Axes flat, fleshy, with a lime-impregnated skin. Deep pink in colour with light banding. Constricted at regular intervals and adorned with black hairs at some constrictions. Forked repeatedly into two or three branches. SIZE: Axes 80 x 5 mm. BIOLOGY: Grows in low-tide pools and gulleys.

156.2 Cylindrical galaxaura *Galaxaura obtusa*

IDENTIFICATION: Axes cylindrical, pulpy, with a tough, lime-impregnated skin, divided by constrictions into segments about 1 cm long. Branching forked. Deep pink in colour. SIZE: Axes 20 cm long x 2–3 mm wide. BIOLOGY: Grows on the sides of pools and gulleys at low-tide level.

156.3 Comb-fan weed *Trematocarpus flabellatus*

IDENTIFICATION: Small, flat, comb-like fans with a distinctive yellow to purple-brown colour. Thin but tough. The flat, narrow axes fork often at close intervals. SIZE: Fans 20–40 mm long, branches 2 mm wide. BIOLOGY: Forms dense growths of overlapping fans on the rims of mid-tide pools and gulleys.

156.4 Dilated gymnogongrus *Gymnogongrus dilatatus*

IDENTIFICATION: Fronds channelled at the base but then flattening out and arched back at the tips. The axes fork regularly and become broader, the lobes wider than 5 mm. The tips of the branches are often truncated and notched and have large ball-like knobs when fertile. Texture crisp and tough. Deep plum-red in colour. SIZE: Axes 15 cm high, more than 5 mm wide. BIOLOGY: The fronds do not form dense clusters but are scattered at the low-tide level.

156.5 Complicated gymnogongrus *Gymnogongrus complicata*

IDENTIFICATION: A tough, upright axis which forks several times, the ultimate branches being ribbed and curled like contorted hands. Red-brown in colour with paler tips. SIZE: Axes 4 x 1–3 mm. BIOLOGY: Occurs in the sublittoral fringe with *Gigartina polycarpa* and *Sarcothalia stiriata*. RELATED SPECIES:

156.6 *Gymnogongrus glomerata* (Namibia to Agulhas) is yellow-brown to purple-brown. The base of its axes is cylindrical, becoming compressed and progressively wider towards the tips of the branches, which form overlapping sands. Fertile fronds bear warty knobs, which are the female spore-containers. Axes about 20–80 mm tall, up to 2–4 mm wide. Occurs along the sublittoral fringe of rocky shores.

156.7 Fine gymnogongrus *Gymnogongrus polyclada*

IDENTIFICATION: Axes narrow and cylindrical, tough and twig-like with pointed tips. The main axis forks repeatedly but also has many short lateral branches that project at an angle of 45 degrees. The tips may proliferate into groups of small branches. Dark purple in colour. SIZE: Axes about 50 x 1 mm. BIOLOGY: Grows on rocks at low-tide level. SIMILAR SPECIES: *Gigartina pistillata* (Cape Point to East London) is also forked and twiggy with many laterals but these stand off the main axis at right-angles, and the branches are compressed and up to 4 mm wide. The tips of the branches taper sharply. Fertile knobs are borne on short laterals.

156.8 Forked gigartina *Gigartina scutellata*

IDENTIFICATION: Succulent twig-like branches that fork at short intervals. Axes compressed but tips swollen with a dent at the apex. Purple-brown in colour. Reproductive structures are small balls borne in groups near the tips of special branches. SIZE: Plants 60 x 1–3 mm. BIOLOGY: Grows in dense stands at low-tide level. Previously called *Gigartina scabiosa*.

156.1 ▲

156.2 ▲

.3 ▲

156.4 ▲

5 ▲   156.7 ▼

156.6 ▲   156.8 ▼

Rhodophyta: 5. Gelatinous Red Algae

Many red algae contain agar, which has good gelling properties. In South Africa, *Gelidium* species and *Gracilaria verrucosa* are harvested commercially for the extraction of their agar for use in food and for microbiological growth-media.

157.1 Saw-edged jelly-weed   *Gelidium pristoides*

IDENTIFICATION: Fronds narrow, flat, with a mid-rib and serrated margin. They branch irregularly and bear small lateral leaflets that are often expanded and paler. SIZE: 60 mm x 5 mm wide. BIOLOGY: A dominant mid-shore alga. Untidy clumps contain juveniles, sporophytes and gametophytes. Often confined to the shells of limpets, barnacles, or reef-worm tubes, out of the reach of grazing limpets. Harvested commercially.

157.2 Abbott's jelly-weed   *Gelidium abbottiorum*

IDENTIFICATION: Finely branched, tough, wiry, maroon in colour. The axis bears lateral branches that have secondary branchlets with tertiary branchlets (pinnate to the third degree). The branches emerge at right-angles to the axis and the secondary branches are often bent. The fertile branches are spatulate and stalked. SIZE: Plants 15 cm x 1 mm. BIOLOGY: In Natal, it grows on vertical rocks, to 2 m depth, that are pounded by the waves. In the Cape it is confined to deep pools. RELATED SPECIES:

157.3 Fern-leafed jelly-weed   *Gelidium pteridifolium* (Port Elizabeth to Durban) forms coarser, neat, fern-like plants, triangular in outline. Dark red-brown in colour. The axes are flattened. Branching is pinnate to the fourth degree; the branchlets get progressively narrower and shorter towards the tip and are not bent. Plants up to 50 cm x 2 mm. Occurs in the low-shore to subtidal.

157.4 Cape jelly-weed   *Gelidium capense* (Cape Peninsula to Durban) is similar but the lateral branches emerge at right-angles to the main axis and then bend towards the tip of the plant; branched to the fifth degree, creating a dense conglomerate appearance. Lower branches often shed. Plant 20 cm x 1 mm. Subtidal.

157.5 Agar-weed   *Gracilaria verrucosa*

IDENTIFICATION: Stringy, red axes with many long cylindrical branches. The holdfast is small and insignificant, and the plant is often detached. SIZE: Up to 50 cm in length; branches 1 mm wide. BIOLOGY: Grows at and below low tide in sheltered sandy bays and lagoons. At Saldanha Bay it washes ashore as tangled masses and is harvested, dried and exported for the extraction of agar.

157.6 Spiny gracilaria   *Gracilaria aculeata*

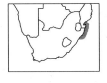

IDENTIFICATION: Stiff, horny, branching bushes, yellow-brown when young, becoming dark reddish-brown when old. The branches arise in whorls, and each branch has swollen zones at regular intervals, which are encircled by leathery spines. SIZE: Axes 10 cm x 3 mm. BIOLOGY: Forms dense growths on the vertical sides of deep mid- to low-tide rocky pools in KwaZulu-Natal.

157.7 Red spirals   *Vidalia serrata*

IDENTIFICATION: Striking plants with a forking stem and elongate forked blades that spiral in a regular manner, with four marginal serrations to each spiral of the blade. Colour red to purple. SIZE: Plant 15 cm long, blades 8–10 mm wide. BIOLOGY: Subtidal in rocky areas. SIMILAR SPECIES: *Gracilaria beckeri* (Cape Agulhas to Durban) also has long, reddish forking blades (10 mm wide) but they bend erratically and do not spiral, and the margins are finely serrated.

157.8 Constricted polyopes   *Polyopes constrictus*

IDENTIFICATION: Axes tough, slender, irregularly constricted with forked branching in all directions. Dark purplish-brown. SIZE: 60 mm long; branches 2 mm wide. BIOLOGY: Grows subtidally on rocks. SIMILAR SPECIES: **157.9 *Prionitis nodifera*** (Natal) has tough constricted axes, coarse, purple-black and knobbly, compressed in parts and contorted; axes maximally 5 mm wide.

157.1 ▲

157.2 ▲

157.3 ▲

157.4 ▲

157.5 ▲   157.7 ▼

157.6 ▲   157.8 ▼   157.9 ▼

Rhodophyta: 6. Spiky & Iridescent Red Algae

158.1 Spiky turf-weed   *Caulacanthus ustulatus*

IDENTIFICATION: A low turf of tangled wiry axes with short spiny branches. Light yellow in colour with maroon tinges. SIZE: Turf 20 mm high, axes 1 mm in diameter. BIOLOGY: A common component of mixed algal turfs on exposed, wave-washed rocks in the Lower Balanoid Zone. They are often found growing on reef-worm tubes, apparently protected there from grazers.

158.2 Elegant net fan   *Martensia elegans*

IDENTIFICATION: Beautiful, soft, iridescent mauve fans; the outer third of each forms a delicate mesh of fine tubes. Warty reproductive bodies occur on the bars of the mesh. SIZE: Fans 20–30 mm. BIOLOGY: Occurs in shallow pools and gulleys at mid-tide level. Usually associated with *Hypnea viridis*, *Dictyota* species and corallines.

158.3 Chylocladia   *Chylocladia capensis*

IDENTIFICATION: Low-growing, cylindrical, succulent axes with constricted segments. The segments are longer than wide. Short branches arise at the joints and are often arranged in whorls. The colour is dark reddish-purple, often dusted with iridescent green. SIZE: Plants 12 mm high x 2 mm wide. BIOLOGY: A common component of intertidal turfs.

158.4 Rosy curled hypnea   *Hypnea rosea*

IDENTIFICATION: A loose tangle of axes with short spiky branches and curling tips. Red in colour. SIZE: Plants 20–30 mm x 1.5 mm. BIOLOGY: Grows epiphytically on other algae, particularly *Gelidium* species at low-tide level, entwining itself with its host plant.

158.5 Iridescent hypnea   *Hypnea viridis*

IDENTIFICATION: Fine axes with short pointed branches that are intertwined into a loosely tangled turf. Iridescent purple-blue in colour. SIZE: Turf is 20 mm high, branches 0.5 mm in diameter. BIOLOGY: Common amid small *Dictyota* species and corallines in mid-tide pools in Natal.

158.6 Green tips   *Hypnea spicifera*

IDENTIFICATION: Creeping rhizomes give rise to stiff, green axes with spiked branches in their upper half. Fertile plants are densely covered with short branchlets. The base of the plant is purplish, giving the only clue that it contains the red pigments diagnostic of the red algae. SIZE: Axes 20 cm x 2 mm. BIOLOGY: In Natal and the Southern Cape it forms a vivid-green band exposed at low spring-tide. It is less common, and a dirty dark-green colour, on the West Coast. Has potential commercial importance for carrageenan extraction.

158.7 Compressed champia   *Champia compressa*

IDENTIFICATION: A dainty plant with short compressed axes with rounded tips and regular segments that are wider than long. They bear lateral branches in one plane with the branches becoming progressively shorter towards the tip. Pinkish-red in colour, often with brilliant iridescent green. SIZE: Axes 25 mm x 3–4 mm. BIOLOGY: Grows subtidally among corallines in clean pools and gulleys.

158.8 Earthworm champia   *Champia lumbricalis*

IDENTIFICATION: Upright axes arise from branching rhizomes and look like stiff earthworms, being cylindrical and divided into segments by septa, which are clearly visible against the light. The axes bear bunches of branches. Colour reddish-brown. SIZE: Axes about 15 cm long x 4 mm wide. BIOLOGY: Grows abundantly on sloping rocks in wave-washed areas at low-tide and shallow subtidal levels, where it is often the dominant seaweed.

158.1 ▲

158.2 ▲

8.3 ▲

3.4 ▲   158.6 ▼

158.5 ▲

158.7 ▲   158.8 ▼

Rhodophyta: 7. Branching Red Algae

Laurencia species have hairy pits at the tips of their branches. *Plocamium* species have curved branches that arise in groups, of two or three, which alternate on the two sides of the main axis (see 159.6d). The branches of *Pterosiphonia* alternate singly up the axes and side branches.

159.1 Flexuose laurencia   *Laurencia flexuosa*

IDENTIFICATION: Crisp, fleshy, pinkish-red plants with narrow, slightly flattened, much-branched axes and truncated tips. The branches arise on both sides, mainly in one plane, giving the plant a neat, uncomplicated appearance. SIZE: Plants about 50 mm tall, branches 2 mm wide. BIOLOGY: Common in rock pools, the Cochlear Zone and subtidal fringe in moderately sheltered areas. RELATED SPECIES: *Laurencia pumila* (East London to Durban) is a small, pink, flattened species, less than 30 mm tall, with a few short branches in all directions. Fertile axes end in stubby clumps of branches.

159.2 Red-tipped laurencia   *Laurencia natalensis*

IDENTIFICATION: Small bunched plants with numerous cylindrical branches in all directions. Green with bright-red tips. SIZE: Fronds 3–4 cm x 1 mm wide. BIOLOGY: Common in rock pools and on sand-covered rocks in the Natal intertidal. RELATED SPECIES: **159.3** *Laurencia glomerata* (Cape Point to Transkei) is uniformly purple-brown. Cylindrical branches and branchlets are densely crowded, giving an outline like an inverted cone. 50 mm tall. Subtidal. SIMILAR SPECIES: *Chondria capensis* (Port Nolloth to Cape Agulhas) has longer axes with well-spaced branches. Tips swollen with bunches of small branchlets ornamented with apical hairs. Purplish-brown.

159.4 Flattened laurencia   *Laurencia complanata*

IDENTIFICATION: Fronds like thick, fleshy, purplish-red feathers. The axes are broad and compressed. There are bunches of short pinnae on the branches, which become swollen when fertile. SIZE: Plants 60–80 mm long, axes about 4 mm wide. BIOLOGY: Grows on rocks subtidally and in pools.

159.5 Coral plocamium   *Plocamium corallorhiza*

IDENTIFICATION: A beautiful deep-pink; iridescent blue-purple underwater. Broad flattened fronds bear claw-like leaflets with marginal teeth, which are themselves serrated. Leaflets arranged in alternating pairs. The lower leaflet is large and simple, and the upper one branched. Reproductive bodies form rosettes of tiny leaflets. SIZE: Fronds 15 cm, claws 10 mm wide. BIOLOGY: Subtidal fringe down to 5 m, often in exposed wave-pounded positions. RELATED SPECIES:
159.6 *Plocamium suhrii* (East London to Zululand) is similar but more delicate, 30 mm long. Claws 5 mm wide, with simple marginal teeth. Reproductive structures form minute axillary branchlets (159.6d). SIMILAR SPECIES: *Portieria hornemannii* (Cape Agulhas to Durban) is flat with alternating branches and curled tips like tiny 'hands'.

159.7 Horny plocamium   *Plocamium cornutum*

IDENTIFICATION: Coarse, untidy, brownish in colour; crowded, cylindrical branchlets in series of not more than two. SIZE: Frond 70 mm, claws about 1 mm wide. BIOLOGY: Abundant at low-tide levels on rocks exposed to heavy surf. RELATED SPECIES:
159.8 *Plocamium rigidum* (central Namibia to Port Elizabeth) is smaller and neater, branches in series of two near the base and three near the tips; 30 mm tall, red in colour. *Plocamium beckeri* (Cape Point to Maputo) is smaller and has branchlets in series of three in all parts of the plant. Reddish-pink in colour.

159.9 Red feather-weed   *Pterosiphonia cloiophylla*

IDENTIFICATION: Small, dark, opaque, brownish-red fern-like axes. The branchlets are narrow and pointed. SIZE: Fronds 40 mm x 1 mm. BIOLOGY: Occurs on wave-swept sublittoral fringe with plocamiums and red-bait.

59.1 ▲

159.2 ▲

59.3 ▲

159.4 ▲

9.5 ▲   159.7 ▼   159.8 ▼

159.6 ▲

159.6d ▲   159.9 ▼

Rhodophyta: 8. Epiphytic & Fine Algae

160.1 Flaccid kelp-weed Carpoblepharis flaccida
IDENTIFICATION: Long, narrow, flattened main axes are haphazardly branched. All the branches have numerous tiny branchlets along each side, creating a frilled margin. Soft and fleshy, purplish-brown in colour. SIZE: Fronds 20 cm long; 2 mm wide. BIOLOGY: A common epiphyte on the kelp *Ecklonia maxima*. *C. minima* grows on *Laminaria*.

160.2 Red ribbons Suhria vittata
IDENTIFICATION: Long ribbon-like fronds with a distinct mid-rib, and a frill of delicate leaflets along the margin. They are often spiralled and have a few branches. Purple-red in colour. SIZE: Fronds about 20 cm long; 10 mm wide. BIOLOGY: Grows on the kelp *Ecklonia maxima*. Can be boiled in water to extract the agar, which is added to flavouring and sugar to make jelly puddings.

160.3 Kelp fern Polysiphonia virgata
IDENTIFICATION: Fine, whip-like axes that branch in all directions. The axes are made up of several rows of microscopic cells, and the reproductive bodies are buried in the swollen ends of the branches. Dark purple-black in colour. SIZE: Axes 20–30 cm. BIOLOGY: Grows epiphytically on the kelp *Ecklonia maxima*.

160.4 Beaded ceramium Ceramium diaphanum-group
IDENTIFICATION: Delicate, fine, branching epiphytes. Each filament has a single row of long, transparent central cells surrounded by bands of small red cortical cells, giving a distinctive beaded appearance just visible to the naked eye. The filaments branch in all directions and have in-curved claw-like tips. Round spore bodies are borne on the branches. SIZE: Axes 20–30 mm x 0.5 mm. BIOLOGY: Epiphytic on other algae, particularly codiums. Several closely related species.

160.5 Flat-fern ceramium Ceramium planum
IDENTIFICATION: Perhaps the most beautiful of the delicate red algae. The axes have neat pinnate branching, creating intricate, flat, fern-like fronds. Purplish-red in colour. SIZE: Fronds 10–20 cm. BIOLOGY: Grows on other algae, notably *Codium* species.

160.6 Cape ceramium Ceramium capense
IDENTIFICATION: Axes whip-like, branching in all directions, usually in whorls. There are large and small branchlets along the axes. Tips form unequal pairs of slightly in-curved claws. Axes opaque, purple, with only a suggestion of beading. SIZE: 50 mm x less than 0.5 mm wide. BIOLOGY: Epiphytic on other algae. RELATED SPECIES: *Ceramium obsoletum* (Port Nolloth to Cape Agulhas) is a larger plant, up to 20 cm, which forks regularly. The branches diverge at a wide angle and have short secondary branches along the inner sides of the forks; tips of branches short. Axes are not beaded but opaque and purple.

160.7 Aristocratic plume-weed Aristothamnion collabens
IDENTIFICATION: Delicate purple-red plants with intricate branching in all directions. The main branches get progressively shorter towards the apex of the plant and are themselves finely branched to form bottlebrush-like fronds. SIZE: 20–30 mm. BIOLOGY: Grows epiphytically on other algae. SIMILAR SPECIES:
160.8 Callithamnion stuposum is also finely branched, plumed and tree-shaped, but the filaments twist together and are slightly sandy. Colour purple to deep-green with iridescing tips. On low-tide rocks, Cape Agulhas to Natal.

160.9 Curl-claw Centroceras clavulatum
IDENTIFICATION: Forms tufts of filaments which are not beaded but jointed. Terminal joints have a ring of microscopic spines. Branches have forked tips that form pairs of strongly in-curled claws. Blackish-brown in colour. SIZE: Fronds usually 40–80 mm. BIOLOGY: Not epiphytic; grows intertidally in damp depressions or pools.

60.1 ▲   160.2 ▲   160.3 ▲

0.4 ▲   160.7 ▼   160.5 ▲   160.8 ▼   160.6 ▲   160.9 ▼

Rhodophyta: 9. Upright Coralline Algae

Upright coralline algae have jointed stems, the segments of which are impregnated with lime as a deterrent to grazers. The type of branching and position and structure of the fertile segments are diagnostic for the genera.

161.1 Finely-forked corallines *Jania* spp.

IDENTIFICATION: Slender, cylindrical axes with symmetrical forked branching; joints are marked by a narrow line. Deep pink to purplish-pink in colour with pale tips. The swollen fertile segments are subterminal with a distal pore (161.1d). SIZE: Axes usually 20–30 mm long; less than 0.5 mm wide. BIOLOGY: Several South African species. Form low turfs in mid-shore rock pools and extensive mats on low-shore flat rocks.

161.2 Nodular coralline *Amphiroa bowerbankii*

IDENTIFICATION: Segments rectangular, flat with thin edges, joints obvious and ornamented with groups of small nodules. Branching forked; ultimate joints have pale rounded tips. Chalky-pink in colour. Surface of segments covered with warty reproductive capsules, each with a central pore (161.2d). SIZE: Axes 50 mm, joints 5–8 mm x 2–4 mm. BIOLOGY: Grows in low-shore pools.

161.3 Horsetail coralline *Amphiroa ephedraea*

IDENTIFICATION: Segments elongate, nearly cylindrical or slightly flattened, with smooth ends. Joints very obvious, black, about 1 mm wide. Branching forked. Purplish-pink in colour. Surface of segments covered with warty reproductive capsules. SIZE: Axes 15 cm long, segments 7–10 mm long x 1–2 mm wide x 1 mm thick. BIOLOGY: Forms large hanging clumps on low-shore and subtidal rocks. RELATED SPECIES:

161.4d *Amphiroa anceps* (Cape Agulhas to Durban) has narrow flattened segments with lateral shoulders at the joints, which are less obvious.

161.5 *Amphiroa capensis* (Cape Peninsula) is much smaller and regularly forked at each joint to form fans about 50 mm long. Segments cylindrical, 3 mm x 1 mm; joints obvious in the centre but narrow at the sides (161.5d).

161.6 Arrowhead coralline *Cheilosporum multifidum*

IDENTIFICATION: Segments arrowhead-shaped with thin joints between them. Branching fairly sparse, forking about every four to ten joints. Segments have irregular toothed edges. Deep pink. Reproductive capsules form swellings on the distal lip of the segment, one on each wing (161.6d). SIZE: Axes 50 mm long, segments 2 mm long x 2–3 mm wide. BIOLOGY: Low shore in rocky pools. RELATED SPECIES:

161.7 *Cheilosporum sagittatum* (Cape Town to Durban) is more delicate, up to 4 cm tall. The segments are smooth, acute arrowheads 1.5 mm long x 1.5 mm wide. Mauve in colour. One fertile swelling to each wing (161.7d).

161.8d *Cheilosporum cultratum* (Cape Point to Maputo) has pink, crescent-shaped segments 1 mm long, 6 mm wide; 1–3 spore capsules per wing.

161.9 Feather corallines *Corallina* spp.

IDENTIFICATION: Branching pinnate (feather-like). There are several species which differ in the widths of their segments. The fertile segments are terminal, club-shaped and may have two horns (161.9d). SIZE: Varies with the species; about 50 mm. BIOLOGY: Often form short, bushy turfs in mid-shore pools.

161.10 Hinged corallines *Arthrocardia* spp.

IDENTIFICATION: Stout axes, oval in section, with up to four branches at a joint; segments are flattened, irregular wedges. Deep pink when not bleached. Fertile fronds fork distally. The pore to the spore container is usually terminal; two lateral branches arise on either side of it (161.10d). SIZE: Varies with the species, about 50-80 mm. BIOLOGY: Close-set tufts grow in rock pools from mid-shore downwards. The several species are difficult to determine.

161.1d

161.2d

161.1 ▲

161.2 ▲

161.4d

161.5d

61.1 ▲

1.3 ▲

161.5 ▲

161.6d

161.7d

161.8d

.6 ▲   161.9 ▼

161.7 ▲   161.10 ▼

161.9d

161.10d

Encrusting Algae

Some seaweeds form flat crusts. Notable are the encrusting corallines which are laden with lime, a putative deterrent to grazers. They are diverse, but accurate identification requires microscopic examination of their tissues and reproductive bodies (conceptacles). Generally, thicker crusts are good competitors, overgrowing thinner ones, but are slower-growing and less tolerant of grazing.

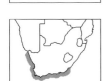

162.1 Ralfsia *Ralfsia verrucosa*

IDENTIFICATION: Forms flat patches or irregular, concentric rings. Brown, grading to khaki; edge pale. SIZE: 1 mm thick; 15 cm wide. BIOLOGY: Found in 'gardens' of the limpet *Patella longicosta* (62.6). Produces polyphenols which deter some herbivores because they inhibit digestion. Replaced by *Ralfsia expansa* in Natal and Moçambique.

162.2 Tar crust *Hildenbrandia lecanellierii*

IDENTIFICATION: Forms irregular, gnarled crusts that lift at the margins. Purple-black. SIZE: 2–10 cm wide; 1–4 mm thick. BIOLOGY: Encrusts crevices low on the shore and in pools. RELATED SPECIES: *Hildenbrandia rubra* (Cape Point to Zululand) forms thin, smooth encrustations in relatively regular circles; resembles red bloodstains.

162.3 Red fan-weed *Peyssonnelia capensis*

IDENTIFICATION: Flat, overlapping, leathery fans; white below and purplish-red above, but may be tinged yellowish or green. Attached to rocks by a central holdfast, but its fans are raised from the substrate. SIZE: Fans about 40 mm. BIOLOGY: Grows on the walls of deep pools or gulleys, down to 20 m. *P. atropurpurea* is dark red-black.

162.4 Cochlear coralline crust *Spongites yendoi*

IDENTIFICATION: Relatively thin. Chalky-white; mauve if shaded. Usually lumpy, or has small, upright knobs. Conceptacles pimple-like and uniporate. SIZE: 0.5 mm thick. BIOLOGY: Abundant on the SW Coast; usually grows on or near the limpet *Patella cochlear* (62.5). SIMILAR SPECIES: *Pneophyllum keatsii* (thin, grey, circular crusts) and *Synarthrophyton eckloniae* (thicker, pink, multiporate) grow on kelps.

162.5 Scrolled coralline crust *Spongites impar*

IDENTIFICATION: Moderately thick. Chalky or beige, margin conspicuously paler. Produces twisted ridges wherever two colonies meet. Surface like an elephant's skin. Conceptacles uniporate. SIZE: 1 mm thick; ridges 10 mm high. BIOLOGY: Mid-shore; common where wave action is strong. Weakly attached and intolerant of grazing. Overgrows *S. yendoi* unless grazed by *Scutellastra cochlear.*

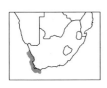

162.6 Velvety coralline crust *Heydrichia woelkerlingii*

IDENTIFICATION: Thick, flat or lobed; surface smooth, glossy, velvety to the touch. Deep purple-pink. Asexual conceptacles uniporate. SIZE: 3–15 mm thick; 5–30 cm diameter. BIOLOGY: A shallow subtidal species. Aggressively overgrows all other crusts, but (as illustrated here) held at bay by the grazing of limpets. Often bored by the polychaetes *Polydora* and *Dodecaceria* (see 21.5).

162.7 Thin coralline crust *Leptophytum foveatum*

IDENTIFICATION: Thin, paint-like, smooth. Browny-pink, with pale squiggles. Edges white. Asexual conceptacles multiporate. SIZE: 0.5 mm thick. BIOLOGY: Forms sheets at or below low tide. Tolerates grazing but susceptible to desiccation. Generates new margins to combat overgrowth. Often associated with the limpet *Patella argenvillei*.

162.8 *Leptophytum acervatum* (Cape Point–Natal) is almost identical but brighter pink; surface often pocked. Grows on pebbles, not rock faces.

162.9 *Leptophytum ferox,* 'packman' (Namaqualand–Moçambique), is 5–10 mm thick, with upright pillars, often capped by slits bordered with 'lips'.

162.10 *Mesophyllum discrepans* (Cape–Natal, low-shore) is purple-brown, white-edged, 1–2 mm thick; it has bumps that erode to form white caps.

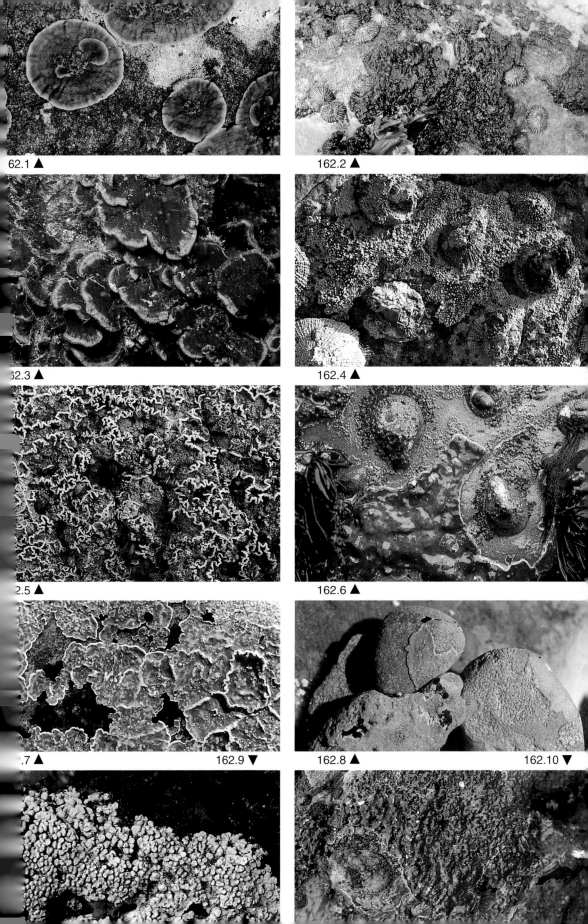

162.1 ▲

162.2 ▲

162.3 ▲

162.4 ▲

162.5 ▲

162.6 ▲

162.7 ▲   162.9 ▼

162.8 ▲   162.10 ▼

Angiospermae: 1. Saltmarsh Plants

Several flowering plants form saltmarshes on estuarine shores, and have adapted to the saline water experienced there. Most are sprawling and low-growing; many are succulent or have reduced leaves to minimise water loss.

163.1 Bulbous marshweed *Triglochin bulbosa*

IDENTIFICATION: A narrow spike of flowers ensheathed by thin upright leaves. Fruit slender, 5–10 mm long. SIZE: 15 cm. BIOLOGY: Common in high-shore saltmarshes at the top of estuaries. RELATED SPECIES:
163.2 *Triglochin striata*: fruit round, 2 mm; leaves splay sideways and then curve up.

163.3 Cord grass *Spartina maritima*

IDENTIFICATION: Forms tufts of narrow, spiky, rough, inrolled leaves; flower head consists of two branches that are pressed together. SIZE: 40 cm. BIOLOGY: Probably an alien species, but is well established in most estuaries on the SW and SE coasts. Forms dense stands just below the high-tide mark.

163.4 Dune slack rush *Juncus kraussii*

IDENTIFICATION: Long, sharply-pointed stems with thin, pointed leaves. Flowers and cylindrical fruits borne in terminal tufts. SIZE: 40 cm. BIOLOGY: Dominates saltmarshes where salinities are low; often extends into dune slacks bordering saltmarshes. One of several species of sedges that border estuaries.

163.5 Oval-leafed saltweed *Halophila ovalis*

IDENTIFICATION: Develops from underground creeping rhizomes that bear slender upright stalks, each with a single, flat, oval leaf. SIZE: 40 mm. BIOLOGY: Found on the lower parts of waterlogged intertidal sandbanks in estuaries, particularly where the water is brackish (*halophila*, 'salt-lover').

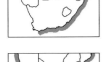

163.6 Soutbossie *Chenolea diffusa*

IDENTIFICATION: Short underground stems with upright branches surrounded by grey-green succulent leaves with a velvety texture. SIZE: 10 cm. BIOLOGY: Grows high on estuarine shores, in sandy areas that lie close to, or just above, the high-tide level.

163.7 Cape eelgrass *Zostera capensis*

IDENTIFICATION: Creeping roots. Leaves narrow, flat; tips rounded, faintly notched. SIZE: Blades 10–30 cm, 1 mm wide. BIOLOGY: Estuarine; binds the sediment, shelters small fish, and yields detrital food. RELATED SPECIES: *Halodule uninervis* (estuarine, St Lucia northwards) is similar, but its leaves have three-pointed tips. Three *Cymodocea* species (common in calm tropical waters) have strap-like leaves 10–15 mm wide.

163.8 Estuarine pondweed *Ruppia maritima*

IDENTIFICATION: Stems and leaves thin, grass-like and tangled (see Pl. 107.4). Flower stalk short and straight (163.8d). SIZE: Stems 1 m long, 1 mm wide. BIOLOGY: Prolific at the heads of estuaries. Always submerged; forms dense mats cursed by boatmen. RELATED SPECIES: **163.9d *Ruppia cirrhosa*:** flower stalk coiled.
163.10d *Potamogeton pectinatus*: flowers clustered on a spike.

163.11 Glasswort samphire *Sarcocornia perennis*

IDENTIFICATION: A sprawling perennial succulent. Stems jointed; no obvious leaves. Flowers tiny, arranged in threes at stem nodes (163.11d). SIZE: 30 mm tall. BIOLOGY: Forms mats low on estuarine saltmarshes. RELATED SPECIES:
163.12 *Sarcocornia littorea* (Namaqualand–Agulhas) is up to 1 m tall, has a thick woody stem, and grows on rocky shores above the high-tide mark.
163.13d *Sarcocornia pillansii* (whole coast) forms shrubs at the upper levels of saltmarshes. Stems thick, flat; each joint ends in a gondola-like tip.
163.14 *Salicornia meyeriana* (Namaqualand–Durban) is an annual and thus has a very short tap-root; forms low bushes with slender, woody branches in the upper zones of estuarine marshes. Flowers arranged in threes (163.14d).

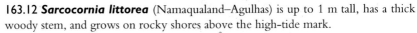

163.1 ▲   163.2 ▲   163.1 ▲

163.3 ▲

163.4 ▲

3.5 ▲   163.11 ▼

163.6 ▲

163.8d   163.9d   163.10d

163.7 ▲   163.14 ▼

163.12 ▲

163.11d   163.13d   163.14d

Angiospermae: 2. Mangroves

Mangroves are trees that grow in saltwater on intertidal mudflats, in estuaries or even on sheltered coasts. Their roots consolidate and trap fine mud; the sediments associated with them are rich in organics but lack oxygen, and are laden with hydrogen sulphide produced by sulphur bacteria. Mangroves support several species of animals that are seldom found elsewhere, including two snails, *Cerithidea* (70.6) and *Terebralia* (70.8) and the mudskipper *Periophthalmus* (132.3). Mangrove roots are shallow and have above-ground pneumatophores to allow them to breathe, exchanging carbon dioxide for oxygen from the air.

164.1 White mangrove   *Avicennia marina*

IDENTIFICATION: Trees of considerable stature, with a grey-white bark and lance-shaped, silvery grey-green leaves (164.1d). Distinctive mats of vertical pneumatophores ('pencil roots') project from the mud beneath them. Fruits egg-shaped with a pointed tip. SIZE: 12 m tall, leaves 65 mm long, fruit 25 mm long. BIOLOGY: The most abundant mangrove in southern Africa, and an early pioneer in the establishment of mangrove swamps, shading and stabilising the sediment with its fine roots, thus providing a 'nursery' for other, less hardy mangroves. Its roots exclude much of the salt in seawater, and its leaves have specialised salt glands that expel salt from the plant. Its leaves contribute substantially to the detritus available as a source of food for animals, and are particularly important for the mangrove crab (41.1).

164.2 Black mangrove   *Bruguiera gymnorrhiza*

IDENTIFICATION: A tall tree with a rough red-brown bark, buttress-roots around the trunk, and knee-like above-ground pneumatophores. Leaves dark green, yellowing with age, smooth and shiny, roughly oval but with a pointed tip. The fruit has a star-shaped basal calyx from which a cigar-shaped hypocotyl projects (164.2d). SIZE: 18 m tall, leaves 12 cm long, fruit 18 cm long. BIOLOGY: Common, but cannot establish new stands of mangroves on its own: grows in the middle of reedbeds or *Avicennia* stands, often outcompeting *Avicennia*. Its roots efficiently prevent most salt from entering its tissues as it draws up water from the surrounding seawater. Its leaves accumulate salt, and the regular loss of leaves may also rid the plant of excess salt. Its mature flowers remain tightly shut until touched by an insect, when they explosively burst open, showering the insect with pollen. Mature fruits drop off and often impale themselves in the mud so that they grow immediately below the parent plant.

164.3 Red mangrove   *Rhizophora mucronata*

IDENTIFICATION: A moderate-sized tree characterised by obvious prop (or 'stilt') roots that arch down from the main stem. Leaves elliptical with an abruptly spiked tip ('mucro'). The fruits have a pear-shaped basal calyx and a long, thin, sharply pointed hypocotyl (164.3d). SIZE: 8 m tall, leaves 10 cm long, fruit 30 cm long. BIOLOGY: Forms thick hedges along the edges of creeks running through the mangrove swamps. This habitat only forms in large, well-established swamps, to which *Rhizophora* is restricted.

164.4 Tagal mangrove   *Ceriops tagal*

IDENTIFICATION: A relatively small tree with light-grey bark, buttress-roots at the base of the trunk, and a small number of 'knee roots' and 'prop roots'. Leaves bright yellow-green, rigidly stiff, oval with a rounded or notched tip (164.4d). Fruit similar to that of *Rhizophora*, with a pear-shaped calyx and a very long, slender hypocotyl. SIZE: 6 m tall, leaves 7 cm, fruit 30 cm. BIOLOGY: Rare in South Africa, only isolated specimens being recorded from Kosi Bay, but becoming common in Moçambique. Grows high on the shore among the landward thickets of *Avicennia*.

164.1 ▲

164.1d ▲

164.2d ▲

164.3 ▼

164.3d ▲

164.4d ▼

Angiospermae: 3. Dune Plants

Several plants are important early colonisers of seashore dunes, stabilising the shifting sands and allowing other plants to establish themselves. Taken to its logical conclusion, this succession can promote coastal shrubs and even dune forests, as happens along the shores of Maputaland. The process is slow, and disturbance can destabilise these dune communities, hence the opposition to mining of coastal dunes (particularly dune forests). Growing in a hostile, desiccating environment with salt spray and shifting sands, most colonists of the beach dunes are low-growing and have tough or succulent leaves. Two European species of grass have been introduced to fix shifting dunes in southern Africa: the sea wheat, *Agropyron distichum*, and marram grass, *Ammophila arenaria*. Both bind loose sands and cut down sand movement by reducing winds close to the ground. However, this can starve the adjacent shore of its natural input of sand, and has led to the disappearance of beaches in some areas (e.g. Arniston).

165.1 Goat's foot   *Ipomoea brasiliensis*

IDENTIFICATION: A sprawling creeper; long runners bear pairs of bilobed leaves (resembling a goat's cloven hoof). Flowers (165.1d) trumpet-shaped, bright purple. SIZE: Leaves 60 mm long; stems many metres long. BIOLOGY: One of the most important early colonisers of seashore dunes. Widely distributed due to the dispersion of its seeds by ocean currents. SIMILAR SPECIES:

165.2 *Canavalia rosea* (northern Natal to Moçambique), a creeping dunecoloniser with trilobed leaves; has small purple flowers resembling those of a pea plant.

165.3 Hottentot's fig   *Carpobrotus sauerae*

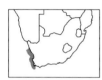

IDENTIFICATION: Sends out long runners with pairs of succulent leaves that are triangular in section. Flowers purple, large and showy, superficially daisy-like. SIZE: Leaves 50 mm long; flowers 70 mm in diameter. BIOLOGY: A coastal creeper on rocky outcrops and dunes that are already partly stabilised. Fruits succulent and edible; may be dried or used to make jams. Actively pollinated by a range of insects. One of several *Carpobrotus* species, some of which are used world-wide to stabilise shifting sands.

165.4 Seeplakkie   *Scaevola plumieri*

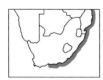

IDENTIFICATION: A small, evergreen, woody shrub with clusters of tough, waxy, oval leaves. Flowers small, white, fan-shaped. Fruits small and berry-like, purple when ripe. SIZE: Leaves 40 mm, shrub 1 m tall. BIOLOGY: Stabilises shifting sands with its extensive roots, building small hillocks. The thick waxy layer on its leaves reduces water loss. SIMILAR SPECIES:

165.5 *Salsola nollothensis* (Port Nolloth northwards) is the most important creator of hummock dunes on the arid Namibian coast. Confronted with water shortage, salt spray, heat and shifting sands, it has a tough, woody stem with pink twigs, and tiny, tightly-packed, silvery leaves.

165.6 Sea pumpkin   *Arctotheca populifolia*

IDENTIFICATION: Stem thick, leaves either simple and oval-shaped or lobed, coated with white hairs that give the leaves a grey colour. Flowers daisy-like with a rim of pale yellow florets. SIZE: Leaves 30 mm wide, flowers 25 mm. BIOLOGY: A common occupant of coastal foredunes which forms small hummocks. The hairs on its leaves are a means of reducing desiccation.

165.7 Sprawling duneweed   *Tetragonia fructicosa*

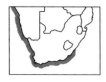

IDENTIFICATION: A sprawling, prostrate plant with a red stem. Leaves succulent, flat, oval. Flowers tiny, with a crown of yellow stamens. Fruits four-winged. SIZE: Leaves 20 mm long; fruits 20 mm. BIOLOGY: An abundant coloniser of foredunes. Its stem, leaves and fruits are covered with turgid, swollen cells that give the plant a glistening appearance and may help reduce water loss. RELATED SPECIES: *Tetragonia decumbens* is similar but has smaller, three-winged fruits.

165.1 ▲

165.1d ▲

165.2 ▲

165.3 ▲

165.6 ▼

165.4 ▲

165.5 ▲

165.7 ▼

Appendix 1 : Regulations for Marine Invertebrates

Species or group	Daily bag limit	Minimum size	Other restrictions
Alikreukel	5	63,5	–
Armadillo	6	–	–
Bloodworm	5	–	Only by hand or suction pump
Clam	8	–	–
Cuttlefish	2	–	By hand or line
East Coast rock lobster	8	65 mm carapace length	By hand, baited hook or approved trap, not from boat, no scuba, not with eggs. (Closed season: 1/11 – 28/2)
Limpets	15	–	Only by hand or with an implement with a blade not exceeding 12 mm in width
Mole crabs	30	–	By hand or triangular trap
Mud crab (*Scylla*)	6	140 mm carapace width	Only by hand, rod and/or line, not with eggs
Mussels	30	–	Only by hand, or with an implement with a blade not exceeding 12 mm in width
Other crabs/ hermit crabs	15	–	Only by hand, rod and/or line
Octopus	2	–	–
Oysters	25	–	Only by hand or with a blade not exceeding 40 mm wide
Pencil bait	20	–	Only by hand or suction pump
Periwinkle/topshell/ turbanshell	50	–	–
Perlemoen/abalone	–	–	Recreational fishery closed until further notice
Polychaete worms/ other marine worms	10	–	Only by hand
Prawn (mud or sand)	50	–	Only by hand or suction pump
Prawn (swimming)	50	–	–
Red-bait	2 kg flesh	–	Only by hand, or with an implement with a blade not exceeding 12 mm wide
West Coast rock lobster	4	80 mm carapace length	Only by hand or with a ring- or scoop-net, no scuba, no diving from boats, not at night, not with eggs or soft shelled. Weekends/holidays only after 1 Feb. Closed season: 1/5 – 15/11
Scallop	10	–	–
Sea cucumber	20	–	–
Sea urchin	20	–	–
Siffie/venus-ear	10	32 mm	Only by hand, or an implement with a blade not over 38 mm wide
Squid/chokka	20	–	–
White mussel	50	35 mm	Only by hand, or an implement with a blade not over 12 mm wide

- Collection of corals, sea fans, pansy shells and giant clams prohibited.
- Permits are required for all coastal resource collection in terms of the Living Marine Resources Act. They may be purchased at post offices.
- These regulations are subject to fairly frequent revision. Watch the press for details or check with Marine and Coastal Management, Dept of Environmental Affairs and Tourism on 021-402 3911.

Appendix 2: Regulations for holders of recreational fishing permits

All anglers are required to purchase an annual recreational angling permit, obtainable at Post Offices at a cost of R50 (short term permits are also available at a reduced fee). Separate permits are required for spearfishing, fishing vessels, use of a cast of throw net, collection of marine aquarium fish and bait collection. To ensure sustainable management of fish stocks angler's catches are restricted by a combination of daily bag limit and minimum size limits for each species, as listed below. Please note that these regulations are subject to quite frequent change – for up to date information contact Department of Environmental Affairs and Tourism, Branch: Marine and Coastal Management (website: deat.gov.za). The following conditions should also be noted:

1. No fish may be sold or offered for sale by the holder of a recreational fishing permit.
2. Minimum sizes are measured from the tip of the snout to the end of the tail and head and tail may not be removed.
3. Species not listed below have a bag limit of ten except sharks, rays, skates or chimaeras (bag limit one) and rockcods (bag limit five).
4. There is an overall cumulative daily bag limit of ten, irrespective of the species caught and provided that this limit does not apply to those species listed here with no bag limit, and to those with a bag limit exceeding ten.
5. There are closed seasons for elf/shad (1 October – 30 November), red steenbras (1 September – 30 November) and galjoen (15 October – end February).

Common name	Minimum size	Common name	Minimum size	Common name	Minimum size
Prohibited species		**Bag limit: 4 per day**		**Bag limit: 10 per day**	
Basking shark	n/a	Carpenter (silverfish)	35 cm	Albacore/longfin tuna	none
Brindle bass	n/a	Elf (shad)	30 cm	Bigeye tuna	3.2 kg
Coelacanth	n/a			Fransmadam (Karel groot oog)	none
Great white shark	n/a	**Bag limit: 5 per day**		Hottentot	22 cm
Natal wrasse	n/a	Baardman (bellman, tasselfish)	40 cm	King mackerel	none
Pipefish & seahorses	n/a	Banded galjoen	none	Pinky (piggy)	7.5 cm
Potato bass	n/a	Billfishes (marlin, sailfish)	none	Queen mackerel	none
Sawfishes	n/a	Blacktail (dassie)	20 cm	Snoek (Cape snoek)	60 cm
Seventy-four	n/a	Blue hottentot	none	Steentjie	none
Whale shark	n/a	Cape knifejaw	none	Strepie (karanteen)	15 cm
		Cape stumpnose	20 cm	Tunas (tunny)	none
Bag limit: 1 per day		Catface rockcod	50 cm	Yellowfin tuna	3.2 kg
Dageraad	40 cm	Dane	none	White stumpnose	25 cm
Englishman	40 cm	Hake (stockfish)	none	Yellowtail	none
Kob caught in estuaries	60 cm	John Brown	none		
& from the shore		Kingfishes	none	**Bag limit: 20 per day**	
[East of Cape Agulhas		Kob from a boat at sea	50 cm	Squid (Chokka)	none
only]		[Cape Agulhas to	(Only 1 kob		
Leopard cat shark	none	Umtamvuna River]	>110 cm/day)	**Bag limit: 50 per day**	
Poenskop (black steenbras/	50 cm	Kob caught from a boat at sea	40 cm	Mullets/harders	none
musselcracker)		[in KwaZulu-Natal]			
Ragged tooth shark	none	Kob [West of Cape Agulhas only]	50 cm	**Bag limit: no limit**	
Red steenbras	60 cm	Large-spot pompano (wave garrick)	none	Anchovies	none
Red stumpnose	30 cm	Natal knifejaw (cuckoo bass)	none	Chub mackerel	none
Scotsman	40 cm	Natal stumpnose (yellowfin bream)	25 cm	Cutlassfish (walla walla)	none
Spotted gulley shark	none	River bream (perch)	25 cm	Garfishes	none
Striped cat shark	none	River snapper (rock salmon)	40 cm	Glassies	none
West coast steenbras	60 cm	Santer (soldier)	30 cm	Halfbeaks	none
White steenbras	60 cm	Slinger	25 cm	Horse mackerel/maasbanker	none
(pignose grunter)		Southern pompano	none	Sardines (pilchard & red-eye)	none
Yellowbelly rockcod	60 cm	Spotted grunter (tiger)	40 cm	Sauries	none
		Springer (ten pounder)	none	Scads	none
Bag limit: 2 per day		Stonebream	none	Wolfherring	none
Bronze bream (bluefish)	30 cm	Swordfish (broadbill)	25 kg		
Galjoen	35 cm	White edged (Captain Fine)			
Garrick (leervis)	70 cm	rockcod	40 cm		
Geelbek (Cape salmon)	60 cm	Zebra (wildeperd)	30 cm		
Roman	30 cm				
White musselcracker	60 cm				
(=cracker)					

Glossary

adipose fatty tissue, particularly that covering part of the eye in some fish

annulated beaded in appearance, or with a series of rings

anterior canal groove projecting from the front of the aperture (opening) of a gastropod shell

aperture opening or mouth (as of a gastropod shell)

avicularium (-a) defensive individual of a bryozoan colony, resembling a bird's head

axil angle between a leaf or branch and the axis from which it originates

axis (-es) main stem or cylinder

balanoid zone middle level of the intertidal zone usually dominated by barnacles

barbel fleshy projection near the mouth used for taste or smell

benthic bottom-dwelling; living in or on the sea floor

body whorl the last turn or coil of a gastropod shell, housing the body

brachidium internal skeleton supporting the lophophore in brachiopods

byssus anchoring threads produced by a gland in the foot of many bivalves

calcareous composed of calcium carbonate or chalk (hence opaque white in appearance)

callus area of thickened shiny shell material on the inner lip of a gastropod shell

canine long, pointed tooth for grasping and piercing

carapace bony or chitinous shield covering part or all of the back of certain animals

cephalothorax fused head and thorax, as in crustaceans

cerata fingerlike processes on the backs of certain nudibranch molluscs (sea slugs)

cheliped crustacean appendage bearing a chela or nipper (e.g. first, clawed legs of crab)

chitin horny organic material forming part of the skeleton of many invertebrates

cirrus (-i) slender flexible appendage; especially claw-like gripping appendage of a feather star or feeding appendage of a barnacle

coenenchyme tissue connecting adjoining polyps in a colony of cnidarians

column trunk or main part of body of a sea anemone

commensal an organism that lives together with another species without harming it

corallite that portion of a coral skeleton housing an individual polyp

dactylozooid individual modified into a tentacle for prey capture in a cnidarian colony

demersal living close to the seabed

detritus particles of decaying plant and animal material, and associated micro-organisms

diatoms unicellular algae with walls impregnated with silica

dichotomous forking into two lobes or branches which are usually of equal size

endemic limited to a particular geographic region

epiphytic growing on the surface of a plant, but not parasitic on it

frond flattened, leaf-like portion of an alga

fusiform spindle-shaped or tapering at both ends

gametophyte sexual or gamete-producing phase of a plant with alternating generations

gastrozooid an individual specialised for feeding in a cnidarian colony

girdle thickened mantle surrounding shell plates of a chiton

gonotheca (-ae) skeletal container of reproductive body of a hydroid

gonozooid reproductive individual in a cnidarian colony

hectocotylus one of the arms of a male cephalopod, specialised to transmit sperm to the female

hermaphrodite an individual that possesses both male and female reproductive organs

holdfast root-, sucker- or disc-like attachment organ of a seaweed

hydrotheca (-ae) cup-like skeletal structure into which a hydroid polyp can withdraw

hypocotyl that part of a seedling projecting below the seed-leaves; elongate in mangroves

intertidal between the high-tide and the low-tide levels on the shore

introvert tubular anterior process which can be protruded or drawn back into the body

lateral line a line along the side of the body of a fish, responsible for detecting sound

ligament elastic horny hinge joining the valves of a bivalve shell; fibrous band of tissue joining two or more movable bones

lophophore horseshoe-shaped organ supporting the tentacles in bryozoans and brachiopods

lunate shaped like the crescent of a new moon

mantle outgrowth of the body wall which lines and secretes the shell of molluscs

manubrium structure hanging beneath the bell of a jellyfish and ending in the mouth

marsupium pouch-like structure in which young complete their development

meristem growing region of a plant

mid-rib central thickening of a leaf or frond of a plant

molar rounded, grinding tooth

mucilage slimy substances produced in cell walls of certain plants

mucus slime, produced by mucous glands

multiporate any structure opening via several pores

muscle scar scar marking the position of muscle attachment to a shell or bone

operculum protective lid which closes the opening of a snail or barnacle's shell or a polychaete's tube, or the gill cover of a fish

opisthosoma hind part of the body, or abdomen, of a spider

osculum (-a) chimney-like opening in a sponge, through which water leaves the body

ovicell structure containing developing eggs of a bryozoan

ovigerous serving to carry eggs

palp sensory appendage originating from the head

papilla (-ae) small, blunt conical projection

paragnath (-es) tiny tooth on the pharynx of certain polychaete worms

parapodium (-a) leg-like locomotory appendage on the body segments of worms

pectoral related to the breast or chest region or limbs, anterior to the belly

pedicel stalk or stem-like attachment of a brachiopod

pedicellaria minute pincer-like structures studding the surface of certain echinoderms

pelagic swimming in the water column of the open sea

pelvic related to the pelvis or hind limbs, posterior to the belly

periostracum outer horny layer of a mollusc shell (often eroded, or shed after death)

perisarc tough outer skeletal covering of the stem of a hydroid

peristomium first true cylindrical segment of a worm, bearing the mouth

pharynx section of the gut immediately behind the mouth; gullet

pinnate having branches on either side of an axis, like a feather

pinnule side- or sub-branch of a plume-like structure

plankton animals (zooplankton) or plants (phytoplankton) which float or drift in the water

pleon abdomen of a crustacean

pleopod abdominal appendage of a crustacean, usually used for swimming

pleotelson rear end of a crustacean; one or more abdominal segments fused to the telson

pneumatophore an above-ground root of a mangrove, used for gas exchange

polyp cylindrical animal, attached at one end, bearing a mouth and tentacles at the other

preoperculum large flat bone lying on top of the operculum or gill cover of fish

proboscis tubular snout or projection of the head, usually extensible

prosoma front part of the body of a spider which bears feeding appendages and the legs

prostomium anterior lobe of the head of a worm, usually bearing sensory appendages

protandrous an individual that is initially male, then becomes female

protogynous an individual that is initially female and later becomes male

protrusible capable of being projected or thrust forward

pseudofaeces material processed by feeding appendages but rejected before being eaten

radial shields pairs of plates on the upper surface of a brittlestar near the base of each arm

radula ribbon-like tongue bearing rows of teeth in a mollusc

raptorial adapted for seizing prey; predatory

ray flexible, soft, jointed cartilaginous element supporting the fin of a fish

resilial ridge band running on the internal surface of a bivalve shell next to the ligament

rhinophore second, sensory pair of tentacles on the top of the head of certain molluscs

rhizome creeping horizontal stem sending out shoots above and roots below

rostrum beak-like process; pointed projection between the eyes of a crustacean

scutes modified fish scales with keels or spines on them

septum (-a) partition separating two cavities or bodies of tissue

sessile attached or stationary

seta (-ae) bristle, particularly on the body of a worm or arthropod

siphon tubular structure through which water is drawn into or expelled from the body

spat newly settled juveniles of bivalve molluscs

spatulate spade-shaped, e.g. the flattened tips of the nippers of some crabs

spherules small spherical bodies found at the top of the column of certain sea anemones

spicule minute spines or granules in the body walls of sponges, soft corals, sea cucumbers

spire coiled portion of a shell above the aperture

sporophyte asexual, spore-producing phase of a plant with alternating generations

statocyst organ of balance consisting of a sac containing granules of lime or sand

stipe stalk- or stem-like portion of an alga

stridulation production of a rasping sound by rubbing two rough surfaces together

substratum (-a) rock or other solid object to which an animal or plant can attach

subtidal below the lowest level on the shore reached by the tides

suture groove that spirals around a gastropod shell, where one whorl meets the next

symbiosis intimate relationship between two organisms which is of mutual benefit

synapticulum (-a) skeletal bar connecting adjacent septa in a coral

telson last segment of the abdomen of a crustacean; often appears as a flattened plate

test shell of a sea urchin, or tough covering of an ascidian or sea squirt

tetrasporophyte asexual algal generation; produces groups of four spores by dividing one spore

thallus body of an alga that is undifferentiated into root, stem or leaves

truncated cut-off or ending abruptly

tubefoot hollow tubular limb typical of echinoderms; operated by internal water pressure

tubercle conical projection or small hump

umbilicus open end of a hollow running up the central axis of a gastropod shell

uniporate structures that open via a single pore, like a salt cellar

uropod posterior abdominal appendage of crustaceans; lies on either side of the telson

utricle swollen, bladder-like tip of a cellular filament in certain green algae

valve one of the pair of shells of a bivalve, or one of the eight plates of a chiton shell

vesicle hollow, bladder-like body usually with fluid

whorl one complete coil of a gastropod shell; or a circle of parts arising from one point

zooecium (-a) the chamber enclosing a single individual or zooid of a bryozoan colony

zooid individual forming part of a colonial organism

zooxanthellae microscopic unicellular algae that are symbiotic in the bodies of some animals

References

GENERAL WORKS

BRANCH, G. and BRANCH, M. 1981. The Living Shores of Southern Africa. C. Struik, Cape Town. 272pp.

BRANCH, M. 1987. Explore the Seashore of South Africa. Struik Publishers, Cape Town. 48pp.

DAY, J.H. 1974. A Guide to Marine Life of South African Shores. A.A. Balkema, Cape Town. 272pp.

GRIFFITHS, C., GRIFFITHS, R. and THORPE, D. 1988. Seashore Life. Struik Pocket Guide Series, Struik Publishers, Cape Town. 64pp.

KALK, M. (Ed.) 1995. A Natural History of Inhaca Island, Mozambique (3rd ed). Witwatersrand University Press, Johannesburg. 395 pp.

LUBKE, R. and DE MOORE, I. (Eds.) 1998. Field Guide to the Eastern and Southern Cape Coasts. University of Cape Town Press in association with Grahamstown Branch, Wildlife and Environment Society of South Africa. 559 pp.

PAYNE, A.I.L., CRAWFORD, R.J.M. and VAN DALSEN, A. 1989. Oceans of Life off Southern Africa. Vlaeberg Publishers, Cape Town. 380pp.

RICHMOND, M.D. 1997. A Guide to the Seashore of Eastern Africa and the Western Indian Ocean Islands. SIDA, Dept for Research Cooperation, SAREL. 448 pp.

IDENTIFICATION GUIDES
Cnidaria

CARLGREN, O. 1938. South African Actinaria and Zoantharia. Kungl. Svensk. Vet.-Akad. Handl. (Series 3) 17(3):1–148.

KRAMP, P.L. 1961. Synopsis of the medusae of the world. J. Mar. Biol. Assoc. U.K. 40:1–469.

MILLARD, N.A.H. 1975. Monograph on the Hydroida of southern Africa. Ann. S. Afr. Mus. 68:1–513.

PAGÈS, S., GILI, J.M. and J. BOUILLIN, 1992. Planktonic cnidarians of the Benguela Current. Scientia Marina 56 (Suppl. 1):1–144.

VERON, J.E.N. 1986. Corals of Australia and the Indo-Pacific. Angus and Robertson, North Ryde, Australia. 644pp.

WILLIAMS, G.C. 1990. The Pennatulacea of southern Africa (Coelenterata, Anthozoa). Ann. S. Afr. Mus. 99:31–119.

WILLIAMS, G.C. 1992. The Alcyonacea of southern Africa. Stoloniferous octocorals and soft corals (Coelenterata, Anthozoa). Ann. S. Afr. Mus. 100: 249–358.

WILLIAMS, G.C. 1992. The Alcyonacea of southern Africa. Gorgonian octocorals (Coelenterata, Anthozoa). Ann. S. Afr. Mus. 101:181–296.

WILLIAMS, G.C. 1993. Coral Reef Octocorals. An Illustrated Guide to the Soft Corals, Sea Fans and Sea Pens Inhabiting the Coral Reefs of Northern Natal. Durban Natural Science Museum, Durban. 64pp.

Unsegmented Worms

WESENBERG-LUND, E. 1963. South African sipunculids and echiuroids from coastal waters. Vidensk. Medd. Dansk. Naturh. Foren. 126:101–146.

Polychaeta

DAY, J.H. 1967. A Monograph on the Polychaeta of Southern Africa. Trustees of the British Museum (Natural History), London. 878pp.

Pycnogonida

BARNARD, K.H. 1954. South African Pycnogonida. Ann. S. Afr. Mus. 41:81–159.

Crustacea

BARNARD, K.H. 1950. Descriptive catalogue of South African decapod Crustacea (crabs and shrimps). Ann. S. Afr. Mus. 38:1–837.

BARNARD, K.H. 1950. Descriptive list of South African stomatopod Crustacea (Mantis shrimps) Ann. S. Afr. Mus. 38:838–864.

BERRY, P.F. 1971. The spiny lobsters (Palinuridae) of the East Coast of southern Africa: distribution and ecological notes. Investl. Rep. Ocean. Res. Inst. S. Afr. 27:1–23.

BODEN, B.P. 1954. The euphausiid crustaceans of southern African waters. Trans. Roy. Soc. S. Afr. 34: 181–234.

DE FREITAS, A.T. 1985. The Penaeoidea of southern Africa 1. The study area and key to the southern African species. Investl. Rep. Ocean. Res. Inst. S. Afr. 56:1–31.

GRIFFITHS, C.L. 1976. Guide to the Benthic Marine Amphipods of Southern Africa. South African Museum, Cape Town. 106pp.

KENSLEY, B. 1972. Shrimps and Prawns of Southern Africa. South African Museum, Cape Town. 65pp.

KENSLEY, B. 1978. Guide to the Marine Isopods of Southern Africa. South African Museum, Cape Town. 173pp.

Brachiopoda

HILLER, N. 1991. The southern African recent brachiopod fauna. In: Brachiopods through Time (Eds. D.I. McKinnon, D.E. Lee and J.D. Campbell), pp. 439–445. A.A. Balkema, Rotterdam.

JACKSON, J.W. 1952. A revision of some South African Brachiopoda, with descriptions of new species. Ann. S. Afr. Mus. 41:1–40.

Mollusca

AUGUSTYN, C.J and SMALE, M.J. 1989. Cephalopods. In: Oceans of Life off Southern Africa (Eds. A.I.L. Payne, R.J.M. Crawford and A.P. van Dalsen), pp. 91–104. Vlaeberg Publishers, Cape Town.

GOSLINER, T. 1987. Nudibranchs of Southern Africa

– A Guide to Opisthobranch Molluscs of Southern Africa. Sea Challengers, Monterey. 136pp.

KAAS, P. and VAN BELLE, R.A. 1985–1991. Monograph of Living Chitons (in 4 volumes). E.J. Brill.

KILBURN, R. and RIPPEY, E. 1982. Sea Shells of Southern Africa. Macmillan South Africa, Johannesburg. 249pp.

LILTVED, W.R. 1989. Cowries and Their Relatives of Southern Africa. A Study of the Southern African Cypraeacean and Velutinacean Gastropod Fauna. Gordon Verhoef, Seacomber Publications. 208pp.

RICHARDS, D. 1981. Shells of Southern Africa – A Concise Guide for Collectors. Struik Publishers, Cape Town. 156pp.

ROELEVELD, M.A. 1972. A review of the Sepiidae (Cephalopoda) of southern Africa. *Ann. S. Afr. Mus.* 59:193–313.

Echinodermata

BALINSKY, J.B. 1957. The Ophiuroidea of Inhaca Island. *Ann. Natal Mus.* 14:1–33

CLARK, A.M. and COURTMAN-STOCK, J. 1976. The Echinoderms of Southern Africa. British Museum (Natural History), London. 277pp.

THANDAR, A.S. 1989. The sclerodactylid holothurians of southern Africa, with erection of one new subfamily and two new genera (Echinodermata: Holothuroidea). *S. Afr. J. Zool.* 24:290–304.

THANDAR, A.S. 1990. The phyllophorid holothurians of southern Africa with the erection of a new genus. *S. Afr. J. Zool.* 25:207–223.

Ascidiacea

MILLAR, R.H. 1955. On a collection of ascidians from South Africa. *Proc. Zool. Soc. Lond.* 125:169–221.

MILLAR, R.H. 1962. Further descriptions of South African ascidians. *Ann. S. Afr. Mus.* 46:113–221.

Fishes

MONNIOT, C., MONNIOT, F., GRIFFITHS, C.L. and SCHLEYER, M. 2001. South African ascidians. *Ann. S. Afr. Mus.* 108:1–141.

COMPAGNO, L.J.V., EBERT, D.A. and SMALE, M.J. 1989. Guide to the Sharks and Rays of Southern Africa. Struik Publishers, Cape Town. 160pp.

HEEMSTRA, P. and HEEMSTRA, E. 2003. Coastal Fishes of South Africa. NISC and SAIAB, Grahamstown. 488 pp.

VAN DER ELST, R. 1988. A Guide to the Common Sea Fishes of Southern Africa (2nd ed). Struik Publishers, Cape Town. 398pp.

VAN DER ELST, R. 1990. Everyone's Guide to Sea Fishes of Southern Africa. Struik Publishers, Cape Town. 112pp.

Birds

GINN, P.J., McILLERON, W.G. and MILSTEIN, P. le S. 1989. The Complete Bird Book of Southern Africa. Struik Winchester, Cape Town. 760 pp.

MACLEAN, G. L. 1993. Robert's Birds of Southern Africa (6th ed). John Voelcker Bird Book Fund, Cape Town. 871 pp.

NEWMAN, K. 2002. Newman's Birds of Southern Africa (8th ed). Struik, London. 512 pp.

SINCLAIR J.C., HOCKEY P.A.R., TARBOTON W.R. 1998. Sasol Birds of Southern Africa (3rd ed.) Struik, Cape Town. 447 pp.

Reptiles & Mammals

BRANCH, W.R. and ROSS, G.J.R. 1988. Marine reptiles and mammals. In: A Field Guide to the Eastern Cape Coast (Eds. R. Lubke, F. Gess, and M. Bruton), pp. 115–130. Grahamstown Centre of the Wildlife Society of Southern Africa.

SKINNER, J.D. and SMITHERS, R.H.N. 1990. The Mammals of the Southern African Subregion. University of Pretoria, Pretoria. 771pp.

Seaweeds

SEAGRIEF, S.C. 1967. The Seaweeds of the Tsitsikamma Coastal National Park. National Parks Board, Pretoria. 147pp.

SEAGRIEF, S.C. 1980. Seaweeds of Maputaland. In: Studies of the Ecology of Maputaland (Eds. M.N. Bruton, and K.H. Cooper), pp. 18–41. Wildlife Society of Southern Africa, Durban.

SIMONS, R.H. 1976. Seaweeds of southern Africa: guidelines for their study and identification. *Fish. Bull. S. Afr.* 7:1–113.

STEGENGA, H., BOLTON, J. and ANDERSON, R. 1997. Seaweeds of the South African West Coast. *Contrib. Bolus Herbarium* 18:1-655.

Salt Marsh, Mangroves & Dune Plants

LUBKE, R.A. and VAN WYK, K. 1998. Terrestrial plants and coastal vegetation. In: Field Guide to the Eastern and Southern Cape Coasts (2nd ed). (Eds. Lubke, R. and De Moore, I.). pp 289–342. University of Cape Town Press.

LUBKE, R.A. and VAN WYK, K. 1998. Estuarine plants. In: Field Guide to the Eastern and Southern Cape Coasts (2nd ed). (Eds. Lubke, R. and De Moore, I.). pp 187–197. University of Cape Town Press.

BERJAK, P., CAMPBELL, G.K., HUCKETT, B.I. and PAMMENTER, N.W. 1977. In: The Mangroves of Southern Africa. Natal Branch of the Wildlife Society of Southern Africa. 73pp.

O'CALLAGHAN, M. 1992. The ecology and identification of the southern African Salicornieae (Chenopodiaceae). *S. Afr. J. Bot.* 58:430–439.

Scientific Index

General Index